Modeling and Simulation of Environmental Systems

This book presents an overview of modeling and simulation of environmental systems via diverse research problems and pertinent case studies. It is divided into four parts covering sustainable water resources modeling, air pollution modeling, Internet of Things (IoT) based applications in environmental systems, and future algorithms and conceptual frameworks in environmental systems. Each of the chapters demonstrate how the models, indicators, and ecological processes could be applied directly in the environmental sub-disciplines. It includes range of concepts and case studies focusing on a holistic management approach at the global level for environmental practitioners.

Features:

- Covers computational approaches as applied to problems of air and water pollution domain.
- Delivers generic methods of modeling with spatio-temporal analyses using soft computation and programming paradigms.
- Includes theoretical aspects of environmental processes with their complexity and programmable mathematical approaches.
- Adopts a realistic approach involving formulas, algorithms, and techniques to establish mathematical models/computations.
- Provides a pathway for real-time implementation of complex modeling problem formulations including case studies.

This book is aimed at researchers, professionals, and graduate students in Environmental Engineering, Computational Engineering/Computer Science, Modeling/Simulation, Environmental Management, Environmental Modeling, and Operations Research.

Modeling and Simulation of Environmental Systems

A Computation Approach

Edited by
Satya Prakash Maurya
Akhilesh Kumar Yadav
Ramesh Singh

CRC Press
Taylor & Francis Group
Boca Raton London New York

CRC Press is an imprint of the
Taylor & Francis Group, an **Informa** business

First edition published 2023
by CRC Press
6000 Broken Sound Parkway NW, Suite 300, Boca Raton, FL 33487-2742

and by CRC Press
4 Park Square, Milton Park, Abingdon, Oxon, OX14 4RN

CRC Press is an imprint of Taylor & Francis Group, LLC

© 2023 selection and editorial matter, Satya Prakash Maurya, Akhilesh Kumar Yadav and Ramesh Singh; individual chapters, the contributors

ISBN: 978-1-032-06698-1 (hbk)
ISBN: 978-1-032-06699-8 (pbk)
ISBN: 978-1-003-20344-5 (ebk)

DOI: 10.1201/9781003203445

Typeset in Times
by SPi Technologies India Pvt Ltd (Straive)

Contents

PART I Water

PART II   Air Pollution

PART III   Internet of Things and Environmental Systems

PART IV   *Future Algorithms in Environmental Systems*

Foreword

Water, air, and soil are interactive elements that constitute the environmental continuum. These elements are fundamental to our way of life, economic well-being, and the health and integrity of the ecosystem. Central to the environmental continuum is water. Because of the growing population, rising standard of living, increasing industrialization, and expanding energy generation, the demand for water to satisfy varied needs is unprecedented. The demand is being further accentuated by pollution abatement, resulting in enormous pressure on water resources throughout the world. In many parts of the world, water is already a scarce commodity and will become so in several other parts in the very near future. Therefore, it requires careful management not only in water scarce regions but everywhere. Water resources management takes on an even greater significance because of the impacts of global warming and climate change. A pressing challenge in water resources management is one of integrating technology with institutional structures and developing tools that will help decision makers when evaluating costs and benefits of water supply not only to the various users but also to the environment and ecosystem. These tools must be able to make a statement on the reliability of supply and demand.

Computational techniques, including soft computing, play an important role in analyzing, modeling, and simulating complex environmental systems. Genetic and other evolutionary algorithms have the capacity to handle large spaces and help with system optimization. Fuzzy approaches help deal with imprecise inputs and states. Machine learning and artificial intelligence techniques have, in general, significantly contributed to the development of computational tools.

This book on modeling and simulation of environmental systems, a timely publication, attempts to introduce different methods that can be effectively integrated to address multi-criteria, multi-objective assessment of complex environmental systems. It contains several chapters that deal with such assessments. The topics included in the book encompass soft computing techniques, computational approaches to data analytics, process dynamics, quantitative and qualitative analytics, artificial intelligence approaches, agent-based modeling, and development of software framework for dynamic adaptive policy pathways. The book will be useful to graduate students, college faculty, and those engaged in research in the field of environmental modeling and simulation approaches.

Vijay P. Singh, Ph.D., D.Sc., D. Eng. (Hon.), Ph.D. (Hon.), D.Sc. (Hon.), P.E., P.H., Hon. D. WRE, Dist. M. ASCE., Hon. M. IWRA, Hon. M. AWRA, Dist. F. AGGS, Academician (GFA)
Distinguished Professor, Regents Professor
Caroline and William N. Lehrer Distinguished Chair in Water Engineering
Texas A&M University
College Station, Texas, USA

Preface

Worldwide, the water governance and environmental management scenario have been undergoing structural changes for more than two decades. Reforms in water sectors management began in the 1990s, but these efforts have been made in bits and pieces depending upon localized scenario/context. The need for integrating various dimensions of planning strategy such as uses and users, and notions of water stress, environmental security for future generations, is a need of today's management—the increasing role of modern soft computational techniques in various disciplines. Motivated academia need to come forward with different approaches for assessment framework, modeling, and simulation of the sustainable environmental management system.

Growing population, expanding urbanization, and industrialization threaten the most vital natural resources, water, and air, for all living things. The impending climate change brings a new level of complexity and challenges in demand and supply chain. Modeling is a better approach to address these issues and remediate problems, but environmental modeling and simulation carry complexities with them as there are various applications. These complexities may include assimilation of parameters, dimension, amount of data, distribution, heterogeneity, and interdependence. However, there may be different purposes for modeling, such as scenario analysis, emergency response, and risk management. Coupling of models, parameter adjustment, and longevity of data may be needed for adaptability in different applications. These demands have led to various current research themes, such as decision support, remote sensing and GIS, adaptive user interfaces, data science, standardization of metadata and system visualizations, workflows for automatic access to distributed resources, and the generic nature of information and simulation systems.

This book presents an overview of modeling and simulation of environmental systems. The book content has been categorized into four sections. The brief subject matter of each section is mentioned.

It creates ample opportunities for modeling and simulation and decision support tools with accuracy and completeness. However, the intervention of other advanced disciplines such as artificial intelligence, machine learning, deep learning, remote sensing and GIS, Internet of Things (IoT), data science and new computer algorithms have great potential in environmental decision-making, prediction, and forecasting. Air and water research dynamics use models dealing with varying scales and resolutions and require new architectures with access to distributed resources. Branch-oriented simulation systems should prove the right software tools to be flexibly adapted to complex environmental system's special structure and data.

This book aims to address the sustainable development goals (SDGs) by applying modern computational techniques and applications for spatio-temporal analysis and simulation modeling solutions for a wider set of environmental problems that are becoming more relevant for environmental engineers and scientists. Hopefully, this book will open the dimensions for explicit programming dynamics to develop decision-making tools to help manage future environmental problems related to air and water.

Satya Prakash Maurya, PhD
Akhilesh Kumar Yadav, PhD
Ramesh Singh, MS

Acknowledgements

First and foremost, praises and thanks to the Almighty for His blessings throughout this project accomplishment and who envisioned us successfully exploring this challenging research domain with a computational perspective. We are also sincerely thankful to CRC Press/Taylor & Francis Group for allowing us to publish this monogram on a contemporary research domain of modeling and simulations.

We would like to express our sincere gratitude to the contributors for submitting the book chapters. Against the proposed chapter's title, the total number of abstracts received was sixty-eight, out of which forty-one were selected for a full chapter contribution. In this book, the project accommodated only twenty-two chapters as per the book's scope with better quality. We appreciate the contribution of authors who have dedicated their valuable time and efforts to writing the assigned manuscripts. The project follows a review process, where the identities of both the authors and reviews are not disclosed to avoid any biased decision. We express our sincere gratitude for valuable guidance with suggestions and comments from Professor Vijay P. Singh (Texas A&M University, Texas, USA), Professor Anil Kumar Tripathi (IITBHU, Varanasi, India), and Professor Dharmendra Kumar Yadav (MNNIT Allahabad, Prayagraj, India), which was invaluable in helping us improve the quality of the manuscript. We are thankful to Mrs. Priyanka for her moral support and appreciate her valuable feedback about field challenges in water resource management applications. We would also like to thank and acknowledge the publication team of CRC Press with the association of Dr. Gagandeep Singh, Ms. Aditi Mittal, and Ms. Divya Muthu for their quick responses and for providing a proper guideline on time. Finally, we acknowledge Mr. Jaydeep Kumar (CISF, HQ, New Delhi, India) and Advocate Dinesh Chandra Gupta (High Court, Allahabad, India), Mr. Suresh Chandra (CRPF, WS HQ, Navi Mumbai, India), and Mr. Uday Pratap Chaudhary for their continuous help and support for the execution of the project.

The project's main aim is to publish high-quality, authentic, relevant and up-to-date research in the relevant field of an environmental system utilizing modeling and simulations based on the basic principle and current development scenario by international organizations (UN-W, EU, WHO, etc.).

Satya Prakash Maurya, PhD
Akhilesh Kumar Yadav, PhD
Ramesh Singh, MS

Editors

Satya Prakash Maurya earned his B. Tech. in Computer Science & Engineering, M. Tech. in GIS & RS. He completed his PhD in Geoinformatics under the research domain of Spatial Decision Support System from Indian Institute of Technology (BHU), Varanasi (India). He has also studied Civil Engineering from IEI (India), Kolkata. He has worked as an Assistant Professor in CSED at PCE, Jaipur (India) with research interest in computational approach in machine learning, data modeling, expert systems, and artificial intelligence with application of urban water modeling and futuristic prediction and soft computing. Presently, he is working as a Project Scientist-I in NSDI funded project at IIT (BHU). He has published a number of research papers and is also a reviewer in a few reputed journals. He is an associate member in professional bodies such as International Association of Engineers, theIRED, USA and IEI, India.

Akhilesh Kumar Yadav earned his B. Tech. in Electronics and Communication Engineering from Chaudhary Charan Singh University, Meerut, India; M. Tech. in Environmental Engineering from Madan Mohan Malaviya Engineering College, Gorakhpur (Dr. APJ AKTU, Lucknow), India; and Doctoral degree with research domain of air pollution from Indian Institute of Technology (Banaras Hindu University), Varanasi, India. His research interests are air pollution, climate change, vulnerability, and human health risk assessments, and GIS applications in air pollution management. He has received the Young Engineer Award by the Institution of Engineers (India), Kolkata in 2017, and published articles in peer-reviewed journals and serves as a reviewer for reputed journals. He is an active member of reputed international professional bodies such as the Institution of Engineers (India), Kolkata; Engineering Council of India, New Delhi; Association of Environmental Analytical Chemistry of India, Mumbai.

Ramesh Singh earned his B.Tech in Civil Engineering with a Gold Medal from IT-BHU, Varanasi and M.S. in Sanitary Engineering from IHE Netherland. He worked as field engineer for 37+ years with Govt. of U.P. (India) and was involved with planning and implementation of various national and foreign aided projects for abatement works of rivers and lakes, and urban and rural water supply and environmental sanitation. He has worked as Visiting Faculty in Civil Engineering at IIT(BHU), Varanasi. Presently, he is engaged as Technical Advisor of Kokusai Kogyo Co. Ltd, Tokyo, Japan, and Directorate Environment and Climate Change, Govt. of Manipur (India). He has published a few research papers and is also a reviewer for reputed journals. He is fellow member of Indian Water Works Association (India), life member of Institution of Engineers (India), Indian Association of Environmental Management and *The Indian Concrete Journal*.

Contributors

Marykutty Abraham
Sathyabama Institute of Science and
Technology
Chennai, India

Ajay Kumar Agrawal
Uttar Pradesh Remote Sensing
Application Centre
Lucknow, India

Aatish Anshuman
Indian Institute of Technology
Bombay
Mumbai, India

Mohammad Aurangzeb Ansari
IBM India Private Limited
Gautam Buddha Nagar, India

Mohammad Irfan Ansari
JMC Projects (India) Limited
Mumbai, India

Ajay Sudhir Bale
CMR University
Bengaluru, India

Rashmi Bhardwaj
Guru Gobind Singh Indraprastha
University, University School of
Basic and Applied Sciences
Dwarka, Delhi, India

Rajesh Biniwale
CSIR National Environmental
Engineering Research Institute
Nagpur, India

Anirban Chakraborty
Indian Institute of Technology
Patna, India

Shivani Dedakia
Marwadi University
Rajkot, India

Shilpa Dongre
Visvesvaraya National Institute of
Technology
Nagpur, India

T. I. Eldho
Indian Institute of Technology Bombay
Mumbai, India

Sahajpreet Kaur Garewal
National Institute of Technology
Raipur, India

Shanky Garg
Guru Gobind Singh Indraprastha
University, University School of
Basic and Applied Sciences
Dwarka, Delhi, India

Rajesh Gupta
Visvesvaraya National Institute of
Technology
Nagpur, India

Rajiv Gupta
Birla Institute of Technology and
Science
Pilani, India

Anand Jayachandran Jolly
CMR University
Bengaluru, India

Rakesh Kadaverugu
CSIR National Environmental
Engineering Research Institute
Nagpur, India

Veena Kashyap
Shoolini University
Solan, India

Anmol Kaur
Government College for Girls
Ludhiana, India

Neha Keriwala
Nirma University
Ahmedabad, India

Mridu Kulwant
The Maharaja Sayajirao University of
    Baroda
Vadodara, India

Akhilesh Kumar
Chitkara University
Patiala, India

Ankit Kumar
Indian Institute of Technology (Banaras
    Hindu University)
Varanasi, India
Dr. Ambedkar Institute of Technology
    for Handicapped
Kanpur, India

Gaurav Kumar
Birla Institute of Technology and
    Science
Pilani, India

Chandrasekhar Matli
National Institute of Technology
Warangal, India

Satya Prakash Maurya
Indian Institute of Technology (Banaras
    Hindu University)
Varanasi, India

Sankaralingam Mohan
Indian Institute of Technology Madras
Chennai, India

Vinay Narayanaswamy
CMR University
Bengaluru, India

Hemanth Kumar Bangalore Naveen
CMR University
Bengaluru, India

Divyashree Neelegowda
CMR University
Bengaluru, India

Chaitanya Nidhi
Rajkiya Engineering College
Azamgarh, India

Govind Pandey
Madan Mohan Malaviya University of
    Technology
Gorakhpur, India

Prerna Pandey
Visvesvaraya National Institute of
    Technology
Nagpur, India

Suchita Pandey
Indian Institute of Technology
    Kharagpur
Kharagpur, India

Anant Patel
Nirma University
Ahmedabad, India

Divya Patel
The Maharaja Sayajirao University of
    Baroda
Vadodara, India

Jayantilal Naginbhai Patel
Sardar Vallabhbhai National Institute of
    Technology
Surat, India

Nilanchal Patel
Birla Institute of Technology
Mesra, Ranchi, India

Prutha Patel
Torrent Pharmaceuticals Ltd.,
Ahmedabad, India

Subbarao Pichuka
National Institute of Technology Andhra
    Pradesh
Tadepalligudem, India

Arunava Poddar
Shoolini University
Solan, India

Om Prakash
Indian Institute of Technology
Patna, India

Baby Chithra Ramasamy
CMR University
Bengaluru, India

Chithra Nelson Rosamma
National Institute of Technology
Calicut, India

Shiv Nath Sharma
Ministry of Jal Shakti, Central Soil and
    Materials Research Station
Department of Water Resources
River Development Ganga
    Rejuvenation (Government of India)
Delhi, India

Saba Shirin
Indian Institute of Technology (Banaras
    Hindu University)
Varanasi, India

Arohi Singh
Pandit Deendayal Energy University
Gandhinagar, India

Rajwinder Singh
Dr. BR Ambedkar National Institute of
    Technology
Jalandhar, India

Ramesh Singh
Indian Institute of Technology (Banaras
    Hindu University)
Varanasi, India

Vijay P. Singh
Texas A&M University
College Station, Texas, USA

Rakesh Kumar Sinha
Indian Institute of Technology
    Bombay
Mumbai, India

Karanvir Singh Sohal
Guru Nanak Dev Engineering College
Ludhiana, India

Jaysukh Chhaganbhai Songara
Sardar Vallabhbhai National Institute of
    Technology
Surat, India

S. Sreedevi
Indian Institute of Technology Bombay
Mumbai, India

Subba Rao Tellagorla
National Institute of Technology Andhra
    Pradesh
Tadepalligudem, India

Ankita Thanki
Marwadi University
Rajkot, India

Arti Thanki
Marwadi University
Rajkot, India

Sashank Thapa
Shoolini University
Solan, India

Subhashish Tiwari
Vignan's Foundation for Science,
    Technology & Research
Vadlamudi, Guntur, India

Avinash D. Vasudeo
Visvesvaraya National Institute of
    Technology
Nagpur, India

Akhilesh Kumar Yadav
Indian Institute of Technology (Banaras
    Hindu University)
Varanasi, India

Part I

Water

1 Computational Models for Water Resource Management
Opportunities and Challenges

Aatish Anshuman and T. I. Eldho

CONTENTS

1.1 INTRODUCTION

Water is a vital resource which is primarily available through surface and groundwater. The availability of surface water is directly dependent on rainfall and evapotranspiration while the availability of groundwater depends on infiltration/percolation through soil media. Surface water is more susceptible to losses and quality deterioration through interaction with the ambient environment. On the other hand, the

DOI: 10.1201/9781003203445-2

effect of pollution and losses due to evaporation is lesser in groundwater. However, groundwater recharge is slow which occurs on a larger time scale. Due to increased dependency on groundwater due to developments in pumping technologies and power availability is causing water table depletion in the major aquifers all over the world. Inadvertent and deliberate disposal of contaminated water and contaminants into the environment is responsible for the deterioration of the quality of surface and groundwater sources. Further, due to unequal distribution of water availability through space and time, it becomes scarce in some regions due to dry periods leading to increased water stress. Moreover, due to climate change, water availability is becoming more and more uncertain (Frederick and Major, 1997). Considering the variability of water availability, the management of water resources is essential to reduce water stress, particularly in scarce areas.

For efficient water resources management (WRM), it is essential to quantify the water resources available in an area, the hydrological processes, and human interventions in the concerned area. These processes may be interrelated and are broadly divided into water inflow, outflow, and storage. Due to the complexity of the water resources system and its management, simple analytical models can't be used.

Over the last few decades, several computational models have been developed for WRM and to approximate the hydrological processes which broadly divided into distributed and conceptual models (Beven, 2001). Distributed models, which are used for detailed understanding or, prediction of hydrological components of the study area, are based on physically based parameters that are representative of characteristics of the study area such as soil hydraulic conductivity, porosity, slope, land use type etc. Few examples of distributed models in surface water prediction are MIKE-SHE (Refsgaard et al., 2007), SHETRAN (Ewen et al., 2000). Similarly, models such as MODFLOW (Harbaugh, 2005), MT3DMS (Zheng and Wang, 1999), RT3D (Clement, 1999), RPCM (Mategaonkar and Eldho, 2011), and MLPG (Swathi and Eldho, 2014) are used for solving the partial differential equations related to groundwater flow and contaminant transport. Conceptual models consider the study area as a single unit and are usually used for the prediction of a single variable. The choice of computational model depends on the requirement of the prediction variable for example streamflow, groundwater level, and groundwater contaminant concentration.

In this chapter, the developments and applications of computational models, especially numerical models, for water resources management is discussed briefly. Further three important numerical/computational models namely FDM, FEM, and meshfree models and their applications for water resources management are elaborated with simple case studies.

1.2 MATHEMATICAL MODELING FOR WRM

Due to the complexity of the water resources system, modeling tools are needed for water resources management (WRM). Among the various modeling techniques, physical models, data-driven models, and computational models are popular. Physical models are expensive and generally done at small scales such as hydraulic structures. Data-driven models based on machine learning techniques are getting popular nowadays. However, these models require a lot of data for training and are not suitable for analyzing different hydrological components. On the other hand, since computational models are physically based, these models can be used for obtaining different components of the water resources systems such as the hydrological cycle. The governing equations used in these models are based on the physical properties of the study area which may be estimated through laboratory studies. In some cases, the study area can be represented using conceptual or lumped parameters (Kunnath-Poovakka and Eldho, 2019). Due to rapid advancement in computational technologies and modeling techniques, mathematical modeling is getting more and more accessible for simulating the hydrological processes for WRM at different scales of the study area.

For WRM, it is essential to understand the hydrological processes in the study area. Mathematical models can be useful to understand different components of the hydrological models. These models can help in the prediction of these components in future which will allow for preparedness for WRM, natural hazards such as floods, drought, depletion of groundwater and contamination, designing of hydraulic structures etc. The uncertainty associated with the climate models can be used as input in these models to provide a range of predictions for dominant hydrological processes. This can help in assessing different approaches for WRM and the consequences.

Although mathematical modeling has many advantages, these models require a mathematical understanding of the governing equations involved in the simulation of the certain hydrological process. Mostly in WRM, we use physically based models using the conservation laws of physics which are generally based on PDE (partial differential equations). For modeling of complex problems which are difficult to solve using analytical solutions, numerical methods such as FDM, FVM, FEM, and meshfree methods are used. These methods may have instability issues depending on the grid/mesh size or, the nodal distance used. The governing equations in the mathematical models are based on physically based or, derived variables that require data related to spatial characteristics of the catchment. The mathematical models often require calibration or, inverse modeling which is done by minimizing the model simulated values and actual values of the variable under consideration such as surface runoff, groundwater level, contaminant concentration etc. However, the estimated model parameters such as hydraulic conductivity, porosity etc. may vary slightly from the field estimated values as the hydrological processes are essentially complex, and the mathematical components may not capture all processes involved.

The essential steps involved in developing computational models for WRM are listed below.

- Identification of the information required for management decisions
- Development of a conceptual model
- Development of a mathematical model
- Development of a numerical model and code
- Code verification
- Model validation
- Model calibration and parameter estimation
- Model applications
- Analysis of model uncertainty and stochastic modeling
- Summary, conclusions, and reporting

Once the mathematical formulation is ready, it is necessary to decide the analysis and solution methodology for the problem concerned. The analysis can be divided into the analytical method, physical method, and computational method. For most of the water resources problems, as the analytical method is not feasible and physical methods or modeling is cumbersome, time consuming, and expensive, mostly preferred option is the computational method.

1.3 COMPUTATIONAL MODELING FOR WRM

In computational modeling, the solution is obtained with the help of some approximate methods using a computer. The solution is only approximate and depends on the type of the approximate method and computer used. Commonly, numerical models are used to obtain the solution in the computational method. Computational methods of analysis apply to a much wider class of mathematical formulations and hence most of the engineering problems can be solved.

The hydrological processes occurring in nature are complex and interdependent. The complexities include irregularity of domain areas, heterogeneity, presence of source/sinks, the uncertainty of the process such as rainfall etc. Analytical modeling for the concerned variable is infeasible considering these complexities. Computational models which replicate the behaviour of the study area for the concerned variable are suitable in the cases. These models serve as essential tools for understanding the water resource system; planning human interventions such as pumping, injection, recharge etc.; predicting the behaviour of the system under different scenarios and communicating key information to stakeholders and decision-makers (Wang and Anderson, 1982).

1.3.1 NUMERICAL MODELS

In the last few decades, a variety of numerical methods such as the method of characteristics, Finite Difference Method (FDM), Finite Volume Method (FVM), Finite Element Method (FEM), Boundary Element Method (BEM), meshfree method etc. have been developed by engineers and scientists for the solution of engineering problems. Depending on the problem to be solved, each of these methods has its advantages and disadvantages and the choice depends upon the complexity of the problem and the investigator's familiarity with the method. Out of the many available numerical methods, finite difference (FDM) and finite element methods (FEM) are the most popular numerical modeling techniques among engineers and scientists. Further, the meshfree methods are the latest development in computational modeling. In this chapter, the application of FDM, FEM, and meshfree methods and opportunities and challenges in using these methods are elaborated with case studies.

1.3.2 DEVELOPMENT OF COMPUTATIONAL MODEL FOR WRM

For the illustration of computer models for application in WRM, groundwater flow and contaminant transport problems in two dimensions are chosen, as

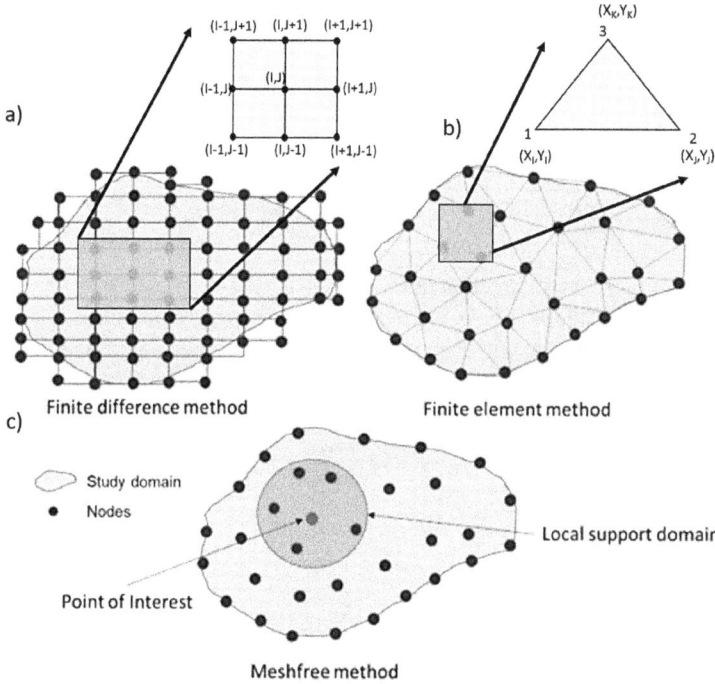

FIGURE 1.1 Discretization of problem domain using FDM, FEM, and meshfree Method.

these problems are more challenging and complex. In groundwater flow and contaminant transport modeling, the governing partial differential equations (PDEs) related to the governing equations and boundary conditions are generally solved using numerical methods such as Finite Difference Method (FDM), Finite Element Method (FEM), Finite Volume Method (FVM), Boundary Element Method (BEM), Analytical Element Method (AEM), meshfree Methods etc. Among these, FDM and FEM are traditionally used for groundwater simulation. In recent times, meshfree methods are also being applied to simulate groundwater flow and contaminant transport (Singh et al., 2016).

Figure 1.1a shows the difference in discretization of the domain in applying these three methods for an irregular domain problem. In the following subsections, these numerical methods for flow and transport processes in groundwater are briefly described.

1.3.3 MATHEMATICAL MODELING OF FLOW AND TRANSPORT IN GROUNDWATER

The governing equation for two-dimensional groundwater flow in a homogeneous isotropic confined aquifer is given by Bear and Cheng (2010).

$$\frac{\partial^2 h}{\partial x^2} + \frac{\partial^2 h}{\partial y^2} = \frac{S}{T}\frac{\partial h}{\partial t} - \frac{Q_w\left(x,y,t\right)}{T} \tag{1.1}$$

Here, h is the groundwater head, S is the storage coefficient, T is the transmissivity coefficient, Q_w is recharge or pumping at the location (x, y). Similarly, the contaminant transport equation in two-dimensional form is given by Bear and Cheng (2010).

$$\frac{\partial C}{\partial t} = D_{xx}\frac{\partial^2 C}{\partial x^2} + D_{yy}\frac{\partial^2 C}{\partial x^2} - V_x\frac{\partial C}{\partial x} - V_y\frac{\partial C}{\partial y} + \frac{q_w C'}{n_e}(x,y,t) \tag{1.2}$$

Here D_{xx} and D_{yy} are dispersion coefficients; V_x and V_y are seepage velocities and n_e is the porosity of the medium. The seepage velocities are obtained from the head distribution in the aquifer. The dispersion coefficients can be computed from dispersivities αL and αT as shown in Eqs. (1.3) and (1.4) (Rastogi, 2007).

$$D_{xx} = \frac{\alpha_L V_x^2 + \alpha_T V_y^2}{V} \tag{1.3a}$$

$$D_{yy} = \frac{\alpha_L V_y^2 + \alpha_T V_x^2}{V} \tag{1.3b}$$

The term C' denotes the concentration of injected solute at the location (x, y) with an injection rate of q_w.

The initial and boundary conditions subjected to the aquifer for flow and transport are as follows (Bear and Cheng, 2010):

$$h(x,y,t=0) = h_i; C(x,y,t=0) = C_i \tag{1.4a}$$

$$h(x,y,t) = h_o; C(x,y,t) = C_o \tag{1.4b}$$

$$T\frac{\partial h}{\partial n} = f_o; \frac{\partial C}{\partial n} = g_0 \tag{1.4c}$$

Here h_i, h_o; C_i; C_o are known head and concentration respectively for Dirichlet boundary; f_o, g_o are known values of fluxes for Neumann boundaries respectively. The term n denotes the direction perpendicular to the boundary.

1.3.4 NUMERICAL MODEL – FINITE DIFFERENCE METHOD

In FDM, the problem domain is represented as grids formed as shown in Figure 1.1a. These nodes may be present either at the centre of the grid or at the intersection of the grid.

The governing equation for flow and transport are approximated through a difference scheme at each node. The difference schemes that can be used are the backward, forward or central difference (Wang and Anderson, 1982). These approximated equations are used to form a system of equations which are solved to compute the concerned state variable head or, contaminant concentration at each node.

Considering groundwater head 'h' as a state variable, the first and second derivatives are computed using the backward, forward, or central difference can be represented as Eqs. (1.5), (1.6) and (1.7) (Wang and Anderson, 1982)

$$\left(\frac{\partial h}{\partial x}\right)_I = \frac{h_I - h_{I-1}}{\Delta x}; \left(\frac{\partial^2 h}{\partial x^2}\right)_I = \frac{h_I - 2h_{I-1} + h_{I-2}}{\Delta x^2} \tag{1.5}$$

$$\left(\frac{\partial h}{\partial x}\right)_I = \frac{h_{I+1} - h_I}{\Delta x}; \left(\frac{\partial^2 h}{\partial x^2}\right)_I = \frac{h_{I+2} - 2h_{I-1} + h_I}{\Delta x^2} \tag{1.6}$$

$$\left(\frac{\partial h}{\partial x}\right)_I = \frac{h_{I+\frac{1}{2}} - h_{I-\frac{1}{2}}}{\Delta x}; \left(\frac{\partial^2 h}{\partial x^2}\right)_I = \frac{h_{I+1} - 2h_I + h_{I-1}}{\Delta x^2} \tag{1.7}$$

Using central difference scheme for spatial discretization and forward difference scheme for temporal discretization, Eq. (1.1) can be represented as follows

$$\frac{h_{I+1,J}^t - 2h_{I,J}^t + h_{I-1,J}^t}{(\Delta x)^2} + \frac{h_{I,J+1}^t - 2h_{I,J}^t + h_{I,J+1}^t}{(\Delta y)^2} = \left(\frac{S}{T}\right)\frac{h_{I,J}^{t+\Delta t} - h_{I,J}^t}{(\Delta t)} - \frac{Q_{wI,J}^t}{T} \tag{1.8}$$

Here, t represents the current time-step and $t + \Delta t$ represents the next time-step. It is to be noted that Eq. (1.8) is written in the explicit form where there is only one unknown term, i.e., $h_{I,J}^{t+\Delta t}$. This formulation is highly unstable and dependent on the grid and time-step size (Rastogi, 2007). The implicit formulation of Eq. (1.1) can be given as follows (Wang and Anderson, 1982)

$$\frac{h_{I+1,J}^{t+\Delta t} - 2h_{I,J}^{t+\Delta t} + h_{I-1,J}^{t+\Delta t}}{\Delta x^2} + \frac{h_{I,J+1}^{t+\Delta t} - 2h_{I,J}^{t+\Delta t} + h_{I,J-1}^{t+\Delta t}}{\Delta y^2} = \frac{S}{T}\frac{h_{I,J}^{t+\Delta t} - h_{I,J}^t}{\Delta t} - \frac{Q_{wI,J}^{t+\Delta t}}{T} \tag{1.9}$$

As it is observed, in implicit formulation there is only one known term i.e., $h_{I,J}^t$. Both formulations can be represented in a matrix form as follows:

$$[K]_{N\times N}\{h\}_{N\times 1}^{t+\Delta t} = [F]_{N\times N}\{h\}_{N\times N}^t + \{Q\}_{N\times 1} \tag{1.10}$$

Here, N represents the total number of nodes. K and F represent the terms associated with heads at $t+\Delta t$ and t time instants respectively. Q represents the vector that contains the source/sink term. Similarly, the governing equation for contaminant transport i.e., Eq. (1.2) can also be discretized.

1.3.5 NUMERICAL MODEL – FINITE ELEMENT METHOD

In FEM, the problem domain is divided into several elements using a mesh. Figure 1.1b shows a triangular element having three nodes. Although the shape of the element is dependent on the user, the triangular element is frequently used due to its flexibility to adapt to irregular geometries to a great extent which is a well-known shortcoming of FDM. Using the element shown in Figure 1.1b, the state variable h is approximated as per Galerkin's method as follows (Desai et al., 2011):

$$\hat{h}(x,y,t) = \sum_{L=1}^{N} h_L(t) N_L(x,y) \tag{1.11}$$

Where h_L is the unknown state variable, N_L is the basis/shape function at node L and N is the total number of nodes in the problem domain.

The concerned state variable is defined over individual elements through the shape function. Using the shape functions, a trial solution is defined for the governing equation at individual elements. Subsequently, a system of equations is formed by setting the weighted residual due to the use of the trial solution to zero. These equations are solved to obtain the state variables at individual elements considering appropriate initial and boundary conditions. Similar to FDM based models, a forward difference is used in time discretization (Desai et al., 2011). Using the weighted residuals and integrating them to zero, the governing equation (1.1) with transmissivities in T_x and T_y in x- and y-directions, can be represented as follows for the domain Ω. For individual element e, the Eq. (1.12) can be written as,

$$\iint_\Omega \left[\frac{\partial}{\partial x}\left(T_x \frac{\partial \hat{h}}{\partial x}\right) + \frac{\partial}{\partial y}\left(T_y \frac{\partial \hat{h}}{\partial y}\right) - Q_w - S\frac{\partial \hat{h}}{\partial t} \right] N_L(x,y)\,dxdy = 0 \tag{1.12}$$

Assembling terms in Eq. (1.13) for an element, it can be represented in a matrix form as follows

$$\sum_e \iint \left(T_x^e \frac{\partial \hat{h}^e}{\partial x} \left\{ \frac{\partial N_L^e}{\partial x} \right\} + T_y^e \times \frac{\partial \hat{h}^e}{\partial y} \left\{ \frac{\partial N_L^e}{\partial y} \right\} \right) dxdy \tag{1.13}$$

$$+ \sum_e \iint \left(S\frac{\partial \hat{h}^e}{\partial t} \right) \{N_L^e\}\,dxdy = \sum_e \iint (Q_w)\{N_L^e\}\,dxdy$$

Here, $\{N_L^e\} = \{N_i\, N_j\, N_k\}$

Where I represent the three nodes in each triangular element namely i, j, and k. G and P matrices are constructed using the conductance and storage terms. F vector represents the recharge term at each element. Aggregating Eq. (1.14) for all elements using the global coordinate system, Eq. (1.14) can be represented as follows,

$$\left[G^{e}\right]\left\{h_{I}^{e}\right\}+\left[P^{e}\right]\left\{\frac{\partial h_{I}^{e}}{\partial t}\right\}=\left\{F^{e}\right\} \tag{1.14}$$

Applying the forward difference scheme for the temporal discretization, Eq. (1.15) can be modified as follows,

$$\left[G\right]\left\{h_{I}\right\}+\left[P\right]\left\{\frac{\partial h_{I}}{\partial t}\right\}=\left\{F\right\} \tag{1.15}$$

The terms in Eq. (1.16) can be rearranged as shown in Eq. (1.10) for solving the head values for the next time-step.

$$\left[G\right]\left\{h_{I}\right\}^{t+\Delta t}+\left[P\right]\left\{\frac{h^{t+\Delta t}-h^{t}}{\Delta t}\right\}=\left\{F\right\} \tag{1.16}$$

A detailed formulation of FEM for groundwater flow and transport problems are given in Wang and Anderson (1982), Rastogi (2007) and (Desai et al., 2011).

1.3.6 NUMERICAL MODEL – MESHFREE RADIAL POINT COLLOCATION METHOD

Meshfree models and are recently being applied to many groundwater-related problems (Mategaonkar and Eldho, 2011; Singh et al., 2016; Anshuman and Eldho 2020). Due to the presence of mesh/grid, the mesh-based methods have disadvantages such as difficulty in representing irregular domain, instabilities in advection and reaction dominant problems such as oscillation and numerical dispersion etc. (Liu and Gu, 2005). In contrast to mesh/grid-based methods, meshfree methods use a set of scattered nodes within the domain without any connectivity to adjacent nodes. Instead, these methods use a local support domain around each node (see Figure 1.1c).

The local support domain can be changed easily depending on the type of problem. For example, for advection dominant problems, the support domain size is increased to stabilize the solution, which is difficult in FDM and FEM based solutions (Singh et al., 2016). Here, a strong form meshfree method named Radial Point Collocation Method (RPCM) is discussed. This method uses Radial Basis Functions (RBFs) for approximation of shape function and its directional derivatives. Few examples of RBFs are given in Table 1.1 (Liu and Gu, 2005).

In Table 1.1, the terms r and d_c represent the Euclidean distance of a node from point of interest i and average nodal distance respectively. In RPCM, the problem

TABLE 1.1
Examples of Radial Basis Functions

Name	Expression	Shape Parameter
Multiquadrics	$(r_i + (\alpha_c d_c)^2)^q$	$\alpha \geq 0, q$
Exponential	$\exp \exp\left[-\alpha_c \left(\dfrac{r_i^2}{d_c}\right)^2\right]$	α
Thin plate spline	r_i^η	η
Logarithmic	$r_i^\eta \log \log r_i$	η

domain is first discretized using nodes. The shape function and its derivatives i.e., ϕ, $\dfrac{\partial \phi}{\partial x}$, $\dfrac{\partial \phi}{\partial y}$, $\dfrac{\partial^2 \phi}{\partial x^2}$, and $\dfrac{\partial^2 \phi}{\partial y^2}$ are computed for each node within its local support domain. These are used to discretize the governing equations in the domain. For example, using the shape function ϕ, the state variable h is presented as follows:

$$h = \sum_i^n \phi_i h_i \tag{1.17}$$

Here n is the number of nodes in the local support domain nodes. In the RPCM, the governing equation (1.1) can be expressed as

$$\theta \frac{T}{S}\left[\frac{\partial^2 \phi}{\partial x^2} + \frac{\partial^2 \phi}{\partial y^2}\right]h^{t+\Delta t} + (1-\theta)\frac{T}{S}\left[\frac{\partial^2 \phi}{\partial x^2} + \frac{\partial^2 \phi}{\partial y^2}\right]h^t = \frac{h^{t+\Delta t} - h^t}{\Delta t} - Qw \tag{1.18}$$

Note that the formulation explained here is known as semi-implicit formulation. Here, θ is a weight coefficient that varies in the range [0, 1]. The numerical scheme is unconditionally stable if the value of θ is higher than 0.5. The terms in Eq. (1.18) can be rearranged and solved as shown in Eq. (1.10) for solving the head values for subsequent time-step. A detailed formulation of RPCM for groundwater flow and transport problems are given in Singh et al. (2016).

1.4 COMPUTATIONAL MODELS FOR WRM

In WRM, computational models are used in simulation of floods, streamflow from catchments, groundwater flow, contaminant transport etc. Depending upon the problem to be solved, we can choose a typical model available in market or, get from open sources.

Each of these computational models has its own advantages and disadvantages and the choice depends upon the complexity of the problem, investigator's familiarity with the method, computational facility, and manpower available. Out of the many available computational models, some of the important computational models for surface water and groundwater management are discussed here.

1.4.1 Models for Surface Water Management

For simulating floods, 1D and 2D hydrodynamic models such as ANUGA MIKE-FLOOD (DHI, 2013), LISFLOOD are used. The 1D hydrodynamic models give the variation of water level in the river, channels etc. while the 2D hydrodynamic models simulate the water depth over the study area for given inflow of water through precipitation. These models can also be coupled to simulate the movement of flood water in a given area (Ghosh et al., 2021). Hydrological models like SWAT (Arnold et al., 2012), VIC (Hengade et al., 2018), SHETRAN (Sreedevi et al., 2019) etc. are used to simulate streamflow at catchment or, basin outlets. Using these models, other hydrological components can be obtained under different scenarios of climate and land use (Madhusoodhanan et al., 2017; Hengade et al., 2018; Sinha and Eldho, 2018). Lumped conceptual models have also been useful in simulating runoff processes with minimal data requirement and provide significantly similar outputs as complex hydrological models (Anshuman et al., 2019). Some of the popular computational models used in surface water management and its applications are briefly described in Table 1.2.

1.4.2 Models for Groundwater Management

For modeling flow and transport processes in groundwater, models such as MODFLOW (Harbaugh, 2005), MT3DMS (Zheng and Wang, 1999), RT3D (Clement, 1999) are used. Apart from that recently meshfree models such as RPCM (Mategaonkar and Eldho 2011; Singh et al., 2016; Anshuman and Eldho, 2020), EFGM (Pathania et al., 2020), MLPG (Swathi and Eldho, 2014) are being applied to groundwater flow and contaminant transport problems. Some of the popular computational models used in groundwater management and its applications are briefly described in Table 1.3.

1.5 CASE STUDIES

To demonstrate the applications of computational models for WRM, two case studies related to groundwater flow and transport are presented here. The FDM, FEM, and RPCM models are used for the case studies and results are compared to illustrate the advantages and their limitations.

1.5.1 Case Study 1

Here a homogeneous confined aquifer of dimensions 600 m × 400 m as shown in Figure 1.2a is selected for demonstration of groundwater flow using FDM, FEM, and RPCM. The depth of the aquifer is 10 m.

The hydraulic conductivity and porosity of the aquifer material are 10 m/d and 0.3 respectively. The east and west boundaries are subjected to Dirichlet boundary conditions with head values of 8 m and 10 m respectively. The north and south boundaries are no-flow boundaries. Two pumping wells located in the aquifer at (175 m, 105 m) and (405 m, 275 m) respectively. The extraction rates from Well 1 and Well 2 are 200 m³/d

TABLE 1.2

Popular Surface Water Management Models

Model Name	Model Description	Model Applications	Remarks
HEC-HMS-RAS	Hydrologic Engineering Centre's Hydrologic Modeling Systems -River Analysis System	Used to model watershed hydrology, flooding, dam break, open-channel flow systems, model hydraulic structures etc.	Distributed 1D/2D models, conceptual models for rainfall-runoff, FDM to do hydraulic routing
MIKE SHE	System Hydrologique Europen	Runoff modeling, soil moisture, irrigation command management	Distributed 2D/3D models using FDM
SWAT	Soil and Water Assessment Tool	Water balance, Sediment, streamflow and non-point source pollution	Semi-distributed conceptual model; Hydrologic model using SCS-CN
HSM	Hydrological Simulation Model	rainfall-runoff model) which uses rainfall and potential evaporation data to simulate the hydrological cycle (surface runoff, percolation)	Physically based, distributed watershed model
TOPMODEL	Topographic Model	simulates hydrologic fluxes of water (infiltration-excess overland flow, saturation overland flow, infiltration, exfiltration, subsurface flow, evapotranspiration, and channel routing) through a watershed	Physically based, distributed watershed model
AGNPS	Agricultural Non-Point Source Pollution Model	evaluating the effect of management decisions impacting water, sediment, and chemical loadings within a watershed system	Lumped semi-distributed model
WEPP	Water Erosion Prediction Project	Integrates hydrology, plant science, hydraulics, and erosion mechanics to predict erosion at the hillslope and watershed scale	distributed parameter, continuous simulation model
HSPF	Hydrological Simulation Prediction-Fortran	Simulate runoff, water balance, and non-point source pollution	Distributed 1D model
WMS	Watershed Modeling Systems	Watershed delineation, modeling, rainfall-runoff, flooding etc.	Include a suite of lumped/distributed models
VIC	Variable Infiltration Capacity model	Macroscale Hydrology Model, water balance, rainfall-runoff	fully distributed, physically based regional-scale hydrology model

TABLE 1.3

Popular Groundwater Management Models

Model Name	Model Description	Model Applications	Remarks
MODFLOW	Modular finite-difference flow model	for simulating and predicting groundwater conditions and groundwater/surface-water interactions	FDM based 2D/3D model
MT3DMS	Modular three-dimensional transport model	for the simulation of advection, dispersion, and chemical reactions of dissolved constituents in groundwater systems	FDM based 2D/3D model
MOC	Method of Characteristics model	MOC Two-dimensional method-of-characteristics groundwater flow and transport model	2D transport model
FEMWATER/ FEFLOW	FEM based flow/transport model	Simulate flow and transport in porous media	2D/3D model
GMS	Groundwater Modeling Systems	A suite of FEM/FDM based models for flow and transport	GIS linked model with GUI; 2D/3D modeling
RT3D	Reactive multispecies Transport in 3-Dimensional groundwater systems	Used for multi-specious reactive transport modeling	Modeling in 2D/3D – FDM based
BIOPLUME		For modeling biodegradation and natural attenuation of contaminants	Use method of characteristics/ FDM; 2D model

and 300 m³/d respectively. The flow in the aquifer is in a steady state i.e., the variation of the head concerning time is negligible. The FDM model is developed in MODFLOW (Harbaugh, 2005) using 2,400 nodes. (see Figure 1.2b).

The COMSOL software is used to develop the FEM based model using 2,109 nodes. It uses refinement near the well location to produce accurate solutions (see Figure 1.2c). In RPCM, the problem domain is represented using 2,501 nodes having $d_c = 10$ m along x- and y- directions. In this study, Multiquadrics RBF (MQ-RBF) is used with $q = 0.98$ and $\alpha = 5$. The results obtained from all models are compared in Figure 1.3 and Table 1.4 respectively.

Visible drawdowns are observed near the pumping wells based on the extraction rates. It is observed that all three models give close results at the observation points described in Table 1.4 with a maximum 2.18% of percentage difference.

Due to the presence of drawdown, the head gradient varies rapidly near the wells which can attract pollutants if any contaminant plume exists near the pumping wells. Modeling such cases can help us taking effective measures to manage water quality and quality in the concerned study area for effective WRM.

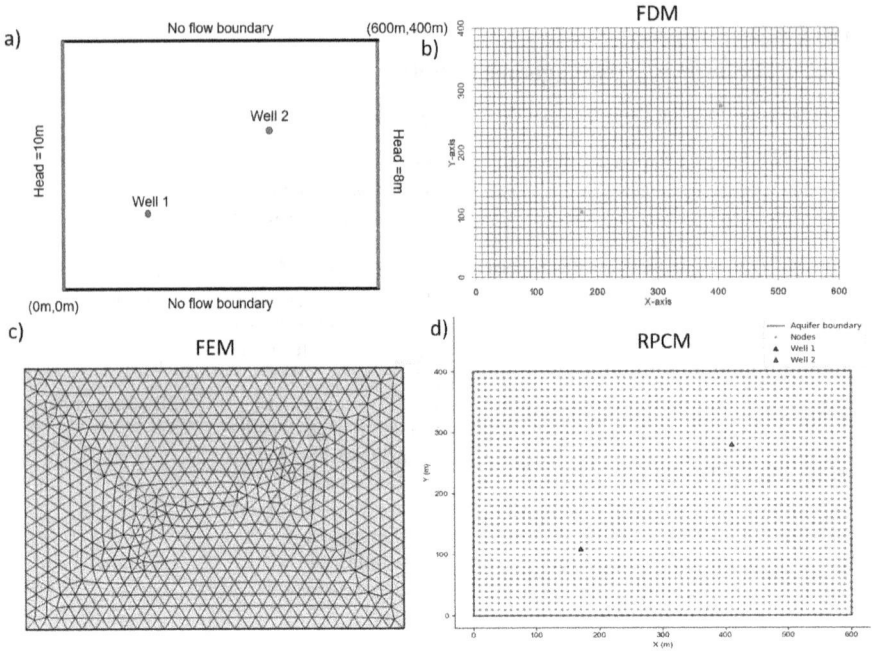

FIGURE 1.2 Problem domain and its discretization of problem domain using FDM, FEM, and RPCM for case study 1.

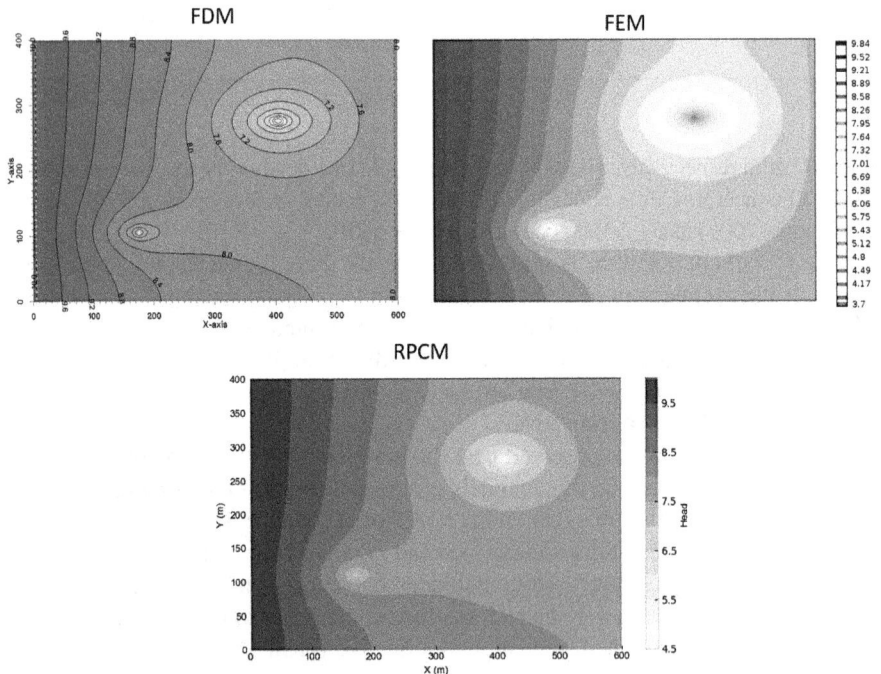

FIGURE 1.3 Head distribution using FDM, FEM, and RPCM for case study 1.

TABLE 1.4

Comparison of Head Values Obtained by Different Models for Case Study 1

Observation Locations	Coordinates	RPCM	FDM	Percentage Difference RPCM & FDM	FEM	Percentage Difference RPCM & FEM
1	(280m, 200m)	7.91	7.90	0.13	7.79	1.54
2	(90m, 300m)	9.29	9.30	0.11	9.18	1.20
3	(200m, 210m)	8.37	8.37	0.00	8.26	1.33
4	(470m, 260m)	6.98	7.00	0.29	7.09	1.55
5	(540m, 100m)	7.96	7.97	0.13	7.79	2.18

1.5.2 CASE STUDY 2

In this case study, a heterogeneous confined aquifer having 3 zones is considered with a depth of 10 m (see Figure 1.4a).

The hydraulic conductivity of these zones are 20 m/d, 5 m/d and 50 m/d. The porosity of the aquifer material is 0.3. The east and west boundaries are Dirichlet boundaries with head values of 10 m and 8 m respectively. The north and south boundaries are no-flow boundaries. A contaminant source was injecting total dissolved solute (TDS) at a part of the west boundary as shown in Figure 1.4a for 10 years. The

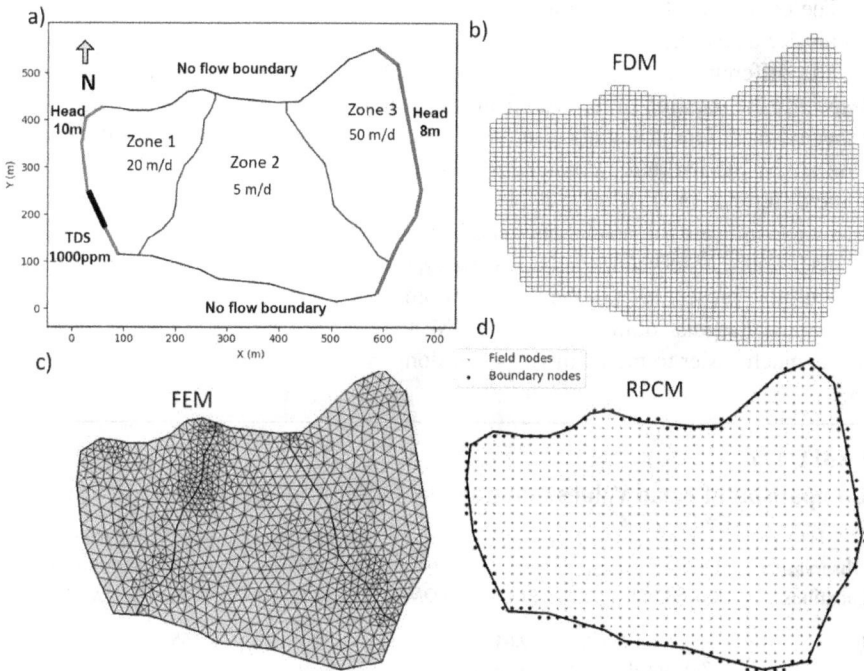

FIGURE 1.4 Problem domain and its discretization of problem domain using FDM, FEM, and RPCM for case study 2.

concentration of a source location is 1000 ppm. The contaminant plume is directed in the west to east direction and measured at four observation wells located at (213.8 m, 1085.2 m), (348.8 m, 1085.2 m), (483.8, 1085.2 m) and (543.8 m, 1085.2 m).

The plume movement can be simulated using a coupled flow and transport model which computes the seepage velocities using head variation and utilizes the same in contaminant transport simulation.

The FDM model for flow and transport are developed using MODFLOW and MT3DMS respectively (Zheng and Wang, 1999). The nodal spacing in MODFLOW-MT3DMS is 15 m (see Figure 1.4).

In grid/mesh-based methods advection dominant problems cause instabilities in the model in the form of artificial oscillations (Zheng and Wang, 1999). To counter that, the MT3DMS model uses operator splitting for simulating advection and dispersion. The Total Variation Diminishing (TVD) scheme is used for simulating advection. The FEM is developed in COMSOL software using average mesh size of 15 m (see Figure 1.4c). The RPCM model is developed with 1,137 nodes with a nodal distance of 15 m in x- and y-directions. The head and concentration values are compared at four observation wells. From Table 1.5, it is observed that the head values obtained by all three models give similar head values (see Figure 1.5).

The maximum value of the percentage difference is 1%. Similarly, the contaminant transport process is simulated. The contaminant transport process is simulated for 10 years are presented in Figure 1.6. It is observed that all three models provide a similar spread of contaminants in the aquifer.

The contaminant concentrations are also compared at four observation wells in Figure 1.7 as mentioned.

The difference in magnitude of concentration is observed at these wells with respect to the location of the well from the source (see Table 1.6). Such difference is expected in different numerical models where heterogeneity is high. In the current case study, the hydraulic conductivity varies sharply from one zone due to which some difference in simulation is observed. However, high R^2 values at the observation wells indicate the results obtained by the models are similar (see Table 1.6).

As it can be observed from the case study, the grid-based FDM model has difficulty inaccurate representation of the irregular domain. While mesh-based FEM, can easily handle the irregular domain. As in RPCM, there is no need for a grid or mesh and hence much easier to model the irregular domains. All three models give satisfactory

---

TABLE 1.5

Comparison of Head Values Obtained by Different Models for Case Study 2

Observation Locations	Coordinates	RPCM	FDM	Percentage Difference RPCM & FDM	FEM	Percentage Difference RPCM & FEM
1	213.8m, 1085.2m	9.64	9.69	0.512	9.65	0.10
2	348.8m,1085.2m	8.88	8.97	1.00	8.85	0.34
3	483.8,1085.2m	8.15	8.12	0.37	8.14	0.12
4	543.8m, 1085.2m	8.05	8.06	0.12	8.05	0.00

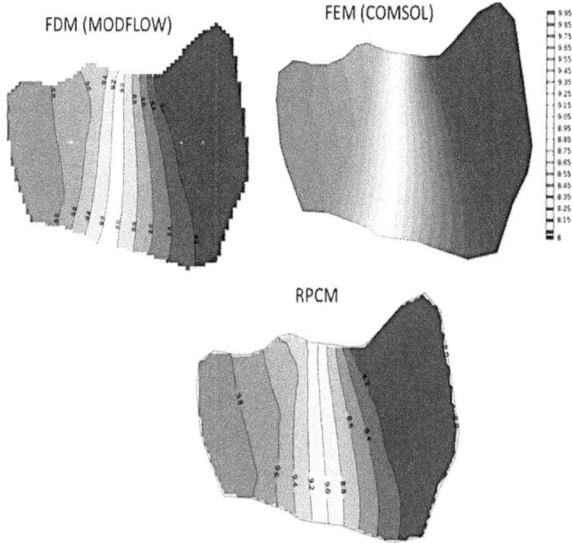

FIGURE 1.5   Head distribution using FDM, FEM, and RPCM for case study 2.

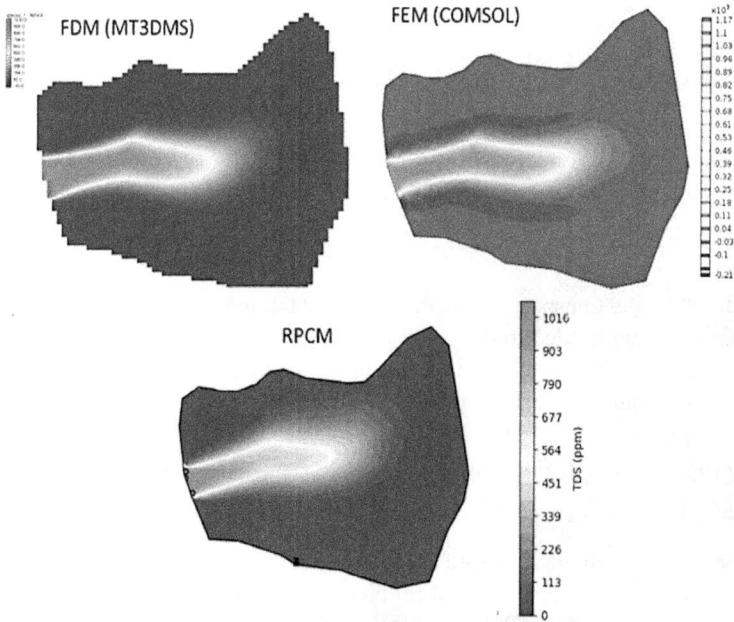

FIGURE 1.6   Concentration contours for TDS after 10 years.

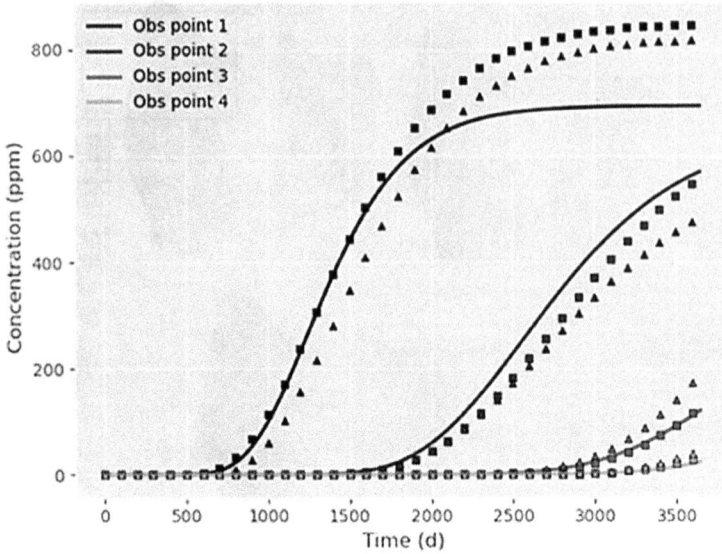

FIGURE 1.7 Breakthrough curves at four observation wells as obtained by FDM (square), FEM (circle) and RPCM (solid line).

TABLE 1.6
Comparison of Breakthrough Curves Obtained by Different Models for Case Study 2

Observation Well	MAE (ppm)		R^2	
	RPCM-FDM	**RPCM-FEM**	**RPCM-FDM**	**RPCM-FEM**
1	62.4	58.8	0.995327	0.984178
2	18.0	31.4	0.997361	0.999167
3	0.6	6.9	0.999077	0.999763
4	0.7	1.9	0.999879	0.99976

results for flow and transport. While modeling contaminant transport using FDM, special techniques are needed to deal with numerical instability. In FEM also, there are issues to deal with highly advective transport problems. The RPCM model could satisfactorily simulate the transport problems more effectively than FDM and FEM.

1.6 OPPORTUNITIES AND CHALLENGES IN COMPUTATIONAL MODELING FOR WRM

With the development of advanced computational models, numerical techniques and fast computers, any complex WRM problem can be solved by computational models. Further, with efficient GUI (graphical user interface) and DSS (decision support systems), any WRM problems can easily be attempted by computational models. The availability of several efficient computational models has opened huge opportunities for engineers and scientists for effective WRM.

The hydrological processes are very complex and mathematical modeling can be done through reasonable assumptions and simplifications. The degree of simplification depends on the accuracy required in simulating the dominant process. Depending on the model selected few parameters are used for calibration which can yield different patterns of prediction of the concerned variable. Data insufficiency is a major challenge while developing a computational model. For example, groundwater modeling requires spatiotemporal information of groundwater levels, and recharge and extraction rates (Zhou and Li, 2011). Data collection can be infeasible in underdeveloped and developing countries which can hinder proper modeling of the hydrological/aquifer systems.

Apart from data issues, computational modeling requires a thorough understanding of the governing equations and model parameters. This can help the modeller to adjust the model parameters within the acceptable range pertaining to the study area. Further, the applicability of the model results for WRM depends on the socioeconomic factors of the study area. A good understanding and communication between the decision-makers and stakeholders are required for the application of policies based on computational modeling results for WRM. Climate change adds another dimension to the challenges involved in computational modeling. The uncertainty in hydrological processes due to climate change is required to be considered in computational modeling in order to provide a possible range of future predictions which can help in making robust decisions in WRM.

Computational models for WRM are rapidly evolving due to advances in numerical techniques and the advent of high-speed modern computers. Though several numerical techniques are developed, still FDM and FEM are most popular. In the last two decades, due to various shortcomings of mesh/grid-based methods, meshfree methods are being developed and applied to various problems related to hydrological modeling. However, due to the unavailability of software, the applicability of these meshfree methods is less accessible compared to mesh/grid-based methods. Further, as per state-of-art research, these methods, especially strong form meshfree methods are complicated to apply in areas subject to stark heterogeneity (Hidayat, 2019). Considering the advantages of the meshfree methods such as simplicity, flexibility and higher stability over grid/mesh-based methods, these methods can be improved for heterogeneous problem domains.

1.7 CONCLUDING REMARKS

For effective WRM, informed decision-making based on predicted water availability under different scenarios is essential. For assessment of water availability, physically based models are essential tools. As computational models are based on physically based mathematical equations, they can assist in predicting the hydrological components under different scenarios. Using this information, the decision-makers can lay out plans for WRM after discussing with various stakeholders.

Most of the computational models are based on a mathematical representation of the study area. The governing equations associated are solved using different numerical techniques such as FDM, FEM, FVM, and recently proposed meshfree methods. In this chapter, we discussed modeling using FDM, FEM, and meshfree RPCM with application to flow and transport simulation in groundwater. The results obtained from these models are similar. Considering the advantages in the modeling process, meshfree methods are used in different groundwater-related problems.

With the advancements in computational models, numerical techniques, fast computers, GUI, artificial intelligence tools, machine learning, and efficient DSS, now any complex WRM problem can be solved using computational models. The recent developments in computational models have opened huge opportunities to Scientists and Engineers for efficient WRM, though still many challenges related to data and better management models are there.

REFERENCES

Anshuman, A., Kunnath-Poovakka, A. and Eldho, T.I., 2019. Towards the use of conceptual models for water resource assessment in Indian tropical watersheds under monsoon-driven climatic conditions. *Environmental Earth Sciences*, 78(9), pp. 1–15.

Anshuman, A. and Eldho, T.I., 2020. Meshfree radial point collocation-based coupled flow and transport model for simulation of multispecies linked first order reactions. *Journal of Contaminant Hydrology*, 229, p. 103582.

Arnold, J.G., Moriasi, D.N., Gassman, P.W., Abbaspour, K.C., White, M.J., Srinivasan, R., Santhi, C., Harmel, R.D., Van Griensven, A., Van Liew, M.W. and Kannan, N., 2012. SWAT: Model use, calibration, and validation. *Transactions of the ASABE*, 55(4), pp. 1491–1508.

Bear, J. and Cheng, A.H.D., 2010. *Modeling groundwater flow and contaminant transport* (Vol. 23). Springer, New York.

Beven, K.J., 2001. *Rainfall-runoff modeling. The primer*. Wiley, New York, p. 360.

Clement, T.P., 1999. *A modular computer code for simulating reactive multispecies transport in 3-dimensional groundwater systems (No. PNNL-11720; EW4010)*. Pacific Northwest National Lab., Richland, WA.

Desai, Y.M., Eldho T.I. and Shah, A.H., 2011. *Finite element method with applications in engineering*. Pearson Education, New Delhi, India.

DHI, 2013, *2013 MIKE FLOOD-1D-2D modeling-user manual*. DHI Group.

Ewen, J., Parkin, G., and O'Connel, P.E., 2000. SHETRAN: Distributed river basin flow and transport modeling system. *Journal of Hydrologic Engineering*, 5(3), pp. 250–258.

Frederick, K.D. and Major, D.C., 1997. Climate change and water resources. *Climatic Change*, 37(1), pp. 7–23.

Ghosh, M., Mohanty, M.P., Kishore, P. and Karmakar, S., 2021. Performance evaluation of potential inland flood management options through a three-way linked hydrodynamic modeling framework for a coastal urban watershed. *Hydrology Research*, 52(1), pp. 61–77.

Harbaugh, A.W., 2005. *MODFLOW-2005, the US Geological Survey modular ground-water model: The groundwater flow process*. US Department of the Interior, US Geological Survey, Reston, VA, pp. 6–A16.

Hidayat, M.I.P., 2019. Meshless local B-spline collocation method for heterogeneous heat conduction problems. *Engineering Analysis with Boundary Elements*, 101, pp. 76–88.

Hengade, N., Eldho, T.I. and Ghosh, S., 2018. Climate change impact assessment of a river basin using CMIP5 climate models and the VIC hydrological model. *Hydrological Sciences Journal*, 63(4), pp. 596–614.

Kunnath-Poovakka, A. and Eldho, T.I., 2019. A comparative study of conceptual rainfall-run-off models GR4J, AWBM and Sacramento at catchments in the upper Godavari River basin, India. *Journal of Earth System Science*, 128(2), p. 33.

Liu, G.R. and Gu, Y.T., 2005. *An introduction to meshfree methods and their programming*. Springer Science & Business Media, London.

Madhusoodhanan, C.G., Sreeja, K.G. and Eldho, T.I., 2017. Assessment of uncertainties in global land cover products for hydroclimate modeling in India. *Water Resources Research*, 53(2), pp. 1713–1734.

Mategaonkar, M. and Eldho, T.I., 2011. Simulation of groundwater flow in unconfined aquifer using meshfree point collocation method. *Engineering Analysis with Boundary Elements*, 35(4), 700–707.

Pathania, T., Eldho, T.I. and Bottacin-Busolin, A., 2020. Optimal design of in-situ bioremediation system using the meshless element-free Galerkin method and particle swarm optimization. *Advances in Water Resources*, 144, p. 103707.

Rastogi, A.K., 2007. *Numerical groundwater hydrology*, Penram Int. Publishing, New Delhi, India.

Refsgaard, J.C., van der Sluijs, J.P., Højberg, A.L. and Vanrolleghem, P.A., 2007. Uncertainty in the environmental modeling process—A framework and guidance. *Environmental Modeling & Software*, 22(11), 1543–1556.

Singh, L.G., Eldho, T.I. and Kumar, A.V., 2016. Coupled groundwater flow and contaminant transport simulation in a confined aquifer using meshfree radial point collocation method (RPCM). *Engineering Analysis with Boundary Elements*, 66, pp.20–33.

Sinha, R.K. and Eldho, T.I., 2018. Effects of historical and projected land use/cover change on runoff and sediment yield in the Netravati river basin, Western Ghats, India, *Environmental Earth Sciences*, 77(3), pp. 1–19.

Sreedevi, S., Eldho, T.I., Madhusoodhanan, C.G. and Jayasankar, T., 2019. Multi-objective sensitivity analysis and model parameterization approach for coupled streamflow and groundwater table depth simulations using SHETRAN in a wet humid tropical catchment. *Journal of Hydrology*, 579, p. 124217.

Swathi, B. and Eldho, T.I., 2014. Groundwater flow simulation in unconfined aquifers using meshless local Petrov–Galerkin method. *Engineering Analysis with Boundary Elements*, 48, pp. 43–52.

Wang, H. and Anderson, M.P., 1982. *Introduction to groundwater modeling finite difference and finite element methods*. W. H. Freeman and Company, New York.

Zheng, C. and Wang, P.P., 1999. *MT3DMS: A modular three-dimensional multispecies transport model for simulation of advection, dispersion, and chemical reactions of contaminants in groundwater systems; documentation and user's guide*. Contract Report SERDP-99-1, US Army Corps of Engineers-Engineer Research and Development Center, p. 220.

Zhou, Y. and Li, W., 2011. A review of regional groundwater flow modeling. *Geoscience Frontiers*, 2(2), 205–214.

2 Applicability of Soft Computational Models for Integrated Water Resource Management

Chithra Nelson Rosamma

CONTENTS

2.1 INTRODUCTION

Stress on water resources is increasing day by day due to natural (climate change) and anthropogenic causes (demographic, industrial uses, land use, and land cover change). The approach of Integrated Water Resources Management (IWRM) is the widely accepted way forward to manage water resources in a sustainable manner, which promotes co-ordinated development and management of water, land, and related resources. This helps to tackle water related challenges such as water scarcity, presence of extreme hydrologic events like flood and drought, water quality deterioration, and soil erosion. A wide variety of modeling approaches were adopted and applied successfully in different parts of the world to tackle these challenges.

DOI: 10.1201/9781003203445-3

Enormous amount of data is required to represent different hydrologic processes meaningfully and their availability is rare in many river basins. Moreover, the quality of data received is poor in many cases.

The hydrologic data required for the IWRM is characterized by complexity, dynamism and non-stationarity and this challenge is being handled in the recent years by developing new scientific approaches for hydrologic modeling and forecasting. Among them soft computing methods based on input–output data has received wide recognition due to their simplicity. Such data-driven approaches involve mathematical equations drawn from an analysis of concurrent input and output time series instead of the physical process in the watershed (Solomatine and Ostfeld, 2008). Artificial Intelligence models have shown promise in modeling non-linear hydrological processes and in handling dynamicity and noise concealed in the data set (Nourani et al., 2014). The limited data availability and distributed nature of the watershed has led to the increased applications of soft computing methods in water resources sector due to their high flexibility to handle complex nonlinear relationship between independent and dependent variables.

Though there were many success stories on application of soft computing methods for IWRM, there is no clear conclusion about the superiority of one model compared to others. Moreover, many shortcomings were visible in many methods and uncertainty associated with them were also more. This has led to the progress of many comparative studies and development of hybrid modeling approaches where the limitation of one model is handled with the help of advantages of another model. Last two decades witnessed enormous amount of research in this area.

In this chapter, an overview on the commonly used soft computing models and their applications for water resources management is presented. Also comparison of different techniques based on review of recent literature is presented. Finally, conclusions and future research direction is presented.

2.2 OVERVIEW OF SOFT COMPUTING METHODS

Fuzzy logic (FL), Artificial Neural Networks (ANN), Support Vector Machine (SVM), Genetic Algorithm (GA) and other deep learning methods are widely applied for solving problems in the field of integrated water resources management. A brief overview of these methods is presented in this section.

2.2.1 ARTIFICIAL NEURAL NETWORK (ANN)

ANN is information processing exemplar, which is constituted by neurons and synapses, having resemblance to the functioning of human brain. The major components of a neuron are Soma (cell body), axons (sends signals), and dendrites (receives signals) and synapse connects an axon to a dendrite. Upon giving a signal, a synapse either excites or inhibits electrical potential and once this electrical potential reaches a threshold, a neuron gets activated, and learning occurs. Similar to this, ANN is constituted by a number of layers of interconnected neurons known as nodes which are the information-processing units in ANN receiving several signals from its input links, each of which has a weight assigned to it, and applies an activation function to transform it into an output signal. The input layer receives the input which will be converted nonlinearly to output by the output layer and all other intermediate layers are called hidden layers. The development of ANN is based on the following rules (Govindaraju 2000):

(1) Information processing occurs at many single elements called nodes.
(2) Signals are passed between nodes through connection links.
(3) Each connection link is associated with a weight representing its strength.
(4) Each node applies a nonlinear transformation called an activation function to its input signal to determine its output signal.

Figure 2.1 presents the block diagram of a neuron. A set of synapses or connecting links is an essential element of an artificial neuron, and this is characterized by a weight. A signal I_j at the input of synapse j connected to neuron k is multiplied by the synaptic weight w_{kj}. The input signals, weighted by the synapses of the neuron are summed up using an adder. Each unit of an ANN receives inputs from the other connected units or from an external source. A weighted sum of the inputs is computed at a given instant of time. For a given external input, the activation dynamics is followed to recall a pattern stored in a network. In order to store a pattern, it is necessary to adjust the weights of the connections in the network and the process of adjusting the weights is referred to as learning. The final set of weight values after the learning process is completed corresponds to the long-term memory function of the network. Learning laws are implementation models of synaptic dynamics. Learning laws describe the weight vector for the ith processing unit at time instant $t+1$, $W_i(t+1)$ in terms of the weight vector at time instant t, $W_i(t)$ as follows:

$$W_i(t+1) = W_i(t) + \Delta W_i(t) \qquad (2.1)$$

where $\Delta W_i(t)$ is the change in the weight vector. There are several learning laws which use only local information for adjusting the weight of the connection between two units. Some commonly used activation functions are linear function ($f(x) = x$), Sigmoid function $\left(f(x) = \dfrac{1}{1+e^{-x}} \right)$, and hyperbolic tangent function $\left(f(x) = \dfrac{e^x - e^{-x}}{e^x + e^{-x}} \right)$; Sigmoid is the most common among these three (Figure 2.2).

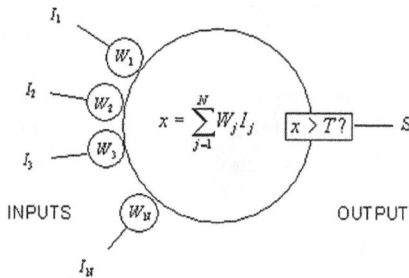

FIGURE 2.1 Block diagram of a neuron.

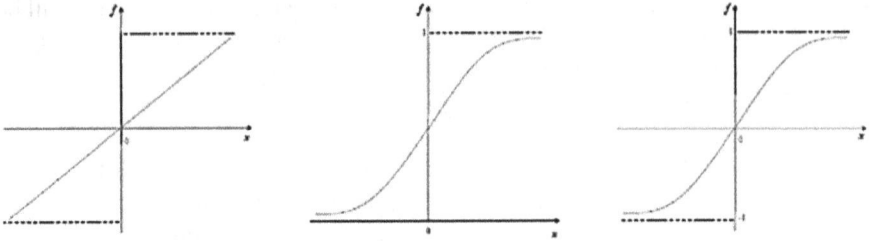

FIGURE 2.2 Activation functions (linear function, sigmoid function and hyperbolic tangent function).

A large number of ANN architectures and algorithms are found in the literature; these include multilayer feed-forward networks, self-organizing feature maps (Kohonen, 1982), Hopfield networks (Hopfield, 1984), counter-propagation networks (Hecht-Nielsen, 1987), radial basis function (RBF) networks, general regression neural network (Specht, 1991), and recurrent ANNs (Elman, 1990). Of these, the most commonly used networks are feed-forward networks and RBF networks (Karunanithi et al., 1994). In a feed-forward ANN, there are layers of processing units, and each layer feeds input to the next layer in a feed-forward mode through a set of connection weights. In back propagation learning, the error from the output layer is propagated backwards to the hidden layers in order to update weights in the hidden layer. The error at the output layer is computed as the difference between the required output and the actual output obtained at each of the output units.

2.2.2 FUZZY LOGIC

Fuzzy logic is a multivalued logic which is derived from fuzzy set theory to deal with approximate reasoning (Huang et al., 2010). As the hydrologic problems often deal with imprecision, vagueness and uncertainty, fuzzy logic-based models may be very well used to handle them (Kambalimath and Deka, 2020). For a comprehensive review of fuzzy logic applications in hydrology and water resources, readers may refer to Kambalimath and Deka (2020).

Important steps involved in fuzzy logic technique are fuzzification, fuzzy inference process, and defuzzification. In the first step fuzzification, crisp data (input and output) is converted into fuzzy set data or membership functions. In the case of binary logic, only 0 and 1 are allowed, whereas, in fuzzy logic any value between 0 and 1 is possible, based on the degree of truth (Figure 2.3). In fuzzy inference process, the membership functions are combined with fuzzy inference tools to obtain fuzzy outputs. In the last step defuzzification, fuzzy output is converted into crisp output based on the associated rules.

2.2.3 GENETIC ALGORITHMS

Genetic algorithm is a supervised machine learning technique. GA is derived based on natural genetics to perform search and optimization procedures. The techniques

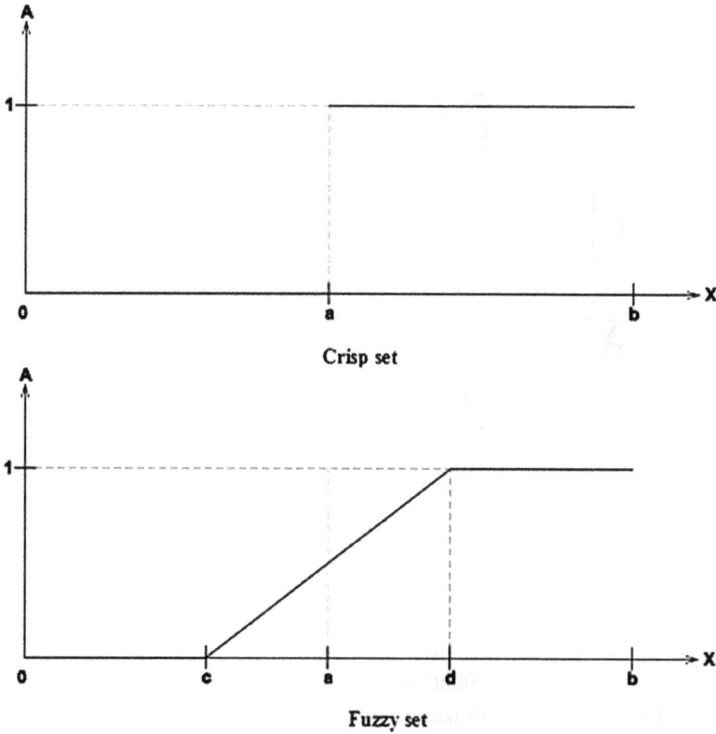

FIGURE 2.3 Characteristics functions of crisp and fuzzy sets.

used in GA are inspired by evolutionary biology such as inheritance, mutation, selection, crossover, and survival of the fittest (Rani et al., 2013). The uniqueness of this evolutionary computation technique compared to other machine learning techniques is its capability to produce explicit mathematical relationships between input and output variables (Herath et al., 2021). GP is considered as a grey box model, whereas SVM and ANN are considered to be black box models. Some of the advantages of GA are their ability to determine global optimum and deal with discontinuous functions (Mohan and Vijayalakshmi, 2009).

The algorithm of GA is based on the concept of survival of the fittest. Initially a population of chromosomes are initialized and fitness of each chromosome is computed. Cross over operator is applied on selected chromosomes based on fitness value to produce an offspring. Mutation operator is applied on offspring to produce new offspring, which is placed in the population. The important operations in GA (selection, crossover, and mutation) are repeated until a new population is completed (Katoch et al., 2021).

2.2.4 SUPPORT VECTOR MACHINE (SVM)

SVMs are very efficient algorithms, based on statistical learning theory, used for classification and regression. Though this method is simple, its predictive accuracy

```
┌─────────────────────────────┐
│          Input data          │
└─────────────────────────────┘
               │
               ▼
┌─────────────────────────────┐
│  Selection of kernel function │◄──┐
└─────────────────────────────┘   │
               │                   │
               ▼                   │
┌─────────────────────────────┐   │
│         SVM Training          │   │
└─────────────────────────────┘   │
               │                   │
               ▼                   │
┌─────────────────────────────┐   │
│       Cross validation        │───┘
└─────────────────────────────┘
               │
               ▼
┌─────────────────────────────┐
│    Selection of best model    │
└─────────────────────────────┘
               │
               ▼
┌─────────────────────────────┐
│         SVM Testing           │
└─────────────────────────────┘
```

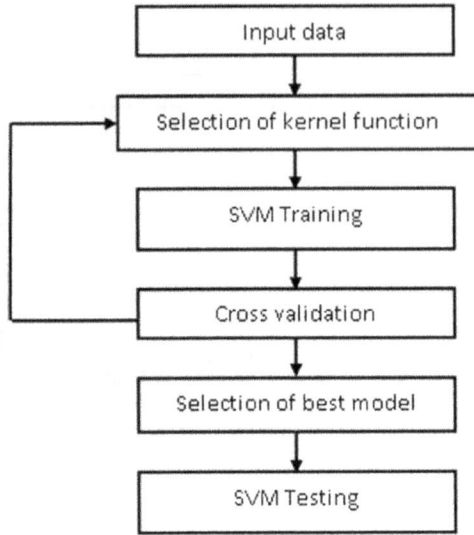

FIGURE 2.4   SVM model development methodology.

is high. In SVMs, original data from input space is mapped to a high dimensional feature space so that the classification become simple. Both complexity and prediction error of the model is minimized utilizing kernel tricks.

In Support Vector Regression, the input space x is mapped to the high dimensional feature space $\Phi(x)$ in a non-linear manner:

$$f(x) = w.\Phi(x) + b \tag{2.2}$$

Where b is the threshold and w is the constant and these are estimated by minimizing the sum of empirical risk and complexity terms. A wide range of kernel functions have been used in the literature and among them linear, radial, and polynomial are the most popular functions. Figure 2.4 presents the methodology for SVM model development.

A detailed review of SVM applications in hydrology especially in the areas of rainfall and runoff modeling, forecasting of streamflow and sediment yield, evaporation and evapotranspiration prediction, lake and reservoir water level prediction, flood forecasting and drought forecasting is found in Raghavendra and Deka (2014).

2.2.5   HYBRID MODELS

There are several recent studies that use hybrid models to improve the accuracy and efficiency of simple models. It is reported that fuzzy hybrid modeling can improve the accuracy of fuzzy system modeling (Kambalimath and Deka, 2020). Several successful applications of hybrid models may be found in literature such as hybrid wavelet-fuzzy (Sahay and Sehgal, 2014) hybrid genetic-fuzzy (Chamani et al., 2013), fuzzy-SVM (He et al., 2014) and Neuro-fuzzy (More and Deka, 2017; Chang et al., 2014); among them ANFIS is the most widely used technique.

Neural networks and fuzzy inference systems are combined in ANFIS. A neural network learning algorithm is used to construct a set of fuzzy if–then rules having membership functions (MFs) based on the stipulated input–output pairs (Muhammad Adnan et al., 2019). Wang et al. (2011) developed a flood simulation model using hybrid GA-WANN model and obtained a very strong relation between rainfall and runoff in arid areas. GA was included in this model to ensure that global optimum is obtained instead of a local optimum, which is a common limitation of ANN.

2.2.5.1   Wavelet Based Hybrid Models

The main limitation of AI models is their inability in handling non-stationary data, which can be resolved with the help of data pre-processing. Hybrid AI-wavelet based pre-processing models are successfully deployed in several hydrologic applications. For a detailed review of such studies, readers may refer to Nourani et al. (2014). They have reported that the use of wavelet-AI models has improved the robustness and accuracy of the model due to the usefulness of wavelet transforms in multi-resolution analysis, de-noising, and edge effect detection and the strong capability of AI methods in optimization and prediction of processes.

Also, Wavelet transform is able to extract nontrivial and potentially useful information from large data sets. Important information that can be drawn from wavelet analysis are understanding multi-temporal scales of hydrologic series, identification of seasonalities and trends, and data de-noising (Sang, 2013).

Two important tasks involved in the wavelet transform based data pre-processing are selection of mother wavelet and level of decomposition. A suitable mother wavelet is chosen based on testing a number of mother wavelets in trial-and-error basis and comparing the similarity in shape between the mother wavelet and the raw time series (Nourani et al., 2014). For time series which have a short memory and short duration transient features, mother wavelets with a compact support form (e.g., Daubechies 1, Haar; and Daubechies 4) are most effective, whereas for the case of time series with long term features, mother wavelets with a wide support form (e.g., Daubechies 2) yield reliable forecasts (Maheswaran and Khosa, 2012).

Nourani et al. (2011) compared the performance of two hybrid wavelet-AI models (WANN, WANFIS) for rainfall runoff modeling and found the performance of WANFIS to be better. The reason may be due to strength of wavelet analysis in extracting dominant frequencies, and fuzzy analysis in handling the uncertainties involved in the relevant phenomena. Alexander et al. (2018) developed hybrid wavelet-ANN model for forecasting of hourly flood stage and compared the performance of this model with that of simple ANN model. Performance of hybrid model was better than simple ANN model and the hybrid model was able to predict the peak flood and time to peak more accurately.

2.3   APPLICATIONS OF SOFT COMPUTING TOOLS FOR IWRM

This section discusses the important applications of various soft computing tools for water resources management. Table 2.1 presents an example list of literatures reviewed for this study and the conclusions derived.

TABLE 2.1
A List of Soft Computing Applications for Water Resources Management

Sl. No.	Methods Used	Water Resources Application	Remarks	Reference
1	Single neural network (SNN) and ensemble neural network (ENN)	Rainfall runoff simulation	ENN performed efficiently than SNN, ENN outperformed rainfall runoff simulation model TANK	Jeong and Kim (2005)
2	ANN, GA	Short term rainfall forecasting,	Hybrid ANN-GA model (GA for selection of suitable input parameters) outperformed MLP neural network	Nasseri et al. (2008)
3	ANFIS	Rainfall runoff modeling	Comparable performance with a physical model, gave a better estimation of peak flow compared to the physical model	
4	ANN, ANFIS	Rainfall runoff modeling	ANN and ANFIS model performed better than a multi-regression model	
5	FL	Rainfall runoff modeling	FL model outperformed the storm water management mode (SWMM) for large rainfall, FL couldn't generate time varying hydrograph	Wang and Altunkaynak (2012)
6	ANN, GA	Rainfall runoff modeling	GA was used for obtaining the weights for neural network. Advantages of this approach were faster training, good accuracy and very high adaptation of nonlinear functional relationships between rainfall and runoff	
7	ANFIS, GP	Ground water level prediction	Performance of GP is better than ANFIS	Fallah-Mehdipour et al. (2013)
8	ANN	Rainfall runoff modeling	Complexity of rainfall runoff model can be reduced by eliminating least significant input variables	Phukoetphim et al. (2014)
9	ANFIS with grid partition (ANFIS-GP) and ANFIS with subtractive clustering (ANFIS-SC), GEP	Forecasting lake level fluctuations	ANFIS-SC models performed better for forecasting 1 to 2 months' lead time, GEP model performed better for 3 months lead time	Sanikhani et al. (2015)
10	ANN (MLP, RBF), SVM	Monthly river flow prediction	Accuracy of MLP and RBF models were better than SVM model, a time index within the inputs of the models increases their performance, uncertainty in the SVM model was less than those in the RBF and MLP models	Ghorbani et al. (2016)
11	ANN, ANFIS, GEP, wavelet-ANN, wavelet-ANFIS, wavelet-GEP	Total dissolved solids	Hybrid wavelet-AI model performance was better than simple AI models, wavelet-GEP was superior to other models	Montaseri et al. (2018)
12	ANN	Drought prediction and forecasting	Results of ANN model was checked with ground observations and government records and found to be accurate	Singh et al. (2021)
13	ANN	Statistical downscaling of monthly mean maximum and minimum temperature	Model performance is very good in terms of prediction accuracy based on coefficient of determination and root mean square error	Chithra et al. (2015)

2.3.1 RAINFALL RUNOFF MODELING

Fuzzy logic has been used for forecasting streamflow and thereby flood in several studies (Deka and Chandramouli, 2005; Toprak et al., 2009; Nayak, 2010). Muhammad Adnan et al. (2019) assessed the accuracy of five soft computing techniques – feedforward neural network (FFNN), the radial basis neural network (RBNN), the generalized regression neural network (GRNN), ANFIS with grid partition (ANFIS-GP) and ANFIS with subtractive clustering (ANFIS-SC), for the prediction of streamflow in a river basin. They found that the performance of RBNN and ANFIS-SC to be better than FFNN, GRNN, and ANFIS-GP models. They also compared the model results with the results from a statistical model seasonal autoregressive integrated moving average (Sarima) and found that the soft computing models outperformed the Sarima model. Havliicek et al. (2013) reported that, for the case of rainfall runoff modeling, GP is more accurate than other machine learning techniques such as ANN in terms of prediction accuracy.

In complex terrains, the relation between various hydrologic processes may be complex and data availability will be very less. In such situations, process based conceptual models will have limited applicability. For such a river basin, Adaptive Regression Spline (MARS), WANN, and simple ANN models were developed and compared for forecasting of runoff by Adamowski et al. (2012). They found the performance of WANN and MARS to be better compared to simple ANN.

Review of several studies on rainfall runoff modeling using AI models showed that the performance of monthly models is better compared to daily models (Nourani et al., 2014). Also they found that the rainfall runoff model performance varies with topographical conditions. For flat terrain, ANN model is preferred whereas for steep mountainous terrain, ANFIS model will be more suitable. Also, for wild terrain wavelet-AI model is preferred and for accurately modeling extreme events, WANFIS model is more suitable.

2.3.2 STATISTICAL DOWNSCALING OF METEOROLOGICAL OBSERVATIONS

Cawley et al. (2003) applied ANN-based approach for statistical downscaling of the extreme events of precipitation in the United Kingdom. They adopted multi-layer perceptron network architecture. The optimal model parameters were determined by gradient descent optimization of a suitable error function (sum of squares error). Chadwick et al. (2011) adopted ANN-based approach to downscale coarse resolution General Circulation Model simulations of temperature and rainfall to regional scale data over Europe. The trained ANN was able to accurately recreate mean values of temperature and precipitation.

It is reported that ANNs are able to distinguish the relation between the input and output variables without explicit physical consideration and they perform well even if the training set of data contain noise and errors due to measurement (Govindaraju, 2000). In order to compensate for numerous changing circumstances ANNs are capable of adapting to solutions over time.

A fuzzy rule-based downscaling technique to estimate the frequency distribution of daily precipitation based on large scale geopotential height was developed by Bardossy et al. (2005). This model was able to capture the predictor-predictand relationship. Valverde et al. (2014) compared the performance of fuzzy downscaling model and neural network based downscaling model to downscale daily precipitation. They concluded that both the models were able to perform better compared to multilinear regression models. Najafi et al. (2011) adopted downscaling based on three approaches – multilinear regression, SVM, and ANFIS. In order to handle the uncertainty due to a single downscaling method, they developed an ensemble of precipitation time series using several downscaling techniques.

2.3.3 WATER QUALITY MANAGEMENT

A fuzzy optimization model was developed for seasonal water quality management addressing the uncertainty in water quality system (Mujumdar and Sasikumar, 2002). The incidence of poor water quality was considered as a fuzzy event and randomness allied with the water quality indicator is connected to this fuzzy event. Oladipo et al. (2021) assessed water quality using fuzzy logic inference and water quality index and inferred the superiority of fuzzy logic inference. Nasiri et al. developed a fuzzy multiple-attribute decision support expert system to recognize the inherent fuzziness in the index making process to compute the water quality index. A study which compared the performance of four artificial intelligence techniques (ANNs, ANFIS with grid partition (ANFIS-GP), ANFIS with subtractive clustering (ANFIS-SC), and gene expression programming (GEP)) to predict total dissolved solids concluded that all the four models could be used successfully, and the performance is the best for GEP model (Ghavidel and Montaseri, 2014).

2.3.4 DROUGHT ASSESSMENT

Several researchers utilized soft computation techniques for drought prediction successfully. Başakın et al. (2021) predicted the metrological drought index, self-calibrated Palmer Drought Severity Index (sc-PDSI) using ANFIS and empirical mode decomposition (EMD)-ANFIS hybrid model. EMD was used as a pre-processing technique to decompose input time series to sub-series and applied to ANFIS models. Performance of EMD-ANFIS model is better than ANFIS model. Another meteorological drought index – standardized precipitation index (SPI) was predicted using co-active neuro fuzzy inference system (CANFIS) with various lead times ranging from 1 to 24 months (Malik et al., 2020). Performance of CANFIS model was better than multilayer perceptron neural network model and multiple linear regression model. SPI forecasts developed for lead times between 1 and 6 months using ANN, WANN, and SVM showed that the performance of WANN to be better than simple ANN and SVM models.

2.3.5 GROUND WATER MODELING

Soft computing methods are utilized for solving complex ground water modeling problems. Several studies used ANN-based models for the prediction of ground water levels (Dash et al., 2010). Natarajan and Sudheer compared the performance of six soft computing methods for ground water level prediction. The methods adopted were ANN, GP, SVM, Extreme Learning Machine (ELM), SVM-QPSO (Quantum Particle Swarm Optimization), SVM-RBF (Radial Basis Function) and the performance of ELM was found to be better compared to other models. Researchers who assessed the suitability of GP for predicting ground water levels concluded that results can be obtained with acceptable accuracy with less computational effort (Natarajan and Sudheer, 2019).

2.4 CONCLUSIONS

Based on the literatures reviewed on the applicability of soft computing techniques for IWRM, the following conclusions were drawn:

- A soft computing method proved to be very useful for a particular study may not perform well in another case. Hence no model can be said as a best model for a particular application.
- Though many studies have discussed about the performance of different models and uncertainty associated with them, there is limited number of studies on quantification of uncertainty. More research is to be conducted in this area.
- The selection of most appropriate model is based on model performance only and uncertainty is not considered for that. More research is to be conducted in the area of criteria for model selection.
- In order to handle the uncertainty associated with soft computing models, it is better to apply a number of models other than relying on a single model.
- Recent decades have witnessed the application of several hybrid models and as a result, model performance has improved considerably. Especially wavelet-based hybrid models performed well with non-stationary data.

REFERENCES

Adamowski, J., H. F. Chan, S. O. Prasher, and V. N. Sharda. 2012. Comparison of multivariate adaptive regression splines with coupled wavelet transform artificial neural networks for runoff forecasting in Himalayan micro-watersheds with limited data. *Journal of Hydroinformatics* 14(3):731–744.

Alexander, A. A., S. G. Thampi, and N. R. Chithra. 2018. Development of hybrid wavelet-ANN model for hourly flood stage forecasting. *ISH Journal of Hydraulic Engineering* 24(2):266–274.

Bardossy, A., I. Bogardi, and I. Matyasovszky. 2005. Fuzzy rule-based downscaling of precipitation. *Theoretical and Applied Climatology* 82:119–129.

Başakın, E. E., Ö. Ekmekcioğlu, and M. Özger. 2021. Drought prediction using hybrid soft-computing methods for semi-arid region. *Modeling Earth Systems and Environment* 7(4):2363–2371.

Belayneh, A., and J. Adamowski. 2012. Standard precipitation index drought forecasting using neural networks, wavelet neural networks, and support vector regression. *Applied Computational Intelligence and Soft Computing* 2012—:1–13.

Cawley, G. C., M. R. Haylock, S. R. Dorling, C. Goodess, and P. D. Jones. 2003. Statistical downscaling with artificial neural networks. *ESANN 2003, 11th European Symposium on Artificial Neural Networks*, Bruges, Belgium, 167–172.

Chadwick, R., E. Coppola, and F. Giorgi. 2011. An artificial neural network technique for downscaling GCM outputs to RCM spatial scale. *Nonlinear Processes in Geophysics* 18:1013–1028.

Chamani, M. R., S. Pourshahabi, and F. Sheikholeslam. 2013. Fuzzy genetic algorithm approach for optimization of surge tanks. *Scientia Iranica* 20(2):278–285.

Chang, F. J., Y. M. Chiang, M. J. Tsai, M. C. Shieh, K. L. Hsu, and S. Sorooshian. 2014. Watershed rainfall forecasting using neuro-fuzzy networks with the assimilation of multi-sensor information. *Journal of Hydrology* 508:374–384.

Chithra, N. R., S. G. Thampi, S. Surapaneni, R. Nannapaneni, A. A. K. Reddy, and J. D. Kumar. 2015. Prediction of the likely impact of climate change on monthly mean maximum and minimum temperature in the Chaliyar river basin, India, using ANN-based models. *Theoretical and Applied Climatology* 121(3):581–590.

Dash, N. B., S. N. Panda, R. Remesan, and N. Sahoo. 2010. Hybrid neural modeling for groundwater level prediction. *Neural Computing and Applications* 19(8):1251–1263.

Deka, P., and V. Chandramouli. 2005. Fuzzy neural network model for hydrologic flow routing. *Journal of Hydrologic Engineering* 10(4):302–314.

Elman, J. L. 1990. Finding structure in time. *Cognitive Science* 14:179–211.

Fallah-Mehdipour, E., O. B. Haddad, and M. A. Mariño. 2013. Prediction and simulation of monthly groundwater levels by genetic programming. *Journal of Hydro-Environment Research* 7(4):253–260.

Ghavidel, S. Z. Z., and M. Montaseri. 2014. Application of different data-driven methods for the prediction of total dissolved solids in the Zarinehroud basin. *Stochastic Environmental Research and Risk Assessment* 28(8):2101–2118.

Ghorbani, M. A., H. A. Zadeh, M. Isazadeh, and O. Terzi. 2016. A comparative study of artificial neural network (MLP, RBF) and support vector machine models for river flow prediction. *Environmental Earth Sciences* 75(6):476.

Govindaraju, R. S. 2000. Artificial neural networks in hydrology. I: Preliminary concepts. *Journal of Hydrologic Engineering* 5:115–123.

He, Z., X. Wen, H. Liu, and J. Du. 2014. A comparative study of the artificial neural network, adaptive neuro-fuzzy inference system and support vector machine for forecasting river flow in the semiarid mountain region. *Journal of Hydrology* 509:379–386.

Hecht-Nielsen, R. 1987. Counterpropagation networks. *Applied Optics* 26:4979–4983.

Havlíček, M., R. Pavelková, J. Frajer, and H. Skokanová. 2013. The long-term development of water bodies in the context of land use: The case of the Kyjovka and Trkmanka River Basins (Czech Republic). Moravian Geographical Reports, 22:39–50.

Herath, H. M. V. V., J. Chadalawada, and V. Babovic. 2021. Hydrologically informed machine learning for rainfall–runoff modeling: towards distributed modeling, *Hydrology and Earth System Sciences*, 25:4373–4401.

Hopfield, J. J. 1984. Neurons with graded response have collective computational properties like those of two-state neurons. *Proceedings of the National Academy of Sciences* 81:3088–3092.

Huang, Y., Y. Lanb, S. J. Thomsona, A. Fang, W. C. Hoffmann, and R. E. Lacey. 2010. Development of soft computing and applications in agricultural and biological engineering. *Computers and Electronics in Agriculture*, 71:107–127.

Jeong, D., and Y. Kim. 2005. Rainfall runoff models using artificial neural networks for ensemble streamflow prediction. *Hydrological Processes* 19:3819–3835.

Kambalimath, S., and P. C. Deka. 2020. A basic review of fuzzy logic applications in hydrology and water resources. *Applied Water Science* 10:191.

Karunanithi, N., W. J. Grenney, D. Whitley, and K. Bovee. 1994. Neural networks for river flow prediction. *Journal of Computing in Civil Engineering* 8:201–220.

Katoch, S., S. S. Chauhan, and V. Kumar. 2021. A review on genetic algorithm: Past, present, and future. *Multimedia Tools and Applications* 80:8091–8126.

Kohonen, T. 1982. Analysis of a simple self-organizing process. *Biological Cybernetics* 44:135–140.

Maheswaran, R., and R. Khosa. 2012. Comparative study of different wavelets for hydrologic forecasting. *Computers & Geosciences* 46:284–295.

Malik, A., A. Kumar, S. Q. Salih, S. Kim, N. W. Kim, Z. M. Yaseen, and V. P. Singh. 2020. Drought index prediction using advanced fuzzy logic model: Regional case study over Kumaon in India. *PLoS One* 15(5):e0233280.

Mohan, S., and D. P. Vijayalakshmi. 2009. Genetic algorithm applications in water resources. *ISH Journal of Hydraulic Engineering* 15(sup1):97–128.

Montaseri, M., S. Z. Z. Ghavidel, and H. Sanikhani. 2018. Water quality variations in different climates of Iran: Toward modeling total dissolved solid using soft computing techniques. *Stochastic Environmental Research and Risk Assessment* 32(8):2253–2273.

More, S. B., and P. C. Deka. 2017. Estimation of saturated hydraulic conductivity using the fuzzy neural network in a semi-arid basin scale for murum soils of India. *ISH Journal of Hydraulic Engineering* 5010:1–7.

Muhammad Adnan, R., X. Yuan, O. Kisi, Y. Yuan, M. Tayyab, and X. Lei. 2019. Application of soft computing models in streamflow forecasting. *Proceedings of the Institution of Civil Engineers-Water Management* 172(3):123–134.

Mujumdar, P. P., and K. Sasikumar. 2002. A fuzzy risk approach for seasonal water quality management of a river system. *Water Resources Research* 38(1):5-1.

Najafi, M. R., H. Moradkhani, and S. A. Wherry. 2011. Statistical downscaling of precipitation using machine learning with optimal predictor selection. *Journal of Hydrologic Engineering* 16(8):650–664.

Nasseri, M., K. Asghari, and M. J. Abedini. 2008. Optimized scenario for rainfall forecasting using genetic algorithm coupled with artificial neural network. *Expert Systems with Applications* 35:1415–1421.

Nayak, P. C. 2010. Explaining internal behavior in a fuzzy if–then rule-based flood-forecasting model. *Journal of Hydrologic Engineering* 15(1):20–28.

Nourani, V., A. H. Baghanam, J. Adamowski, and O. Kisi. 2014. Applications of hybrid wavelet–artificial intelligence models in hydrology: A review. *Journal of Hydrology* 514:358–377.

Nourani, V., Ö. Kisi, and M. Komasi. 2011. Two hybrid artificial intelligence approaches for modeling rainfall runoff process. *Journal of Hydrology* 402(1–2):41–59.

Oladipo, J. O., A. S. Akinwumiju, O. S. Aboyeji, and A. A. Adelodun. 2021. Comparison between fuzzy logic and water quality index methods: A case of water quality assessment in Ikare community, Southwestern Nigeria. *Environmental Challenges* 3:100038.

Phukoetphim, P., A. Shamseldin, and B. Melville. 2014. Knowledge extraction from artificial neural networks for rainfall runoff model combination systems. *Journal of Hydrologic Engineering* 19(7):1422–1429.

Raghavendra, N. S., and P. C. Deka. 2014. Support vector machine applications in the field of hydrology: A review. *Applied Soft Computing* 19:372–386.

Rani, D., S. K. Jain, D. K. Srivastava, and M. Perumal. 2013. Genetic algorithms and their applications to water resources systems. *Metaheuristics in Water, Geotechnical and Transport Engineering* 43–78.

Sahay, R. R., and V. Sehgal. 2014. Wavelet-ANFIS models for forecasting monsoon flows a case study of the Gandak River (India). *Water Resources* 41:574–582.

Sang, Y. F. 2013. A review on the applications of wavelet transform in hydrology time series analysis. *Atmospheric Research* 122:8–15.

Sanikhani, H., O. Kisi, H. Kiafar, and S. Z. Z. Ghavidel. 2015. Comparison of different data-driven approaches for modeling lake level fluctuations: The case of Manyas and Tuz Lakes (Turkey). *Water Resources Management* 29(5):1557–1574.

Singh, T. P., P. Nandimath, V. Kumbhar, S. Das, and P. Barne. 2021. Drought risk assessment and prediction using artificial intelligence over the southern Maharashtra state of India. *Modeling Earth Systems and Environment* 7:2005–2013.

Solomatine, D. P., and A. Ostfeld. 2008. Data-driven modeling: Some past experiences and new approaches. *Journal of Hydroinformatics* 10(1):3–22.

Specht, D. F. 1991. A general regression neural network. *IEEE Transactions on Neural Networks* 2:568–576.

Toprak, Z. F., E. Eris, N. Agiralioglu, H. K. Cigizoglu, L. Yilmaz, H. Aksoy, H. G. Coskun, G. Andic, and U. Alganci. 2009. Modeling monthly mean flow in a poorly gauged basin by fuzzy logic. *Clean–Soil, Air, Water* 37(7):555–564.

Valverde, M. C., E. Araujo, and H. C. Velho. 2014. Neural network and fuzzy logic statistical downscaling of atmospheric circulation-type specific weather pattern for rainfall forecasting. *Applied Soft Computing* 22:681–694.

Wang, K., and A. Altunkaynak. 2012. Comparative case study of rainfall runoff modeling between SWMM and fuzzy logic approach. *Journal of Hydrologic Engineering* 17(2):283–291.

Wang, Y., H. Wang, X. Lei, Y. Jiang, and X. Song. 2011. Flood simulation using parallel genetic algorithm integrated wavelet neural networks. *Neurocomputing* 74(17):2734–2744.

3 Computational Models for Exchange of Water between Ground Water and Surface Water Resources over a Sub-Basin

Subba Rao Tellagorla and Subbarao Pichuka

CONTENTS

DOI: 10.1201/9781003203445-4

39

3.1   INTRODUCTION

The groundwater availability varies spatio-temporally and largely depending on the geological formations (soil properties) and climatic conditions (Liu et al., 2007; Sciuto and Diekkrüger, 2010; Famiglietti, 2014; Chunn et al., 2019). Moreover, the recent studies have also confirmed the declining trends in the fresh water (surface and subsurface) resources (Winter et al., 1998; Famiglietti, 2014; Seo et al., 2018). Therefore, the detailed scientific study on the Ground Water and Surface Water (GW–SW) interactions is a need-of-the-hour for the effective management of the available freshwater resources.

The planning and management of the water resources in an integrated manner has drawn scientific attention to achieve sustainable development. The impact assessment has become one of the major parts in the management of the water resource systems. In this regard, computer-based modeling techniques gain more popularity among the various methodologies used for modeling the GW–SW interactions. The model performance significantly varies based on the consideration of the heterogeneous nature of the parameter.

A plethora of studies has focused on modeling the basin-scale GW–SW interactions in several river basins around the world. For instance, Bailey et al. (2016) have estimated the spatio-temporal variation of the GW–SW interactions in Upper Klamath basin in Oregon, USA. Hence the scientific understating of the connections and related components are essential not only for the assessment studies and also for framing the management strategies (Scibek et al., 2007; Chunn et al., 2019).

The computational models (single or coupled) are developed to solve various surface and subsurface flow processes. The integrated models are developed to solve both processes simultaneously in a single package. Some models are developed for either surface or subsurface systems and then coupled with another model for solving the overall interactions. The level of coupling among the various models and the recent trends in the modeling techniques are also illustrated in this chapter.

3.2   FLOW PROCESSES

The components of the hydrologic system are broadly categorized into three subsystems namely atmospheric, surface, and subsurface systems. The surface and the subsurface systems are connected mainly through infiltration and subsurface flow. A part of the precipitated water infiltrates into the subsurface and forms the groundwater storage. The stored water reaches the surface through the process called base flow or seepage (Rumsey et al., 2015; Karamage et al., 2018). These processes vary spatio-temporally over a basin based on the variations in precipitation (Hu et al., 2009).

3.2.1   SURFACE RUNOFF

The part of the precipitated water flows over the surface and forms surface runoff. The amount of surface flow depends on the climatic conditions and various surface and subsurface parameters (topography, waterways, stream gauging, aquifer thickness, soil properties, hydraulic conductivity, specific storage etc.) of the region

(Scibek et al., 2007; Hu et al., 2009; Prabhakar and Tiwari, 2015; Saha et al., 2017; Chunn et al., 2019). The amount of the surface runoff and the impact of the mentioned parameters on runoff can be analyzed using advanced models (Brunner and Simmons, 2012; Guay et al., 2013; Barthel and Banzhaf, 2016; Muma et al., 2016; Karamage, et al., 2018; Seo et al., 2018). These models compute the surface runoff based on certain governing equations. The commonly used governing equations for estimation of the surface runoff are presented as follows:

3.2.1.1   Governing Equations

The St. Venant equation is being used widely for the surface flow. The simplified depth-averaged equations (Sleigh and Goodwill, 2000) are given below:

The conservation of mass equation:

$$v\frac{\partial A}{\partial x} + A\frac{\partial v}{\partial x} + b\frac{\partial h}{\partial t} = 0 \tag{3.1}$$

The momentum equation:

$$g\frac{\partial h}{\partial x} + v\frac{\partial v}{\partial x} + \frac{\partial v}{\partial t} = g\left(S_i - S_f\right) \tag{3.2}$$

where:

A = cross sectional area of the channel
v = mean flow velocity
h = depth of flow
g = acceleration due to gravity
t = time
S_i = bed slope of the channel
S_f = friction slope of the channel

The governing equations have been derived based on certain assumptions and cannot be solved using explicit methods. Hence various numerical methods are evolved to solve implicitly. The selection of a particular numerical method is depending on various factors such as data availability, methodological approach, and objective of the problem (see Section 3.3).

3.2.2   INFILTRATION

Infiltration is defined as the downward entry of water into the soil and separates precipitation into surface runoff and groundwater recharge (through soil layers). The cracks and macropores on the ground surface are generally assumed to be neglected in mathematical conceptualization to minimize the complexity. At basin scale, the infiltration rate is largely affected by the soil layers based on their physical properties. However, at the local scale, the hydraulic properties, vertical soil profile, soil moisture distribution, and rainfall dictate the infiltration rate. The vertical profile of the subsurface system can be divided into (i) near-surface saturated zone, (ii) transmission zone of unsaturated flow, (iii) wetting zone (where moisture decreases with depth), and (iv) saturation zone. To model the downward movement of water, the soil moisture distribution is approximated using various approaches. The piston-type wetting front movement approach developed by Green and Ampt (1911) is the most widely adopted model for infiltration modeling.

3.2.2.1 Governing Equations

Vertical movement of unsaturated flow can be described by combining Darcy's law with the continuity principle. The resulting partial differential equation is known as Richard's equation (Richards, 1931). Since Richard's equation does not have any closed form of analytical solution, it must be solved numerically for complex practical situations. Different methods are available for estimating the infiltration and are listed in Table 3.1. The comprehensive details of these methods are presented in Todd and Mays (2004).

3.2.2.2 Boundary Conditions

The portion below the wetting front is assumed as unsaturated and the soil is having initial moisture content. The soil portion above the wetting front is having saturated soil moisture content. The bottom impermeable base and the water table are two vertical boundary conditions considered in the subsurface system. It is necessary to consider the water table level because it varies with climate, Land Use Land Cover (LULC) and topography (Todd and Mays, 2004; Liu et al., 2007; Scibek et al., 2007; Chen et al., 2020). During the infiltration process, the wetting front progresses downward until it reaches a point, where the infiltration ceases, at an impermeable base. Hence, the depth and slope of the confining layers are to be considered for modeling the interactions.

TABLE 3.1
Details of Various Methods Used to Estimate the Rate of Infiltration in a Soil Medium

Method Name	Cumulative Infiltration (F_t)	Infiltration Rate (f_t)	Variables
Green–Ampt	$\psi \Delta\theta \ln\left(1 + \dfrac{F_t}{\psi \Delta\theta}\right) + Kt$	$K\left(\dfrac{\psi\Delta\theta}{F_t} + 1\right)$	K – hydraulic conductivity at t Ψ – wetting front suction head $\Delta\theta$ – change in moisture content
Horton's equation	$f_c t + (f_0 - f_c)(1 - e^{-kt})/k$	$f_c + (f_0 - f_c)e^{-kt}$	f_0 – initial infiltration rate f_c – constant infiltration rate K – Decay constant
Soil Conservation Service- Curve Number (SCS-CN)	$\dfrac{S(P_t - I_a)}{P_t - I_a + S}$	$\dfrac{S^2 \dfrac{dP_t}{dt}}{\left(P_t - I_a + S\right)^2}$	S – potential maximum retention I_a – initial abstraction P_t – rainfall at time t
Philip's equation	$St^{0.5} + kt$	$0.5\, St^{-0.5} + K$	S – Sorptivity

3.2.3   BASE FLOW

The outflow of groundwater from subsurface to surface takes place through the process known as base flow. This forms the groundwater discharge to the streams, rivers, lakes, and water ponds in the low stage situations. Estimation of the base flow is crucial for water resources management. It can be estimated using various methods i.e., graphical hydrograph separation, recession curve method, base flow index, etc., are few to list (Winter et al., 1998; Rumsey et al., 2015). The volume of base flow is depending on the topography, climate, and basin characteristics (Winter et al., 1998; Khan et al., 2019).

3.2.4   GROUND WATER–SURFACE WATER (GW–SW) INTERACTIONS

The GW–SW interactions take place either through the entrance of the surface water to the ground or baseflow of the water from the subsurface to the surface or a combination of both. These two are interdependent (Winter et al., 1998; Liu et al., 2007; Scibek et al., 2007; Khan et al., 2019). The interactions between surface and subsurface can be studied in two spatial scales: local and basin (Sophocleous, 2002; Barthel and Banzhaf, 2016). The local-scale interactions take place in a zone near the bottom bed of the water body, called as hyporheic zone. The water in surface water bodies first interacts with the immediate bottom layer of the bed. These localized interactions are mainly controlled by the stream bed properties and hydraulic conductivity (Sophocleous, 2002; Khan et al., 2019). These small interactions are superimposing on the overall interactions of the surface water body.

The basin-scale interaction isa major concern of hydrologic systems modeling. Both the flow process and their parameters vary over the spatial scale which leads to complexity in understanding the interaction. The spatio-temporal variations like distribution of precipitation and hydraulic conductivity; surface water and groundwater table levels, etc., can be extensively represented in the basin-scale interactions (Sophocleous, 2002). The climate, LULC, landscapes, water bodies, aquifer configurations and anthropogenic activities are also some of the key factors which influence the basin-scale interactions (Sophocleous, 2002; Scibek et al., 2007; Hu et al., 2009; Saha et al., 2017; Chunn et al., 2019; Chen et al., 2020; Haque et al., 2021).

3.3   COMPUTATIONAL MODELS

The computational models use the computer program to simulate complex flow processes with the help of numerical methods. These are developed based on the scientific knowledge of hydrological processes. The modeling involves sequential steps of objective identification, data collection, conceptual framework, checking the consistency of available data, selection of suitable numerical model, calibration, validation, and application. Among these, the data collection and the methodological approach are found to be most crucial. The details of the methodological framework and the data collection are discussed in subsequent sections.

3.3.1 METHODOLOGICAL FRAMEWORK

The flow processes of GW–SW interactions are multifaceted in basin scale. The scientific knowledge of these complex physical processes can be helpful to develop an efficient model. The modeling concept starts with objective or problem identification. The objective of the model determines the type and amount of data needed for modeling a system. The data sources necessary for GW–SW interactions are presented in Table 3.2.

The following step is to select/develop a conceptual model. The conceptual model is a representation of the interactions between flow processes. Next, the relationships between various components of the flow system like surface flow, infiltration, and base flow are specified while developing the mathematical equations. Further, these mathematical equations are solved using numerical methods and computer programs. The historical data are used to calibrate and subsequently validate the model. The basic methodological modeling framework is shown in Figure 3.1.

3.3.2 DATA REQUIREMENT

The computational models are range from simple to complex. The simple models require minimum input parameters (lesser data) compared with the complex models. The data availability and its accuracy are the two major constraints associated with the model selection (Liu et al., 2007; Guay et al., 2013). Generally, the model is developed for a specific purpose based on a methodological framework with certain assumptions. The data requirement mainly depends on the overall framework of the model. For instance, spatial variability of the hydrogeological properties can be considered as single or distributed values over the basin. In the lumped case, the average value is required whereas in the distributed case the parameters at each point in the basin are required as input for the model. Further, the available data may not always fit into the model (as direct input) and requires processing. For example, topographic data of a basin needs to be extracted from the Digital Elevation Model. The different types of input data required for modeling the GW–SW interactions along with their sources are outlined in Table 3.2.

TABLE 3.2
Model Input Data Requirements and Their Sources

Data Category	Features	Sources
Surface	The boundaries, elevations, topography, waterways, stream gauging, LULC, dams, reservoirs, lakes, streams	Topography maps, Digital Elevation Models, Bathmetric maps, land use maps, state agencies, vegetation maps
Subsurface	Aquifer thickness, soil properties, hydraulic conductivity, specific storage, specific yield, hydraulic head, geology, aquifer confining details etc.	Geological maps, remote sensing, state agencies, groundwater monitoring databases, catchment management authorities
Meteorological	Precipitation, evaporation, temperature, moisture content, solar radiation, wind speed, relative humidity etc.	Meteorological departments, remote sensing data

---

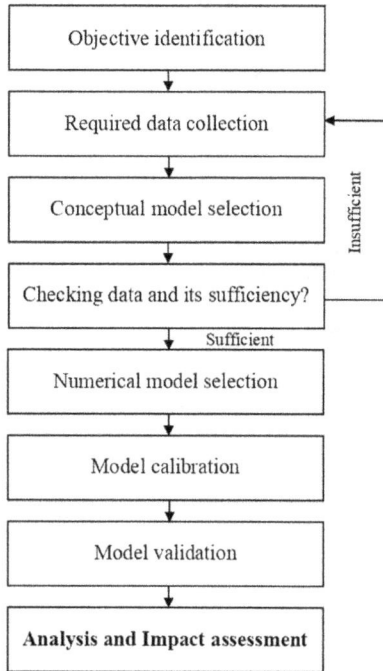

A flowchart with the following boxes connected top to bottom by arrows:

Objective identification
↓
Required data collection ←
↓
Conceptual model selection
↓
Checking data and its sufficiency? ── Insufficient (returns to Required data collection)
↓ Sufficient
Numerical model selection
↓
Model calibration
↓
Model validation
↓
Analysis and Impact assessment

FIGURE 3.1   The general methodological modeling framework.

In addition to the above sources, the data can also be collected from the published literature, government/project reports, survey reports, consultancy reports, etc.

Some of the notable online sources of the data are:

- The United Nations' agency Food and Agriculture Organization is having the World soil maps. (http://www.fao.org/).
- United States Geological Survey (USGS) provides geo-spatial data. (https://earthexplorer.usgs.gov).
- The Soil & Water Assessment Tool (SWAT) platform is having data for usage on its website (https://swat.tamu.edu).
- The state governments are also maintaining the data. For example, Government of India is maintaining DEM data (https://bhuvan.nrsc.gov.in/), hydrologic data (https://indiawris.gov.in/) and meteorological data (http://dsp.imdpune.gov.in/).

3.3.3   CLASSIFICATION OF GW–SW MODELS

Each flow process consists of several parameters which generally vary with different spatio-temporal scales. For instance, the precipitation is not uniform (i.e., magnitude is not the same) over the entire basin and the performance of a model is largely depends on the consideration of this variation (Liu et al., 2007; Hu et al., 2009; Pichuka and Maity, 2017; Karamage et al., 2018). Further, the overall GW–SW interaction contains various flow processes. All those processes can be solved in one model (fully coupled) or multiple models (loosely coupled). In view of the above complexities, the model classifications are presented in this section to distinguish the applicability of various models.

3.3.3.1 Deterministic and Stochastic Models

The models are broadly classified into deterministic or stochastic models based on the presence of randomness in the parameter data (Macian-Sorribes et al., 2017; Haque et al., 2021). The stochastic approach considers the parameter randomness, and such a model predicts the outcome with a certain level of uncertainty. In the case of deterministic approach, the model always produces the same outcome with the same set of input parameters. The deterministic models are sometimes referred to as process-based models (Haque et al., 2021). They are divided into the following three types:

- Empirical models in which mathematical equations are used to assess the processes.
- Lumped conceptual models: the averaged value of the parameter over an area is considered for the modeling.
- Distributed-physically based models: the spatial variability of the parameter is considered over an area for the modeling.

The distributed-physically based models are widely used due to their accurate representation of the physical and hydrological processes (Abbott and Refsgaard, 2012).

3.3.3.2 Coupled Models

The integrated models have drawn attention in recent years as they consider the characteristics of parameters like flow processes, climate, geology, geomorphologic, LULC and human factors (Brunner and Simmons, 2012; Guay et al., 2013; Barthel and Banzhaf, 2016; Chunn et al., 2019; Haque et al., 2021; Zhu et al., 2012). The movement of water in each of the surface and subsurface (saturated and unsaturated zones) systems can be solved in one coupled model or multiple separate models and there after combine the results. Therefore, based on the level of GW–SW integration, the models are classified as fully coupled and loosely coupled models.

3.3.3.3 Fully Coupled Models

The surface and subsurface flow processes are generally interdependent. The models which represent both surface and subsurface systems in a single package can be classified as the fully coupled model (Sciuto and Diekkrüger, 2010; Brunner and Simmons, 2012; Guay et al., 2013; Barthel and Banzhaf, 2016; Muma et al., 2016; Seo et al., 2018; Haque et al., 2021; Zhu et al., 2012). These models consider the continuous flow of water in the surface and subsurface systems. Since all the processes are solved in a single model the requirement of the additional packages can be reduced. Some of the commonly used fully coupled models and their applications in various studies are presented in Table 3.3.

TABLE 3.3
Applications of Various Fully Coupled and Loosely Coupled Models

Model Type	Model	Usage Purpose
Fully coupled models	HydroGeosphere	Influence of the spatial variation of the soil properties in Wüstebach river basin, Germany (Sciuto and Diekkrüger, 2010)
	ParFlow	Impact of landscape heterogeneity on groundwater availability in Amazon basin, USA.
	CATHY	Assessment artificial subsurface drainage effects on flow in a sub-watershed of the Beaurivage river (Muma et al., 2016).
	Penn state Integrated Hydrologic Model (PIHM)	Estimation of the streamflow and groundwater levels in Haw River Basin located in North Carolina (Seo et al., 2018)
Loosely Coupled models	SWAT for surface system and MODFLOW for subsurface system.	Assessment of spatio-temporal patterns of ground water discharge to River in a semi-arid region in Sprague Watershed in Upper Klamath Basin in Oregon, USA (Bailey et al., 2016)
	Florida Institute of Phosphate Research – Hydrological Model (FIPR-FHM) is an integration of MODFLOW for subsurface system and HSPF for surface system.	Estimation of the groundwater budget in Big Lost River Basin in Idaho, USA (Said et al., 2005)
	MIKE SHE for subsurface flow and MIKE 11 for surface flow.	Estimation of groundwater response to surface flow and topology in Tarim Basin, China (Liu et al., 2007)

3.3.3.4 Loosely Couple Models

The loosely coupled models model the GW–SW flow processes separately using two or more models and then inter-exchange the results (Said et al., 2005; Liu et al., 2007; Barthel and Banzhaf, 2016; Chunn et al., 2019). One model is used for the simulation of either surface or subsurface systems. Then the simulated result of that model is used as input for another model to simulate the other system. Some of the commonly used loosely coupled models along with their usage in various studies are presented in Table 3.3.

3.3.4 CHALLENGES AND OPPORTUNITIES

Although both fully coupled and loosely coupled models are useful to simulate the GW–SW interaction, they have their own set of advantages and limitations. The specific challenges and opportunities pertain to each of the model categories (fully/loosely coupled) are illustrated in the following paragraphs.

Some of the challenges and opportunities of the fully coupled models are as follows:

- Fully coupled models can model the continuous movement of water between surface and subsurface in a composite manner (Guay et al., 2013). Hence these are found to be capable of providing solutions for most of the water resource problems (Brunner and Simmons, 2012).
- Consideration of a large number of parameters in a single model increases the complexity, uncertainty and takes longer computational time (Barthel and Banzhaf, 2016; Haque et al., 2021). It is a major challenge associated with the fully coupled models.
- The lateral flow occurring over an unsaturated zone may affect the final outcome. Therefore, care must be taken by the modeller while modeling the interactions in such cases (Zhu et al., 2012).

Some of the opportunities and limitations of the loosely coupled models are as follows:

- These models study the movement of water mainly either in surface or subsurface system by simplifying the movement (through suitable assumptions) in another system (Guay et al., 2013; Barthel and Banzhaf, 2016).
- In contrast with the fully coupled models, the loosely coupled models are computationally less intensive and often complete the analysis quickly (Schilling et al., 2019).
- These models are unable to consider the feedbacks from the surface when modeling the subsurface system and vice-a-versa (Guay et al., 2013).

3.4 APPLICATIONS OF COMPUTATIONAL MODELS IN IMPACT ASSESSMENT

The computational models are not only used for understanding/estimating the GW–SW interactions but also used for the impact assessment studies. Some of the computational model studies and their findings are presented here for illustrative purposes. The models used for assessing the impact of urbanization on groundwater fluctuations. It is concluded that the vegetation cover is directly proportional to the groundwater in a region. Prabhakar and Tiwari (2015) noticed that the increase in land development activities and urbanization intensify the impervious land which further leads to a reduction in groundwater recharge. The change in climate impacts the surface and subsurface water resources and thereby changes the peak time of surface runoff (Scibek et al., 2007). The recent studies confirm the changes in hydroclimatic variables, e.g., temperature (rising trend) and precipitation (falling trend), which in turn reduces the mean contribution of groundwater to streamflow (Saha et al., 2017). Intensive groundwater withdrawal has an immediate impact on the river flow than the climate change (Chunn et al., 2019).

3.5 RECENT TRENDS IN MODELING TECHNIQUES

Several advanced techniques have been developed in recent years to achieve a noble goal of sustainable development of water resources. Some of the recent models related to GW–SW interactions are detailed below.

- A trained artificial neural network model is developed for the simulation and a genetic algorithm is used for the optimization to minimize the shortages of the water demand in irrigation systems (Safavi et al., 2010).
- The optimization can be either in a deterministic or stochastic approach. A Combined Surface-Groundwater Stochastic Dual Dynamic Programming (CGS-SDDP) is developed for stochastic optimization of water resource systems to frame the management strategies (Macian-Sorribes et al., 2017).
- The groundwater level can be estimated with the surface water fluctuations through the AI approach (Cobaner et al., 2016). The AI is used to develop a relationship among groundwater level, lake water level, evaporation, and perception.
- Various machine learning techniques are developed to estimate the groundwater spring potential. Groundwater spring maps have been constructed in a region using the Kernel Logistic Regression (KLR), Random Forest (RF), and Alternating Decision Tree (ADTree) models in China (Chen et al., 2020, Al-Fugara et al., 2020).
- The coupled model which is an integration of Muskingum method for open channel flow and machine learning method for the groundwater flow, is used to ascertain the effect of ecological water replenishment on dynamic change in surface and groundwater levels in a river (Sun et al., 2021).
- There are many other studies which focusses on modeling the GW–SW interactions over various parts of the world (Cobaner et al., 2016; Chunn et al., 2019; Joseph et al., 2021).

3.6 SUMMARY

This chapter reviews various numerical and computational models used for modeling the GW–SW interactions. The generalized methodological approach for a computational model and various steps involved are clearly described. The opportunities and complexities associated with those models are demonstrated. Different components involved in surface and subsurface systems are also discussed. Next, several influencing factors and their impact on the GW–SW interaction are illustrated. The input data required for developing a model and its consistency along with the available sources are also presented. Finally, the classification of existing models is well explained along with the recent trends that evolved in the modeling techniques. The comparative study of the available models to ascertain the performance and applicability for a specific basin can be kept as the future scope of this study.

REFERENCES

Abbott, M. B., & Refsgaard, J. C. (Eds.). (2012). *Distributed hydrological modeling* (Vol. 22). Springer Science & Business Media. Kluwer Academic, Boston.

Al-Fugara, A. K., Pourghasemi, H. R., Al-Shabeeb, A. R., Habib, M., Al-Adamat, R., Al-Amoush, H., & Collins, A. L. (2020). A comparison of machine learning models for the mapping of groundwater spring potential. *Environmental Earth Sciences, 79,* 1–19.

Bailey, R. T., Wible, T. C., Arabi, M., Records, R. M., & Ditty, J. (2016). Assessing regional-scale spatio-temporal patterns of groundwater–surface water interactions using a coupled SWAT–MODFLOW model. *Hydrological Processes, 30*(23), 4420–4433.

Barthel, R., & Banzhaf, S. (2016). Groundwater and surface water interaction at the regional-scale–A review with focus on regional integrated models. *Water Resources Management, 30*(1), 1–32.

Brunner, P., & Simmons, C. T. (2012). HydroGeoSphere: A fully integrated, physically based hydrological model. *Groundwater, 50*(2), 170–176.

Chen, W., Li, Y., Tsangaratos, P., Shahabi, H., Ilia, I., Xue, W., & Bian, H. (2020). Groundwater spring potential mapping using artificial intelligence approach based on kernel logistic regression, random forest, and alternating decision tree models. *Applied Sciences, 10*(2), 425.

Chunn, D., Faramarzi, M., Smerdon, B., & Alessi, D. S. (2019). Application of an integrated SWAT–MODFLOW model to evaluate potential impacts of climate change and water withdrawals on groundwater–surface water interactions in West-Central Alberta. *Water, 11*(1), 110.

Cobaner, M., Babayigit, B., & Dogan, A. (2016). Estimation of groundwater levels with surface observations via genetic programming. *Journal-American Water Works Association, 108*(6), E335–E348.

Famiglietti, J. S. (2014). The global groundwater crisis. *Nature Climate Change, 4*(11), 945–948.

Green, W. H., & Ampt, G. A. (1911). Studies on Soil Physics. *The Journal of Agricultural Science, 4*(1), 1–24.

Guay, C., Nastev, M., Paniconi, C., & Sulis, M. (2013). Comparison of two modeling approaches for groundwater–surface water interactions. *Hydrological Processes, 27*(16), 2258–2270.

Haque, A., Salama, A., Lo, K., & Wu, P. (2021). Surface and groundwater interactions: A review of coupling strategies in detailed domain models. *Hydrology, 8*(1), 35.

Hu, H. R., Mao, X. L., & Liang, L. (2009). Temporal and spatial variations of extreme precipitation events of flood season over Sichuan Basin in last 50 years. *Acta Geographica Sinica, 64*(3), 278–288.

Joseph, N., Preetha, P. P., & Narasimhan, B. (2021). Assessment of environmental flow requirements using a coupled surface water-groundwater model and a flow health tool: A case study of Son river in the Ganga basin. *Ecological Indicators, 121,* 107110.

Karamage, F., Liu, Y., Fan, X., Francis Justine, M., Wu, G., Liu, Y., & Wang, R. (2018). Spatial relationship between precipitation and runoff in Africa. *Hydrology and Earth System Sciences, Discussions,* 1–27.

Khan, H. H., Khan, A., Senapathi, V., Prasanna, M. V., & Chung, S. Y. (2019). Groundwater and surface water interaction. In Venkatramanan, S. (Ed.). *GIS and geostatistical techniques for groundwater science* (pp. 197–207).

Liu, H. L., Chen, X., Bao, A. M., & Wang, L. (2007). Investigation of groundwater response to overland flow and topography using a coupled MIKE SHE/MIKE 11 modeling system for an arid watershed. *Journal of Hydrology, 347*(3–4), 448–459.

Macian-Sorribes, H., Tilmant, A., & Pulido-Velazquez, M. (2017). Improving operating policies of large-scale surface-groundwater systems through stochastic programming. *Water Resources Research, 53*(2), 1407–1423.

Muma, M., Rousseau, A. N., & Gumiere, S. J. (2016). Assessment of the impact of subsurface agricultural drainage on soil water storage and flows of a small watershed. *Water*, *8*(8), 326.

Pichuka, S., & Maity, R. (2017). Spatio-temporal downscaling of projected precipitation in the 21st century: Indication of a wetter monsoon over the Upper Mahanadi Basin, India. *Hydrological Sciences Journal*, *62*(3), 467–482.

Prabhakar, A., & Tiwari, H. (2015). Land use and land cover effect on groundwater storage. *Modeling Earth Systems and Environment*, *1*(4), 1–10.

Richards, L. A. (1931). Capillary conduction of liquids through porous mediums. *Physics*, *1*(5), 318–333.

Rumsey, C. A., Miller, M. P., Susong, D. D., Tillman, F. D., & Anning, D. W. (2015). Regional scale estimates of baseflow and factors influencing baseflow in the Upper Colorado River Basin. *Journal of Hydrology: Regional Studies*, *4*, 91–107.

Safavi, H. R., Darzi, F., & Mariño, M. A. (2010). Simulation-optimization modeling of conjunctive use of surface water and groundwater. *Water Resources Management*, *24*(10), 1965–1988.

Saha, G. C., Li, J., Thring, R. W., Hirshfield, F., & Paul, S. S. (2017). Temporal dynamics of groundwater–surface water interaction under the effects of climate change: A case study in the Kiskatinaw River Watershed, Canada. *Journal of Hydrology*, *551*, 440–452.

Said, A., Stevens, D. K., & Sehlke, G. (2005). Estimating water budget in a regional aquifer using HSPF-MODFLOW integrated model. *Journal of the American Water Resources Association*, *41*(1), 55–66.

Schilling, O. S., Park, Y. J., Therrien, R., & Nagare, R. M. (2019). Integrated surface and subsurface hydrological modeling with snowmelt and pore water freeze–thaw. *Groundwater*, *57*(1), 63–74.

Scibek, J., Allen, D. M., Cannon, A. J., & Whitfield, P. H. (2007). Groundwater–surface water interaction under scenarios of climate change using a high-resolution transient groundwater model. *Journal of Hydrology*, *333*(2–4), 165–181.

Sciuto, G., & Diekkrüger, B. (2010). Influence of soil heterogeneity and spatial discretization on catchment water balance modeling. *Vadose Zone Journal*, *9*(4), 955–969.

Seo, S. B., Mahinthakumar, G., Sankara Subramanian, A., & Kumar, M. (2018). Conjunctive management of surface water and groundwater resources under drought conditions using a fully coupled hydrological model. *Journal of Water Resources Planning and Management*, *144*(9), 04018060.

Sleigh, P. A., & Goodwill, I. M. (2000). *The St Venant equations*. School of Civil Engineering: University of Leeds, England.

Sophocleous, M. (2002). Interactions between groundwater and surface water: The state of the science. *Hydrogeology Journal*, *10*(1), 52–67.

Sun, K., Hu, L., Guo, J., Yang, Z., Zhai, Y., & Zhang, S. (2021). Enhancing the understanding of hydrological responses induced by ecological water replenishment using improved machine learning models: A case study in Yongding River. *Science of The Total Environment*, *768*, 145489.

Todd, D. K., & Mays, L. W. (2004). *Groundwater hydrology*. John Wiley & Sons, New Jersey, US

Winter, T. C., Harvey, J. W., Franke, O. L., & Alley, W. M. (1998). Report on *Ground water and surface water: A single resource*. Circular 1139, US Geological Survey, Colorado, USA.

Zhu, Y., Shi, L., Lin, L., Yang, J., & Ye, M. (2012). A fully coupled numerical modeling for regional unsaturated–saturated water flow. *Journal of Hydrology*, *475*, 188–203.

4 Computational and Field Approach to Assess Artificial Recharge of Groundwater

Marykutty Abraham and Sankaralingam Mohan

CONTENTS

4.1 INTRODUCTION

Pumping requirements from aquifers are increasing due to an increase in water demand because of population growth and higher standard of living. An indiscriminate extraction of aquifers can cause decline in physical and chemical quality of water, saline water intrusion in coastal aquifers, and land subsidence.

DOI: 10.1201/9781003203445-5

Sometimes environmental impacts may not be readily identifiable in the beginning. It is imperative that a hydrological equilibrium between the water pumped and the water recharged is achieved by reducing pumping or by increasing recharge. Reducing pumping is not practical due to the ever-increasing demand. Thus, enhancement of recharge may be more advantageous when compared to other options. Replenishment of groundwater can be achieved after analyzing the possibility and feasibility of artificial recharging using various recharge configurations.

Groundwater recharge can be implemented in an area of permeable formation where feeding is done directly or by changing the natural conditions of the region to reuse it again. This improves the quality of the water also.

For selection of site and suitable type of artificial recharge structure, it is necessary to have the detailed knowledge on hydrogeological and hydrological aspects of the study area. The parameters and features to be considered while selection of site include: inflow and outflow, specific yield, storativity, porosity, hydraulic conductivity, hydraulic boundaries, water balance studies, depth and type of aquifer, natural recharge, and transmissivity. To evaluate the performance of the artificial recharge structures with respect to the above-mentioned parameters, an effective numerical model is required to establish guidelines for managing groundwater resources systems in an optimal manner. Artificial recharge effects could be simulated with the recent breakthroughs in groundwater flow and transport models.

The frequently used recharge estimation method is water balance models (Xu and Chen, 2005; Ladekarl et al., 2005; Rushton et al., 2006; Richey, et al., 2015; Dhungel and Fiedler, 2016). Determination of groundwater recharge and the recharge coefficient through case studies was carried out by several researchers (Yeh et al., 2007; Adeleke et al., 2015; Han et al., 2017; Subramanian and Abraham, 2019; Salem et al., 2019; Nolte et al., 2021). From hard rock towards alluvial areas in India, groundwater recharge range between 5% and 20% of rainfall (Athavale et al., 1992). Natural recharging was estimated to be 19% of rainfall in the area by Mohan and Abraham (2010). Using MODFLOW, numerical models were developed to mimic aquifer response in diverse hydrogeological settings (Lubczynski and Gurwin 2005; Aish and Smedt, 2006; Szucs et al., 2009; Wang et al., 2008; Panagopoulos, 2012; Chitsazan and Movahedian, 2015; Pramada et al., 2018; Abraham and Mohan, 2019; Sithara et al., 2020; Sundararajan and Sankaran, 2020).

Based on the literature review carried out, it was found that not many computational and field methods have been carried out to find the efficacy of artificial recharge structures.

For sustainability of water resources artificial recharge is an essential infrastructure component for a semi-arid region. Groundwater in cities is under threat due unplanned urbanization, industrial development, and improper waste management. This chapter explains the applicability of groundwater modeling for sustainable management of groundwater resources through artificial recharge.

4.2 STUDY AREA

The study area is part of a sub-basin of Vellar river, situated on the western side of the Cudallore aquifer system. In research conducted by the Geological Survey of India, the region was designated as favourable recharge region (Sukhija et al., 1996). A geo-electrical resistivity survey was conducted, and maps of iso-resistivity lines were prepared to detect the changes in bedrock and aquifer depth. The geological section was interpreted on the basis of electrical resistivity survey and favourable locations were identified. A village named Nadiyapattu was selected as the favourable location for construction of recharge structures. Nadiyapattu is located between the latitudes of 11°39′ 36″ and 11°45′ 00″ N and the longitudes of 79°24′ 12″ and 79° 24′ 48″ E. Location plan of the recharge arrangements is given in Figure 4.1.

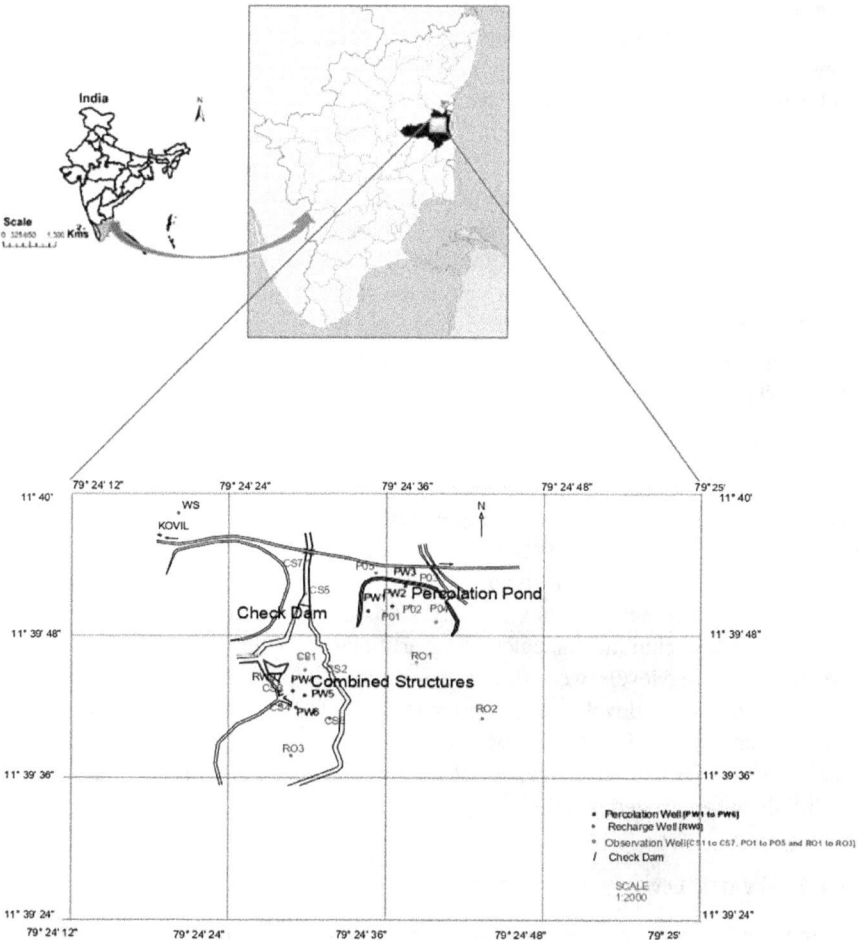

FIGURE 4.1   Location plan of the recharge arrangements.

The resistivity survey conducted at Nadiyapattu, showed that from the ground level, the top formation consists of 18 m to 24 m thick unconsolidated lateritic sandstone, followed by a 50 m to 55 m thick mottled clay with thin sand bands with a thickness of 2 m to 4 m, followed by a 30 m to 40 m layer of sand. The area receives an average annual rainfall of 1000 mm, of which 60% is brought in by North-East monsoon during October to December, while the South-West monsoon from June to September brings 25%. The weather in the area is hot and humid, with temperatures usually ranging from 33°C to 25°C.

4.2.1 ARTIFICIAL RECHARGE STRUCTURES

The artificial recharge schemes can be implemented after studying the effect of recharge structures in offsetting the deficit condition, hence avoiding potential negative consequences such as water level drop and water quality variance. Thus, for augmenting groundwater levels, various recharge structures such as check dams, recharge well, percolation pond, and percolation wells were constructed at Nadiyapattu village in Cuddalore basin individually, and also in combination, to study their effectiveness in recharging the aquifer. Infiltration study was conducted at 19 locations using double ring infiltrometer to identify suitable recharge locations and the constant infiltration rate was obtained as 440.85 mm/hr.

Locations were identified in such a way that all the different recharge structures could be constructed independently in different locations. For the combined effect of these structures a separate location within the sub-basin was identified in which all the artificial recharge structures were constructed (Check dam, Percolation pond, and percolation wells). Fifteen observation wells were constructed to monitor water level fluctuations.

4.3 METHODOLOGY

Aquifer recharge assessment is a complex undertaking that necessitates the use of a variety of approaches. Natural recharge alone is not sufficient to nullify the effect of pumping, and the ability of artificial recharge scheme to replenish the aquifer must be evaluated. Different models were used to assess the recharge caused by various structures. The technique for calculating artificial recharge is divided into several phases. The water levels were drawn using data obtained from observation wells to analyse the water level changes prior to and after the construction of recharge arrangements. The effectiveness was also determined using the basin scale water balance approach. The effectiveness of various arrangements for both quantity and quality fluctuations were studied using numerical models.

4.3.1 WATER LEVEL FLUCTUATIONS

There were 15 observation wells to study the groundwater level improvement. Daily water levels were observed for one preceding year and more than two years following the construction of the structures to gain a better understanding on the recharge phenomenon. Water levels were observed two times daily and the average daily water levels were calculated.

4.3.2 MASS BALANCE STUDY

Water balance study gives a numerical estimate of a structure's contribution to recharge. Small reservoirs and ponds are frequently shallow, with high surface area to volume ratios which can lead to quite substantial evaporation losses that can outweigh groundwater recharge contributions.

The reduction in water levels over time in a recharge basin is an indicator of the recharge structure's performance. The fall in pond water level during periods when there is no inflow and outflow is either due to recharge or evaporation. After deducting open pan evaporation rates, this is transformed into groundwater recharge. The losses due to seepage, abstraction, and other factors can also be taken into account and the water balance for a reservoir can be found (Neumann et al., 2004).

4.3.3 NUMERICAL GROUNDWATER MODELING

A model should be regarded as a rough approximation rather than an exact replica of field conditions. Mathematical models can evaluate, predict, and compare various scenarios of hydrogeological systems in a simpler way. The more complicated equations that explain groundwater flow are solved using numerical models. The numerical solutions to the governing flow equation allow for fewer restrictive assumptions, allowing the model to more closely mirror natural conditions. The effectiveness of individual structures as well as various combinations of structures was further studied using numerical models. Numerical models were developed using MODFLOW (Harbaugh, 2000). The graphical user interface for the study was Groundwater Modeling Systems (GMS) (EMRL, 2002).

4.3.3.1 Flow Model (MODFLOW)

For flow modeling, MODFLOW software was employed to depict the processes in the system, which has the following mathematical equation.

$$\frac{\partial}{\partial x}\left(k_{xx}\frac{\partial h}{\partial x}\right) + \frac{\partial}{\partial y}\left(k_{yy}\frac{\partial h}{\partial y}\right) + \frac{\partial}{\partial z}\left(k_{zz}\frac{\partial h}{\partial z}\right) + -w = S_s\frac{\partial h}{\partial t}$$

where

h: hydraulic head
x, y, z: coordinates of the principal components of the hydraulic conductivity
    $K_{xx}, K_{yy}, \& K_{zz}$
w: volumetric flux per unit volume and is referred to as a source or sink term
K_{xx}, K_{yy}, K_{zz}: hydraulic conductivities in the principal directions
S_S: specific storage

The equations were solved using finite-difference technique with a block-centred formulation. Defining model domain, defining boundary conditions, assigning parameters, calibration are important steps in the application of numerical groundwater models. The model construction is divided into three steps: conceptual model design, 3D finite difference grid development, and MODFLOW simulation.

The hydraulic characteristics of the aquifer were averaged over the cell area to use a finite-difference approximation. Each cell's basic differential flow equation is replaced by an algebraic equation, resulting in a flow field. Through an iterative procedure, the algebraic equations were numerically solved, and at the node, the average head value for a cell is calculated. The hydraulic properties are believed to be constant over these cells, which are the smallest volumetric units.

4.4   RESULTS AND DISCUSSIONS

Artificial recharge is not only a good way to store water, but it also allows for better management of the water that is available with improvement on water quality. The recharge structures were constructed in April 2004 and the rains from July 2004 were collected in them for recharging the groundwater aquifer. Since September 2003, daily water levels have been observed. Water level fluctuations, mass balance, analytical model, and numerical models were used to estimate the quantity recharged following the installation of artificial recharge structures. Even though this study was carried out during the years 2004 to 2006, the results will be useful for water managers to improve the groundwater situation.

4.4.1   WATER LEVEL FLUCTUATIONS

The recharging pattern was investigated by plotting water levels and rainfall near the recharge structures (Figure 4.2). The image shows that following the building of recharge structures, the water level rose by 2 to 3 metres overall.

4.4.2   MASS BALANCE

The water balance was assessed by tracking the changes in pond water level over time for percolation pond and check dam. After deducting pan evaporation rates, these data were converted into groundwater recharge rates during times without direct abstraction and rainfall.

Individual structures' responses were assessed using data collected every two days. Within the recharge system, recharging was expected to be spread equally at rates stated in $m^3/d/m^2$. The impact of the storage is derived by assessing the recharge per unit ponding area. Percolation ponds and check dams were subjected to this basin-scale volume balance investigation.

4.4.2.1   Percolation Pond

Water in the pond area was measured on alternate days, and the monthly data was compiled and used to analyse the response of the structure (Table 4.1). The percolation

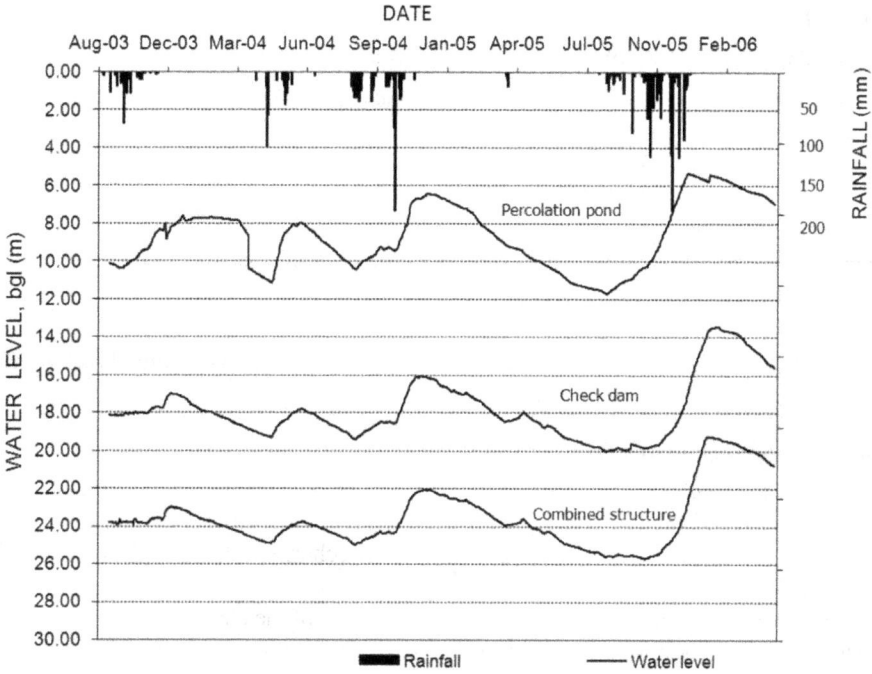

FIGURE 4.2 Water level variation in recharge structure areas.

TABLE 4.1
Effect of Percolation Pond on Recharge

Month	Area m²	Volume m³	Change in Volume m³	Evaporation m³	Recharge m³	Recharge mm
Sep-04	16735.75	46764.93	683.38	41.20	642.17	38.31
Oct-04	13180.64	33313.91	756.93	55.73	701.20	52.90
Nov-04	18899.51	56634.37	561.12	43.63	517.49	27.38
Dec-04	17485.67	47724.97	1140.89	42.51	1098.39	62.82
Jan-05	15572.58	36546.30	704.14	56.99	647.15	41.56
Feb-05	11980.86	24393.34	1086.92	64.30	1022.62	85.35
Mar-05	6750.02	11153.91	549.39	47.66	501.73	74.33
Apr-05	5174.82	6728.12	237.54	35.08	202.46	39.12
May-05	4529.26	4628.66	238.65	40.52	198.13	43.75
Jun-05	3184.98	2428.19	127.03	32.41	94.61	29.71
Jul-05	2680.58	1483.46	86.94	22.43	64.51	24.07
Aug-05	4629.05	2969.30	147.78	32.84	114.94	24.83
Sep-05	15899.24	12839.67	634.00	124.09	509.91	32.07
Oct-05	20919.51	29881.06	605.37	118.61	486.76	23.27
Nov-05	38299.08	134032.50	1773.65	117.96	1655.69	43.23
Dec-05	43010.36	164342.01	2366.07	180.64	2185.43	50.81
Jan-06	41092.82	137739.49	2165.66	213.34	1952.32	47.51
Feb-06	37822.14	103500.29	2162.17	225.04	1937.13	51.22

TABLE 4.2
Recharge Capacity of Structures by Mass Balance

S. No.	Recharge Structure	Monthly Recharge per m² of Ponding Area
1	Percolation pond along with 3 percolation wells	0.69 m³
2	Check dam	0.30 m³
3	Check dam with 1 recharge well	0.54 m³

pond with three percolation wells could recharge 0.69 m³ of water/month/m² of the ponding area.

4.4.2.2 Check Dams

The response of check dam and check dam with one recharge well were evaluated in the similar way. The effectiveness of check dam was obtained as 0.30 m³/month/m² of ponding area and that of check dam with one recharge well was 0.54 m³/month/m² of pond area.

According to mass balance study, around 80–90% of the change in the volume of water was penetrated to the aquifer, with just 10–20% was evaporated. The percolation pond with percolation wells had the maximum effect compared to the other structures in recharging the aquifer (Table 4.2).

4.4.3 NUMERICAL GROUNDWATER MODELING

4.4.3.1 Flow Model

Based on the established water balance components and hydrogeological regime, numerical models were developed to simulate groundwater heads. Incorporating observed and inferred hydrologic data for the years 2003–2005, groundwater flow model was created using the MODFLOW. Observed water levels during September 2003 were taken as initial condition. Average monthly data from September 2003 to December 2004 was used for model calibration, and the data from January 2005 to February 2006 was used for validation.

4.4.3.2 Model Construction

Defining the model domain, boundary conditions, parameter assignment, and calibration require careful consideration when creating numerical groundwater models. Model building included three steps: conceptual model design, 3-D finite difference grid generation, and execution of MODFLOW. The base map was first imported, followed by the determination of system boundaries and assignment of boundary conditions. Constant head boundaries were assigned, and stresses were defined.

The modeling domain was considered as a single layer (80 m above the MSL to 45 m below MSL), with an area of 1450 m by 1100 m with 32 rows and 39 columns comprising 1,248 cells, and 24 wells were used to obtain the layer elevation.

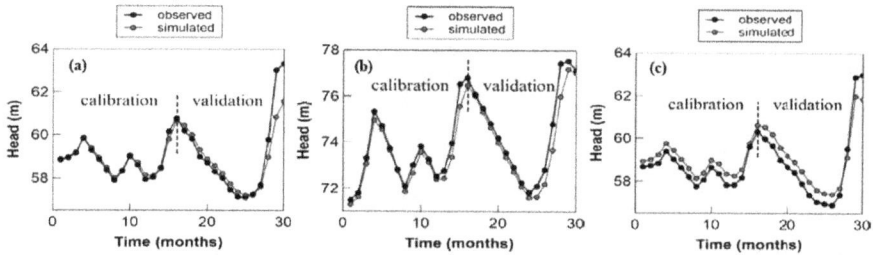

FIGURE 4.3 Observed and simulated heads – (a) check dam area; (b) percolation pond area; (c) combined structure area.

4.4.3.3 Initial Conditions, Boundary Conditions, and Stresses

The initial condition was chosen to be the water levels in the monitoring wells in September 2003. For the whole modelled region, boundary criteria were defined. Dirichlet boundaries were specified for the model. The eastern limit ranged from 69 metres north to 68 metres south. The western border was defined by a series of constant head cells ranging from 58 metres in the north to 54 metres in the south. The top was taken as recharge boundary with a recharge rate of 20% of rainfall. Evapotranspiration was computed from Penman-Monteith method.

Hydraulic conductivity was assigned as per the infiltration study. Vertical hydraulic conductivity was set as 1/10 of horizontal conductivity value. Hence the daily hydraulic conductivity values in the x, y, and z directions were respectively assigned as 10 m, 10 m, and 1 m. It was expected that the specific yield was 0.2. Near the southern border of the area, there was just one agricultural pumping well.

Thirty stress periods were created from September 2003 to February 2006. Pre-conditioned Conjugate Gradient technique was used as the solution (PCG). The model was executed in transient state.

4.4.3.4 Model Calibration and Validation

Monthly average water levels from September 2003 to December 2004 and from January 2005 to February 2006 were used for calibration and validation respectively. 15 observation wells were used for the calibration. During calibration, hydraulic conductivity was changed within a tolerable range and determined to be 12 m/day.

Figures 4.3a–c show the comparison of observed and simulated heads (i.e., water table elevations from MSL) which matched well. The observed heads on monthly basis and simulated heads on ten days interval are used for the comparison. The mean error, mean absolute error, and root mean square error during the calibration period were 0.13 m, 0.71 m, and 0.87 m respectively, indicating satisfactory calibration.

There was significant rise in water levels in the percolation pond area as well as in the combined structures area within two years of recharge. The average increase in the peak season due to artificial recharge was in the range of 2 m to 4 m.

4.4.3.5 Performance Evaluation of Individual Structures

The spatial and temporal changes in head for individual structures in the study region were studied using the calibrated numerical model. The distances were computed

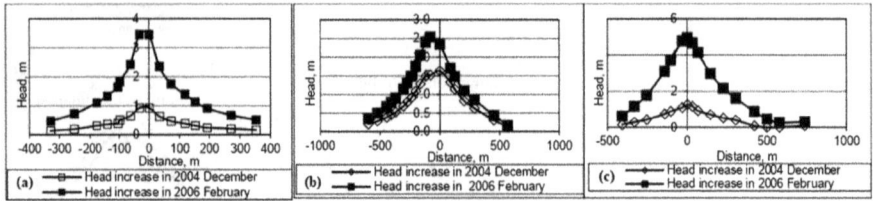

FIGURE 4.4   (a) Increase in head due to a check dam; (b) Increase in head due to percolation pond with percolation wells; (c) Increase in head due to combined structure arrangement.

TABLE 4.3
Maximum Head Increase for Structures with Two Years of Recharge

Sl. No	Recharge Structure	Ponding Area	Mount in Head from the Centre of the Structure	Radius of Influence
1	Check dam	1300 m^2	3.46 m	400 m
2	Percolation pond with 3 percolation wells	15000 m^2	2.54 m	600 m
3	Combined structure (Check dam, Pond, 3 Percolation wells)	15000 m^2	4.7 m	500 m

from the centre point of the structure. The influence region of the percolation pond was extending up to the boundary of the study region. The recharge mound obtained with recharge due to the structures in December 2004 and February 2006 were compared individually with that of the condition without recharge structure which was obtained from the calibrated numerical model. Figures 4.4a–4.4c show the increase in head, due to various structures from the centre of the structure. From the figures it is clear that the increase in head due to check dam during 2004 and 2006 are 0.93 m and 3.46 m respectively. The rise in head near percolation pond area was 1.63 m and 2.54 m respectively during the same period. The rise in head was 1.27 m and 4.7 m respectively during December 2004 and February 2006 at combined structure area.

The influence zones for check dam, percolation pond and combined structure from the centre point of the structure were 400 m, 600 m, and 500 m respectively, with the highest rise in water level of 3.46 m, 2.54 m, and 4.7 m in two years of artificial recharge (Table 4.3).

4.5   CONCLUSIONS

The study analyzed the viability of a variety of artificial recharge schemes. Effectiveness of artificial recharge structures was assessed using different methods such as water level fluctuations, water balance, and numerical models. Water balance method was used to quantify the efficacy of ponding structures. According to water balance study, 80–90 % of the water in the structure was recharged, whereas only 10–20% was evaporated. Numerical models for flow were constructed using

MODFLOW to establish a tool for evaluating the long-term performance of artificial recharge systems.

Within two years of recharging, the water levels at the centre of check dam, percolation pond, and combined structure rose to 3.81 m, 2.48 m, and 3.95 m, respectively. Artificial recharge, according to the study, is a realistic alternative for the augmentation of groundwater resources. The model can be effectively used to determine the recharge pattern in aquifers and may also be used as a planning tool for artificial recharge projects.

One of the main disadvantages of this study is that it was carried out during the years 2003 to 2006. Now, the land use and river characteristics might have changed in the study area. Hence the recharge values may vary if recent data were to be included.

REFERENCES

Abraham, M., Mohan, S., 2019. Effectiveness of Check Dam and Pond with Percolation Wells for Artificial Groundwater Recharge Using Groundwater Models, *Water Science and Technology-Water Supply*, 19(7), 2107–2115.

Adeleke, O., et al., 2015. Estimation of Groundwater Recharges in Odeda Local Government Area, Ogun State, Nigeria Using Empirical Formulae. *Challenges*, 6(2), 271–281.

Aish, A., Smedt, F., 2006. Modeling of a Groundwater Mound Resulting from Artificial Recharge in the Gaza Strip, Palestine. *Water for Life in the Middle East*, 2, 779–787.

Athavale, R.N., Rangarajan, R., Muralitharan, D., 1992. Measurement of Natural Recharge in India. *Journal of Geological Society of India*, 39(3), 235–244.

Chitsazan, M., Movahedian, A., 2015. Evaluation of Artificial Recharge on Groundwater Using MODFLOW Model (Case Study: Gotvand Plain-Iran). *Journal of Geoscience and Environment Protection*, 3(5), 122–132.

Dhungel, R., Fiedler, F., 2016 Water Balance to Recharge Calculation: Implications for Watershed Management Using Systems Dynamics Approach, *Hydrology*, 3(13), 1–19.

EMRL (Environmental Modeling Research Laboratory), 2002. Groundwater Modeling System (GMS) Version 4.0 Tutorial Documents, Vol. 1–4, Environmental Modeling Research Laboratory, Brigham Young University, Provo, UT.

Han, D., et al., 2017. Alterations to Groundwater Recharge due to Anthropogenic Landscape Change. *Journal of Hydrology*, 554, 545–557.

Harbaugh, A.W., 2000. MODFLOW-2000, the U.S. Geological Survey Modular Ground-Water Model – User Guide to Modularization Concepts and the Groundwater Flow Process. U.S. Geological Survey Open-File Report 00-92, 121.

Ladekarl, U.L., et al., 2005. Groundwater Recharge and Evapotranspiration for Two Natural Ecosystems Covered with Oak and Heather. *Journal of Hydrology*, 300(1–4), 76–99.

Lubczynski, M.W., Gurwin, J., 2005. Integration of Various Data Sources for Transient Groundwater Modeling with Spatio-Temporally Variable Fluxes-Sardon Study Case, Spain. *Journal of Hydrology*, 306(1–4), 71–96.

Mohan, S., Abraham, M., 2010. Derivations of Simple Site-Specific Recharge-Precipitation Relationships: A Case Study from the Cuddalore Basin, India. *Environmental Geosciences*, 17(1), 37–44.

Neumann, I., Barker, J., MacDonald, D., Gale, I., 2004. Numerical approaches for approximating technical effectiveness of artificial recharge structures, Groundwater Systems and Water Quality Programme Commissioned Report CR/04/265N, British Geological Survey, Nottingham, UK.

Nolte, A., et al., (2021). Hydrological Modeling for Assessing Spatio-Temporal Groundwater Recharge Variations in the Water-Stressed Amathole Water Supply System, Eastern Cape, South Africa, *Hydrological Processes*, 35, e14264. https://doi.org/10.1002/hyp.14264

Panagopoulos, G, 2012. Application of MODFLOW for Simulating Groundwater Flow in the Trifilia Karst Aquifer, Greece. *Environmental Earth Sciences*, 67, 1877–1889.

Pramada, S.K., Minnu, K.P., Roshni, T., 2018. Insight into Seawater Intrusion due to Pumping: A Case Study of Ernakulam Coast, India. *ISH Journal of Hydraulic Engineering*. https://doi.org/10.1080/09715010.2018.1553642

Richey, A.S., et al., 2015. Quantifying Renewable Groundwater Stress with GRACE. *Water Resources Research*, 51, 5217–5238.

Rushton, K.R., et al., 2006. Improved Soil Moisture Balance Methodology for Recharge Estimation. *Journal of Hydrology*, 318(1–4), 379–399.

Salem, A., Dezső, J., El-Rawy, M., 2019. Assessment of Groundwater Recharge, Evaporation, and Runoff in the Drava Basin in Hungary with the WetSpass Model. *Hydrology*, 6(1), 23.

Sithara, S., Pramada, S.K., Thampi, S.G., 2020. Impact of Projected Climate Change on Seawater Intrusion on a Regional Coastal Aquifer. *Journal of Earth System Science*, 129, 218. https://doi.org/10.1007/s12040-020-01485-y

Subramanian, S.T., Abraham, M., 2019. Assessment of Natural Groundwater Recharge for North Chennai Aquifer. *Environmental Geosciences*, 26(2), 1–10.

Sukhija, B. S., Nagabhushanam, Reddy P. D.V., 1996. Groundwater Recharge In Semi-Arid Regions of India: An Overview of Results Obtained Using Tracers, *Hydrogeology Journal*, (4), 50–71.

Sundararajan, N., Sankaran, S., 2020. Groundwater Modeling of Musi Basin Hyderabad, India: A Case Study. *Applied Water Science*, 10, 14. https://doi.org/10.1007/s13201-019-1048-z

Szucs, P., Madarasz, T., Civan F., 2009. Remediating Over-Produced and Contaminated Aquifers by Artificial Recharge from Surface Waters. *Environmental Modeling & Assessment*, 14(4), 511–520.

Wang, S., et al., 2008. Application of MODFLOW and Geographic Information System to Groundwater Flow Simulation in North China Plain, China. *Environmental Geology*, 55, 1449–1462.

Xu, C.Y., Chen, D., 2005. Comparison of Seven Models for Estimation of Evapotranspiration and Groundwater Recharge Using Lysimeter Measurement Data in Germany. *Hydrological Process*, 19, 3717–3734.

Yeh, H.F., Lee, C.H., Chen, J.F., 2007. Estimation of Groundwater Recharge Using Water Balance Model. *Water Resources*, 34(2), 153–162.

5 Multi-Objective Optimization in Water Resource Management

Rashmi Bhardwaj and Shanky Garg

CONTENTS

DOI: 10.1201/9781003203445-6

5.1 INTRODUCTION

As we all know, water is one of the most precious resources which directly affect our life. No matter how many other resources we have but we can't ignore the importance of water, especially in this era because of so many factors that affect both the quality and quantity of these resources. This is not only because of natural factors, but there is also a range of manmade activities behind this which eventually lead to the crisis of clear and unpolluted water in comparison with the demand. The natural factors include precipitation and sudden change in environmental conditions, and the manmade factors include population growth, and consumption of food, which requires plenty of water. According to the NITI Aayog report of June 2018, there are around 600 million people facing high water-scarcity. Moreover, because of inadequate access to water, around 200,000 people died every year and the country's demand is expected to increase rapidly by 2030 which eventually leads to a 6% decrease in GDP and by 2050 the requirement for water is likely to be increased to 1,180 BCM, and availability is just 695 BCM. So, there is an utmost requirement to take immediate action to conserve water, not only for the current generation, but also for the future.

There is a need to bridge the gap between sustainability and water resource. Therefore, the role of optimization techniques and their mathematical models play an important role in environmental management from past decades (Mishra, 2020). Optimization is an amalgam of mathematics, statistics, and computer science which enables the decision-maker to analyse the problem and then after modeling arrive at a conclusion. The most commonly used model is the linear programming model in which both the objective functions and their constraints are linear (Clyde, 1971; Singh et al., 2013). But there are some cases where we have a non-linear model where either the objectives or constraints are non-linear (Alothaimeen and Arditi, 2019; Papadatos et al., 2002). Pareto Front is one of them where we take two objective functions at a time (Rojas et al., 2015). These multi-optimization techniques play a vital role in the management of natural resources, during project management even in adverse conditions (Xevi and Khan, 2006; Gonzalez et al., 2018). A GRASP is another method used earlier to find solutions along the Pareto front (Higgins et al., 2008). The increase in literature review regarding advances in mathematical programming techniques for the management of water resources under uncertainty like changes in the weather pattern, economic development and therefore help in using these techniques efficiently (Archibald and Marshall, 2018). If we want to achieve our goals, then there is a need to stop the overutilization of these resources. (Damle, 2006). But the application of optimization techniques is difficult in water resource management due to different objectives and uncertainty in the behaviour of resource input (Datta and Harikrishna, 2005). There are many software systems actively working for to provide the optimal solution (Watts et al., 2009). However, sometimes we try to make decisions based on many conflicting objectives, and this kind of situation is called multi-objective optimization (Lahdelma et al., 2000; Vanderpooten 1989; 1990). Earlier Adaptive management is being used for learning and management of natural resources, but it has its own limitations (Ehrgott, 2006; Williams, 2011). Moreover, each and every sector require a quick decision-making approach to grow business (Beria et al., 2012; Kiker et al., 2005; Marler and Arora, 2010; Mendoza and Martins, 2006).

This chapter deals mainly with the integration of applied mathematics with water resource management and will formulate a multi-objective model keeping in mind all the objectives and considering each objective with different weights with their limitations and will try to consider most of the factors that affect this resource management and optimize the model with the Posteriori technique. We will illustrate this through a case study based on Delhi. In this chapter, Section 5.2 shows the basics and methodology of the process and the mathematical formulation of all methods, and why the weighing method is preferred over all other methods. In Section 5.3 the process of implementing the Weighing Method is given. In Section 5.4 we explain the results, using a case study based on Delhi water resource management. In Section 5.5, conclusion and future scope are given, and in Section 5.6, references for the articles which were read during this chapter.

5.2   MATHEMATICAL OPTIMIZATION

An optimization model is an important characteristic to solve many problems which may be a business-oriented problem, or maybe a resource allocation problem. It mainly consists of an objective function, a set of constraints on the objective functions, and decision variables.

Objective Function: An objective function is a goal that the decision-maker want to achieve. There may or may not have constraints associated with it. In case of training of the personnel, the authorities want that the cost incurred in the training of their employees to be minimal.

Constraints: Constraints are the restrictions on the goals or objective functions. In case of training of the personnel, budget is the major limitations or constraints for this problem.

Decision Variables: A variable that is controlled by a decision-maker hand also helps in deciding the output. In case of training personnel, a decision variable can be the number of employees who takes participate in the training.

The problem can be linear or non-linear based upon the objective functions or constraints.

Linear Programming Problem: A mathematical modeling in which objective functions and constraints both are in linear form as shown in **Formulation 5.1**. The areas where LPP has been used include transportation and resource allocation.

Non-Linear Programming Problem: A mathematical modeling in which either objective functions or constraints or both are in non-linear form. There are many applications of NLP in the oil industry, including the power network model.

This chapter deals mainly with the multi-objective optimization model. **A multi-Objective Optimization problem** is a part of MCDM Problems in which we consider many different conflicting objectives with their respective constraints and trying to optimize each and every objective function simultaneously. The problem can be linear or non-linear. Here we focus more on linear model. The standard form of MOLP can be seen in Formulation **5.2**. There does not exist a single optimal solution for all objective functions instead of that it gives an infinite number of non-inferior solutions. So, the main idea behind considering the multi-objective programming is that we are trying to out a set of Pareto optimal solutions and basically trying to establish a relationship between all those objectives. There are many methods to solve the Multi-Objective Optimization Model. But here in this chapter, we try to focus on the **vector optimization techniques**, preparation of a model, what are the important objectives that should be taken care of, and what are the limitations behind this and finally prepare a model using the **Posteriori method**.

There is a huge reason behind choosing only the Posteriori technique. There are many methods to solve like **No Preference Methods** in which the views of the decision-makers are not considered, and as there is no role of decision-maker, it is his choice to accept or reject the solution. Because of this, there is a high chance of rejection occurring due to unfavourable solutions. Similarly, in the case of the **Method of global criteria**, we have to minimize the distance between the objective region, which is feasible, and the individual optimal, as Seen in **Formulation 5.3**. The main problem is that here we considered each objective function equally important which is not true in the real-life scenario because if we are considering conflicting objective functions then there should be different priorities for different objectives based on the person's perceptions. This is the main reason to choose the Posteriori method over all other methods. As seen in Table 5.1, there is a significant difference between all other methods.

Basically, the **Posteriori method** – also known as Method of Generation of Non-Inferior Solution – is the method in which Pareto optimal set is generated and then the set of Pareto solution is presented in front of the decision-maker and then it's the decision-maker choice to choose the most preferred among all the alternatives. In the

TABLE 5.1
Comparison Table of all Methods of Vector-Optimization

S. No.	Method Name	Description
1.	No Preference Method	Decision-Maker views are not taking into consideration. Chances of Error are relatively high in this method.
2.	Global Criteria Method	Distance between some point of reference and the objective reason which is feasible is minimized. There are no weights associated with it.
3.	Weighing Method	In this, we assign weights to the objective function and after finding the optimal solution most preferred will be chosen.
4.	Epsilon-Constraint Method	In this only one objective is optimized at a time and all other objective functions are treating as a constraint by giving a lower or upper bound value based on min or max type respectively.
5.	Apriori Method	In this the expectations or preferences are known before executing the techniques and this will create a problem as sometimes decision-makers do not know how realistic their preferences are.
6.	Lexico-Graphic Ordering	In this, each objective function is taken one by one according to its priority and if any of the objective functions has a unique value, then that value will be considered optimal for other objectives functions too. In this loss of generality is there.

Posteriori method Weighing method is used in which each objective has its own importance so different weights will be given to each objective function according to our priority and convert the multiple-objective functions into a single objective and then minimize the single objective for finding the optimal allocation. As seen in **Formulation 5.4 and Formulation 5.5**.

The mathematical formulation of the above methods are as follows:

5.2.1 LINEAR PROGRAMMING MODEL

Let us consider there are m variables and n constraints:

$$\max\left\{w^T z\right\} \tag{5.1}$$

subject to
$Bz \le d$
$z \ge 0$

Here z is a set of decision variables. w^T is a cost vector. B matrix of constraint coefficients & d is constraint level.

5.2.2 SIMPLE MULTI-OBJECTIVE MODEL

$$\min\left\{w_1\left(x\right), w_2\left(x\right), w_3\left(x\right) \ldots w_n\left(x\right)\right\} \tag{5.2}$$

subject to
$x \in S$

Here consider individual optimal $W_1^*, W_2^*, W_3^*, \ldots, W_n^*$

5.2.3 GLOBAL CRITERIA METHOD

So, the formulation of the method of global criteria from **(5.2)** can be written as:

$$\min\left(\sum_{J=1}^{n}\left|w_i\left(x\right) - W_i^*\right|^p\right)^{1/p} \tag{5.3}$$

subject to
$x \in S$

Here, $w_i(x)$ is the individual objective function value.

$W_i(x)$ is the individual objective function optimal value.
x refers to a vector of decision variables that belong to the feasible region set
    S and $1 \le p < \infty$.

5.2.4   POSTERIORI METHOD (WEIGHING METHOD)

The formulation of the Weighing Problem can be written as:

$$\min\{w_1(x), w_2(x), w_3(x)\ldots w_n(x)\} \tag{5.4}$$

subject to

$$x \in S$$
$$\lambda_i \geq 0$$

$$\sum_{J=1}^{n} \lambda = 1_i$$
$$xj \geq 0$$

After introducing weights to the objective function. Now the formulation becomes:

$$\min \sum_{J=1}^{n} \lambda_i w_i(x) \tag{5.5}$$

subject to

$$x \in S$$
$$\lambda_i \geq 0$$

$$\sum_{J=1}^{n} \lambda = 1_i$$
$$x \geq 0$$

Here, $w_i(x)$ is the individual objective function expression.

λ_i represents the weight vector.
x refers to a vector of decision variables that belong to the feasible region set S.

5.3   STEPS TO IMPLEMENT POSTERIORI TECHNIQUE

STEP 1: Understand the problem and decide the decision variables, objectives that you wish to fulfil, and constraints of the problem.
STEP 2: Change all the objective functions to the same type.
STEP 3: Analyse each objective function and assign weight to them according to their importance.
STEP 4: Merge all the objective functions into a single objective function.
STEP 5: Weight value should be greater than or equal to zero and the sum of all weight values should be equal to 1.
STEP 6: Solve the problem using the linear/non-linear approach using a single merged objective with a set of constraints and default weight constraints.

5.3.1 PROBLEM DEFINITION

As management of water resources is our main concern. So, here, basically, our aim is to formulate a model with multiple objectives which plays an important role and some constraints like budget constraint, demand and supply constraint are also associated with those objective functions.

This chapter considers six main objective functions, which are conflicting in nature, and play important roles in the management of water resource. Each objective function and the reason behind choosing it is explained in detail.

5.3.2 OBJECTIVE FUNCTIONS

5.3.2.1 Maximization of Revenue

As for the adequate supply of clean and fresh water, there is a need for investment which can be achieved through the revenue generated from water and the commercial activities which mainly depend on water. Water resources can be monetized from the various agricultural and industrial activities. The profit from these revenues will be used in the betterment of the quality and supply of these resources. Moreover, there is a need for a financing strategy that is sustainable and also ensures a pure supply of water and sanitation. In developing countries, many authorities of water depend on the funding of the public for their investment in extensions of networks. And even do not bear their maintenance and operation cost so revenue maximization is an important criterion for the water resource and the same can be formulated as:

$$\max \text{Revenue}\,(\text{RE}) = \sum_j R_j x_j$$

5.3.2.2 Minimization of Overutilization

There is nothing in this world that is available free throughout our life. One day everything will vanish whether it is renewable or not. Overutilization of freshwater has been increased which in turn increases the pressure on the availability of water which directly and indirectly affects the quality of the environment. There are many factors which contribute a major role in overutilization of water like the products which are sugar-based, and meat products demand more consumption of water and mainly due to increase in population, growth of an economy, and demand of agricultural products. So, our main actions are to suppress these activities which lead to the overutilization of water and the formulation of this objective function is as follows:

$$\min \text{Overutilization}\,(\text{OU}) = \sum_j x_j + \sum_j W_j x_j$$

5.3.2.3 Minimization of Cost

As we all know, the cost and revenue move in parallel. If the cost of providing safe and clean water increases, then in turn revenue decreases. There are many cost factors that unnecessarily increase the cost. First, there is a need to identify the factors that directly or indirectly increase the cost of providing good quality of water. The formulation of this model is as follows:

$$\min \text{Cost}(C) = \sum_j C_j x_j$$

5.3.2.4 Minimization of Pollution

As we all know the earth is made up of around 2/3 of water, so it should be our primary responsibility to conserve water as much as possible. According to the reports of UN many deaths occur because of contaminated water rather than nature or because of some wars. Moreover, it not only affects our health but also our environment and the economy. There are many factors that are responsible for the contamination of water like wastes of industries. Moreover, increasing population also indirectly affects this as they dump their household waste into water. Similarly, untreated wastewater and sewage generation also play a vital role in increasing pollution related to water. The formulation of this objective function is as follows:

$$\min \text{Pollution}(P) = \sum_j P_j x_j$$

5.3.2.5 Maximization of Treatment Plants

As the water pollution increases due to reasons mentioned in above objective. There is a need to control this pollution and moreover, the requirement of the treatment plans which treat the untreated water like sewage water, polluted water from industries is necessary which make the water fit for drinking and for other purposes. So, it is necessary to know how many plants are required in each state or district to treat those effluents and the capacity of CETP plants to treat the water for drinking purposes. The formulation of this objective function is as follows:

$$\max \text{No.of CETP}(\text{Common Effluent Treatment Plans}) = \sum_j D_j N_j$$

5.3.2.6 Minimize Waste Generation

There is a need to minimize the waste generated in water as if more waste is generated it eventually leads to more water pollution which in turn requires more treatment plants which increases the cost of treatment. The formulation of this objective function is as follows:

$$\min \text{Sewage Generation}(SE) = \sum_j Gp_j x_j$$

Notations used in all the six objective functions are as follows:

R_j is the revenue generated by the all the activities in the jth district
x_j is the allocation of water to the jth district.

C_j represents the cost involved in supplying clean water to the jth district.
W_j represents the wastage of the water in the jth district.
P_j represents the percentage of impurity present in the jth district.
D_j represents the capacity of CETP plants to treat water in the jth district.
N_j represents the number of CETP in j^{th} district.
Gp_j represents the percentage of sewage production in j^{th} district.

Each objective function has its own limitations. So here, we are describing some constraints based on the above objective functions.

5.3.3 CONSTRAINTS

5.3.3.1 Supply Constraint
This constraint explains the limitations on the supply of groundwater resources.

$$\Sigma x_j \leq A \tag{5.6}$$

$$x_j \leq a_j \tag{5.7}$$

Here **Formulation 5.6** represents that the sum of allocation to the water in all the districts should be equal to the total water availability in a particular state. Also, **Formulation 5.7** represents that the water allocation in a particular district should be less than or equal to the water availability in that district.

5.3.3.2 Demand Constraint
This constraint put some limitations on the demand side as the supply of water is not infinite.

$$\Sigma x_j = Y \tag{5.8}$$

Here **Formulation 5.8** represents that the water demand should be equal to the total demand in a particular state.

5.3.3.3 Future Needs
This constraint basically takes care of the future demand.

$$P_j \leq x_j + y_j - Y_j \tag{5.9}$$

Here **Formulation 5.9** represents that expected demand in the future should be less than or equal to the available resources.

5.3.3.4 Budget Constraint
This constraint represents the limitations on the budget of the government in treating wastewater, setting up plants, and supplying clean water to the people.

$$\sum_j C_j x_j \leq B. \tag{5.10}$$

Here **Formulation 5.10** represents that the total cost incurred during treatment of water should be less than or equal to the budget available with the government.

5.3.3.5 Max. No. of CEP Plants

This constraint represents the number of treatment plants in a particular district.

$$\sum_j D_j N_j \leq Cp \tag{5.11}$$

Here, **Formulation 5.11** represents that the waste treated by number of plants in a particular district should be less than or equal to the total available capacity of plants to treat wastewater in the state.

Notations used in all the constraints:

x_j represents the water allocation in the jth district.
y_j represents the water allocation in future in the jth district.
Y_j represents the current water demand in the jth district.
a_j represents the water available in the jth district.
Y represents the demand in a particular state.
P_j represents the projected demand in a particular district.
Cj represents the cost involved in treating wastewater.
N_j represents the number of effluent plants in the jth district of a particular state.
D_j represents the maximum capacity of CETP plants to treat water in the jth district.
A represents the total groundwater availability in the state.
B represents the total budget available with the state.
Cp represents the total capacity of the state to treat wastewater.

5.4 RESULTS

5.4.1 FINAL MODEL FORMULATION

After combining all the six objective functions to a single objective and the constraints based on this and assigning the appropriate weight values to each objective function, the final formulation is as follows:

$$\min \left\{ W_1 \left(hx_1\right) + W_2 \left(hx_2\right) + W_3 \left(hx_3\right) + W_4 \left(hx_4\right) + W_5 \left(hx_5\right) + W_6 \left(hx_6\right) \right\}$$

subject to
$\Sigma x_j \leq A$
$x_j \leq a_j$
$\Sigma x_j = Y$
$P_j \leq x_j + y_j - Y_j$

$$\sum_j C_j x_j \leq B$$

$$\sum_j D_j N_j \leq Cp$$

$$\sum_{j=1}^{n} W_j = 1$$

$W_j \geq 0\ j = 1,2,3....,6$
$N_j \geq 0$
$x_j \geq 0$

From Figures 5.1 and 5.2 we can say that all objectives are conflicting in nature and can be optimized only when the other objective function compromises.

(1) Here we convert each objective function to minimization type and assign different weights to them which is represented by W_j.
(2) The constraints from 5.11 to 5.19 represent the constraints described above.
(3) The summation of all the weight values should be equal to 1.
(4) Weight values should be greater than or equal to 0.

5.4.2 Case Study Based on the Allocation of Water in State Delhi with the Available Data

Delhi, the capital of India and also the Union Territory is one of the most densely populated state which is surrounded by the Haryana and Uttar Pradesh and ranked as second-most productive area. Because of huge population and, according to WHO report most polluted city in the world poses a major problem in providing resources in adequate quantity to its population and this continue in the case of water supply

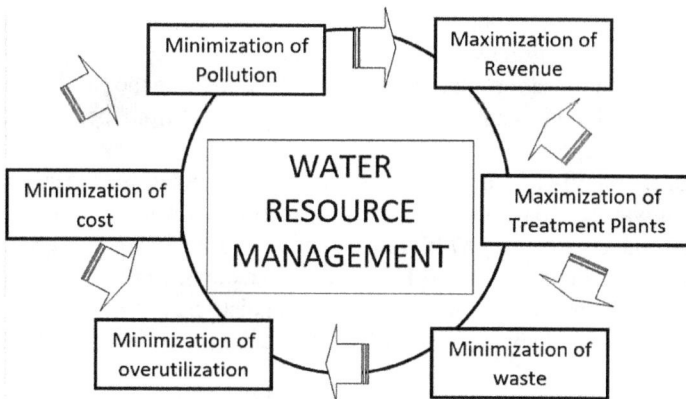

FIGURE 5.1   Resource management.

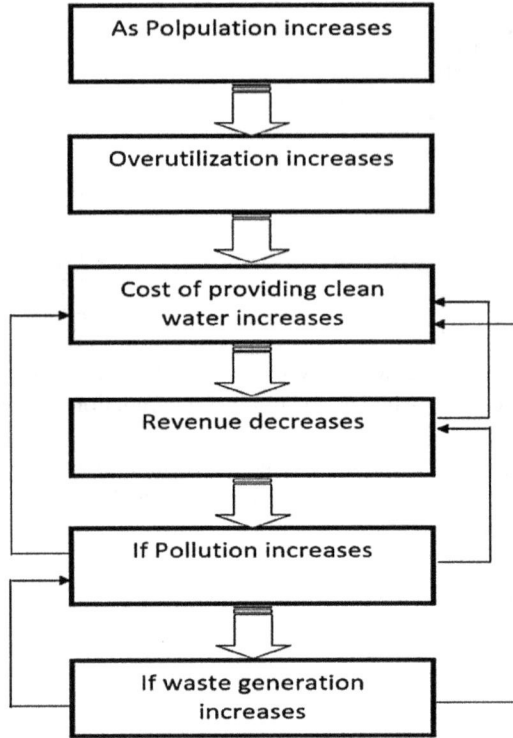

FIGURE 5.2 Optimization process.

and sanitation. This requires optimal solutions to tackle this problem and to allocate the resources properly.

Delhi is divided into 11 districts, with each district different in its own way, and with a different speciality.

In this case study we are considering the Delhi state with nine districts that are North-West Delhi, South Delhi, West Delhi, South-West Delhi, North-East Delhi, East Delhi, North Delhi, Central Delhi, and New Delhi with their respective population.

From the graph we can see that the highest population is in the North-West district and lowest population is in New Delhi. Data is taken from population census. According to the reports of Planning Department, Economic Survey of India, 2019–2020 on an average 141.55 litres of water per day is required but the supply is only 104 litres of water per day on an average. Expenditure on water is around 27.64 billion and income is around 25.87 billion, and nearly 40% of water is wasted for non-profitable purposes and around 40% in pipeline leakages. Here we are considering proportions on the basis of population of different district.

Let's solve the single objective function of revenue using the simple LPP solving technique:

$$\max \text{Revenue} \left(\text{RE} \right) = \sum_j R_j x_j$$

subject to

$\Sigma x_j \leq A \ x_j \leq a_j$

$\Sigma x_j = Y$

$$\sum_j C_j x_j < = B$$

Table 5.2 shows the average revenue calculated based on the population of each district and total revenue generated from water. Data is taken from CGWB (2019–2020), Ground Water Report.

Average demand and average supply are calculated from the overall supply and demand given.

By solving the problem, we get the following result as shown in Figure 5.3. Similarly, we solve cost function & overutilization function with respect to budget, supply and demand constraints and get the following result as Figure 5.4.

Similarly, we can solve objective functions of individually too and now we solve it using the Posteriori method of vector optimization using different weights. Here we are considering three objective functions: maximize revenue, minimization of overutilization, and minimization of cost with supply, demand, and budget constraint with weights 0.7, 0.2 & 0.1 respectively. From Figure 5.3 in which we take revenue as objective function, and its value is ₹6,669.87 crore and allocations are shown in the figure.

Similarly, objective function value for cost function is ₹278.67 million and for overutilization of water function is 1,001.681 mgd. From Figure 5.5 we can see that there are different allocations when we are considering three objective functions at a time i.e., revenue, cost and overutilization and get the optimal value of the combine function of 3,602.26, which is far better than the value when we solve individually.

Moreover, we can change the weight according to our priority and make the allocations more optimal by adding more data if available. The details of recent studies are also given (Table 5.3).

TABLE 5.2
Average Revenue

District Name	Revenue from Water (in Crores ₹) CGWB (2019–2020)
NW Delhi	62.60
South Delhi	46.77
West Delhi	43.54
SW Delhi	39.26
NE Delhi	38.28
East Delhi	29.26
North Delhi	15.20
Central Delhi	9.97
New Delhi	2.43

Allocations(Revenue in %)

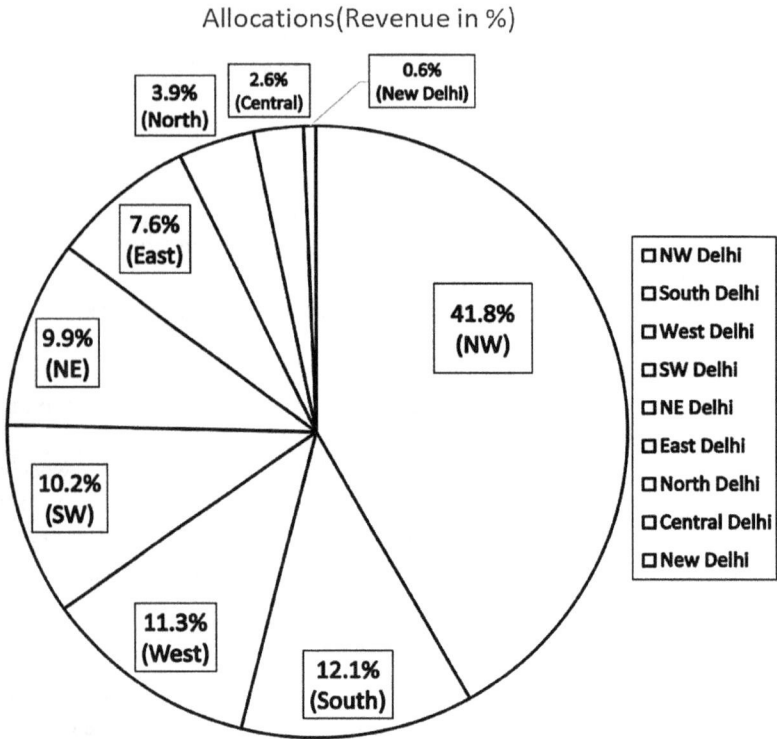

FIGURE 5.3   Allocations for revenue function.

5.5 CONCLUSIONS

The chapter deals with these six multi-objective functions and five constraints by keeping in mind the fact that 62,000 MLD of sewage is generated by urban areas while the treatment is only 37% of its which is a lot less, and moreover out of 816 STP plants, only around 500 are working, and 70% of sewage generated is left untreated, so there is a need to decrease the sewage production especially in urban areas because of increasing population, and that's why more resources are needed. This is the main reason of overutilization in the urban areas and consequently this would lead to an increase in the number of STP plants to reduce the quantity of untreated sewage but establishing these plants will incur a heavy cost, and also maintenance will require more efforts, and this cost will be covered from the revenue generated from water so there is a need to make a balance between these objectives. Moreover, if over-utilization increases then at some point in time these resources get depleted, and consequently supply is less than demand which eventually degrade the quality of water, and which again leads to an increase in the cost of providing clean water and parallelly decreases revenue. So, a slight change in a single objective will lead to a drastic change in other objective functions. Moreover, in optimizing these functions we face many challenges like there is a limitation on the budget part. Government has

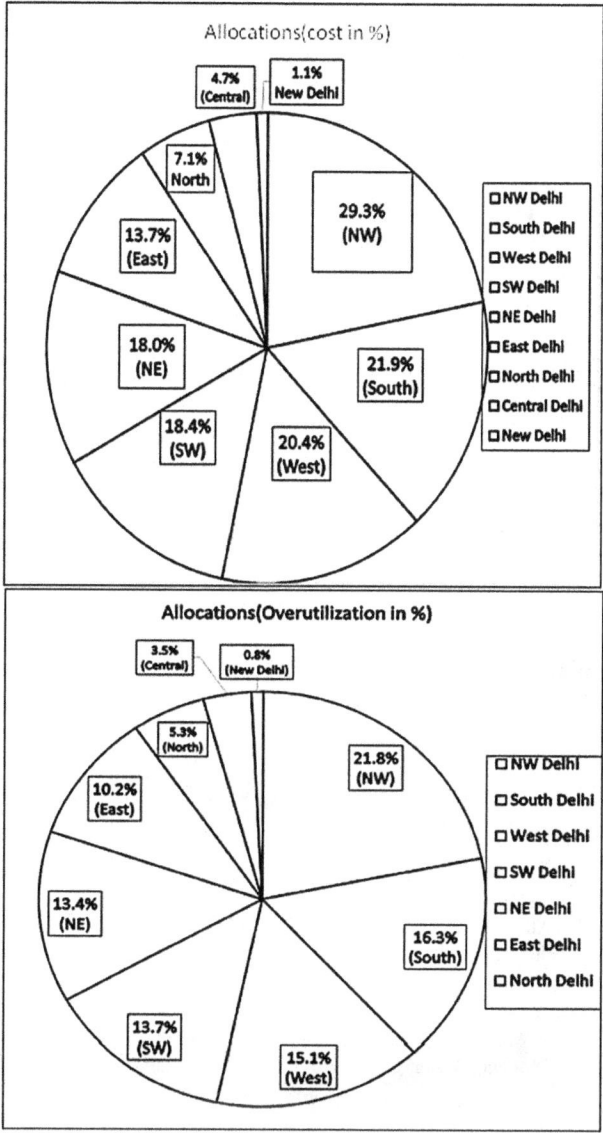

FIGURE 5.4   Allocations for cost function & overutilization function.

its own budget to install these STP plants and in implementing the policies. Demand and Supply are also not instantaneous. Moreover, if we succeed in conserving water then there is a chance that we can conserve the water resource for our future generation too. This technique is useful for the government, NGOs, for making the policy in advance. If so that the optimal utilization of resources be planned well in advance and then effectively utilize keeping in mind the future generation.

Water Allocation

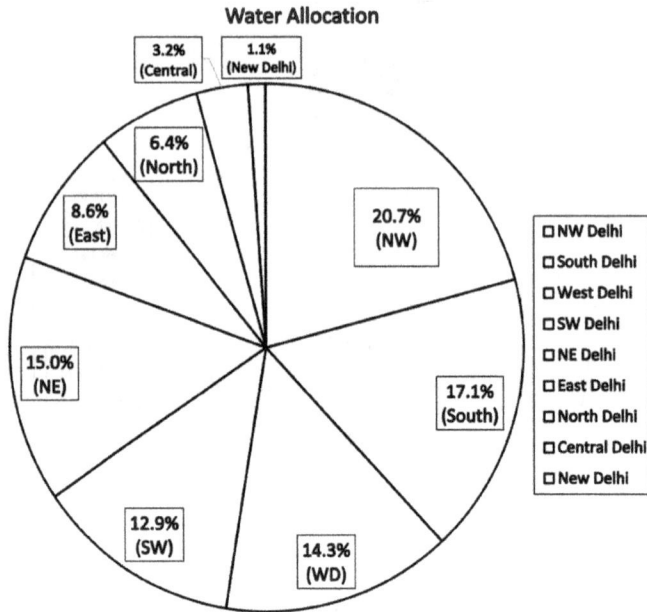

FIGURE 5.5   Water allocation (combination of 3 objectives).

TABLE 5.3
Water Management

Authors & Year	Past Research	Methodology Used
Calvin G. Clyde (1971)	Application of Operations Research Techniques for Allocation of Water Resources in Utah: Utah water research Laboratory	• A single Objective function i.e., Cost Function is taken by the Researcher. • In this the simple Linear Programming Approach is used to minimize the cost of ground water.
Jaume Freire G., Christopher A. Decker et al. (2018)	A Linear Programming Approach to Water Allocation during a Drought – Water	• Different Multi-Objective functions are taken simultaneously in terms of water allocations and Employment. • In this multiple-objective linear programming input/output method is used to allocate water resources during Drought condition.
Ma Guadalupe R., Fabricio R. et al. (2015)	Multi-objective optimization for designing and operating more sustainable water management systems for a city in Mexico – AIChE	• Two objective functions i.e., revenue and water consumption are taken. • In this the methodology used by researcher is the Pareto front Technique i.e., by taking two objective functions at a time.
V. Dample	Sectoral Allocation and Pricing of Groundwater	• In this, two factors are considered i.e., the cost and pricing factors involved in the management of water resources.

(*Continued*)

TABLE 5.3 (CONTINUED)

Authors & Year	Past Research	Methodology Used
Bithin D., Harikrishna V. (2005)	Optimization Applications in Water Resources Systems Engineering	• In this, the theoretical view of the integration of civil engineering with Optimization in water resource management is explained and Uncertainty of water is taken into consideration.
Mishra, Mukesh Kumar (2020)	Application of Operational Research in Sustainable Environmental management and Climate Change: ECONSTOR	• In this paper, the importance of operation research in managing the environment and how climate affects the resources is explained.
Thomas W. Archibald, Sarah E. Marshall (2018)	Review of Mathematical Programming Applications in Water Resource Management Under Uncertainty: Springer	• In this paper, Mathematical Models on the management of water resources under uncertainty is developed. • Stochastic Process and Dynamic Programming is used in this.
Bhardwaj R., Garg S. (2021)	Multi-Objective Optimization Techniques in Water Resource Management	• In this, 6 different objectives are taken. • Posteriori method (Weighing Method) is used to formulate mathematical model.

5.6 FUTURE SCOPE

This multi-objective optimization model can be used by the government to allocate water resources in the state/district by feeding the appropriate data into the model and by giving the priorities in terms of weights to each objective function. Moreover, as this is a linear model because both the objective function and its constraints are linear in nature, but it can be extended to non-linear optimization model.

ACKNOWLEDGEMENTS

We are sincerely thankful to Guru Gobind Singh Indraprastha University for this research.

REFERENCES

Alothaimeen I, Arditi D. 2019. "Overview of Multi-Objective Optimization Approaches in Construction Product Management." *Multicriteria Optimization-Pareto-Optimality and Threshold-Optimality*. London, UK: IntechOpen. https://doi.org/10.5772/intechopen.88185

Archibald TW, Marshall SE. 2018. Review of Mathematical Programming Applications in Water Resource Management under Uncertainty. *Environmental Modeling and Assessment* 23: 753–777. https://doi.org/10.1007/s10666-018-9628-0

Beria P, Maltese I, Mariotti I. 2012. Multicriteria versus Cost Benefit Analysis: a Comparative Perspective in the Assessment of Sustainable Mobility. *European Transport Research Review* 4(3): 137–152.

CGWB. 2019–2020. *Ground Water Year Book, NCT, Delhi*. Central Ground Water Board, Ministry of Jal Shakti, Government of India. http://cgwb.gov.in/Regions/NCT/GWYB_2019-2020_Final.pdf

Clyde, CG. 1971. *Application of Operations Research Techniques for Allocation of Water Resources in Utah*. Utah Water Research Laboratory, 1–83. https://digitalcommons.usu.edu/water_rep/531

Damle V. 2006. *Sectoral Allocation and Pricing of Groundwater*. http://117.252.14.250:8080/jspui/handle/123456789/4612

Datta B, Harikrishna V. 2005. *Optimization Applications in Water Resources Systems Engineering*, 57–63. https://www.researchgate.net/publication/242115216_Optimization_Applications_in_Water_Resources_Systems_Engineering

Ehrgott M. 2006. *Multicriteria Optimization*. Springer Science & Business Media. Boston, UK.

Gonzalez JF, Decker CA, Hall JM. 2018. A Linear Programming Approach to Water Allocation during a Drought. *Water* 10: 1–14.

Higgins AJ, Hajkowicz S, Bui E. 2008. A Multi-Objective Model for Environmental Investment Decision Making. *Computers & Operations Research* 35(1): 253–266.

Kiker GA, Bridges TS, Varghese A, Seager TP, Linkov I. 2005. Application of Multicriteria Decision Analysis in Environmental Decision Making. *Integrated Environmental Assessment and Management* 1(2): 95–108.

Lahdelma R, Salminen P, Hokkanen J. 2000. Using Multicriteria Methods in Environmental Planning and Management. *Environmental Management* 26(6): 595–605.

Marler RT, Arora JS. 2010. The Weighted Sum Method for Multi-Objective Optimization: New Insights. *Structural and Multidisciplinary Optimization* 41(6): 853–862.

Mendoza G, Martins H. 2006. Multi-Criteria Decision Analysis in Natural Resource Management: A Critical Review of Methods and New Modeling Paradigms. *Forest Ecology and Management* 230(1): 1–22.

Mishra M. 2020. Application of Operational Research in Sustainable Environmental Management and Climate Change. *ECONSTOR*, 1–16. http://hdl.handle.net/10419/215782

Papadatos A, Berger AM, Pratt JE, Barbano DM. 2002. A Non-linear Programming Optimization Model to Maximize net revenue in Cheese Manufacture. *Journal of Dairy Science* 85(11): 2768–2785.

Rojas MG, Napoles F, Ponce JM, Serna M, Guillen G, Jimenez L. 2015. Multi-Objective Optimization for Designing and Operating More Sustainable Water Management Systems for a City in Mexico. *AIChE Journal* 61(8): 2428–2446.

Singh RK, Varma SP, Kumar A. 2013. Application of Linear Programming techniques in Personnel Management. *IOSR Journal of Mathematics* 8(1): 45–48.

Vanderpooten D. 1989. The Interactive Approach in MCDA: A Technical Framework and Some Basic Conceptions. *Mathematical and Computer Modeling* 12: 1213–1220.

Vanderpooten D. 1990. "Multiobjective Programming: Basic Concepts and Approaches." *Book Chapter: Stochastic Versus Fuzzy Approaches to Multiobjective Mathematical Programming under Uncertainty*. Editors: S-Y Huang, J Teghem. Springer Science & Business Media, 7–22.

Watts ME, Ball IR, Possingham HP. 2009. Marxan with Zones: Software for Optimal Conservation-Based Land-and Sea-Use Zoning. *Environmental Modeling & Software* 24(12): 1513–1521.

Williams BK. 2011. Adaptive management of natural resources: framework and issues. *Journal of Environmental Management* 92(5): 346–1353.

Xevi E, Khan S. 2006. A Multi-Objective Optimization Approach to Water Management. *Journal of Environmental Management* 77(4): 269–277.

6 Tools in Decision-Making of Allocation of Non-Traditional Resources for Sustainable Water Development

Sahajpreet Kaur Garewal and
Avinash D. Vasudeo

CONTENTS

6.1   INTRODUCTION

Ground water is a ubiquitous natural resource which needs great attention due to increasing urban sprawl, overexploitation, and quality degradation. Surface and groundwater contamination is a major issue reported worldwide (Singh, 2014). The adulteration of groundwater is due to ingress of certain contaminants, which make its use restricted or sometime impossible. The groundwater quality degradation is

DOI: 10.1201/9781003203445-7

a concern due to its hazardous effect on human life and surrounding atmosphere (Samake et al., 2011). With the increasing consciousness regarding the importance of groundwater resources, various attempts have been made as a measure to diminish, eliminate, and prevent groundwater contamination.

To protect the groundwater from contamination, it is necessary to take effective measure at regional level. Groundwater vulnerability mapping can identify the region less or more susceptible to contamination and helps in making effective policies (Jang et al., 2015). Various methods has been used for groundwater vulnerability assessment in last few decades, out of which DRASTIC has been successfully implemented worldwide including China (Huan et al., 2012; Wu et al., 2016), India (Umar et al., 2009; Sinha et al., 2016), Iran (Javadi et al., 2011; Neshat et al., 2013; Barzegar et al., 2015; Sadeghfam et al., 2016), and many more.

The DRASTIC method (Aller et al., 1987) uses selected intrinsic parameters, assigned rates, and weights for groundwater vulnerability assessment. This, however, by some opinion (Sadeghfam et al., 2016), is a limitation of DRASTIC, as the changing geology and hydrology are not considered. But if some modification is to be done to the DRASTIC method, a detailed study of the area litho logs, stratigraphy, and data of monitoring wells are required, which would help in better understanding the area under study, and of any groundwater vulnerability assessment. The various modification has been applied to the standard DRASTIC to reduce its subjectivity and to enhance the result in their respective study (Leone et al., 2009; Sener and Davraz 2013; Sadeghfam et al., 2016).

In Nagpur City, due to increasing urbanization and careless disposal of untreated city waste, groundwater contamination has grown to be a serious environmental issue (Pujari et al., 2007; CGWB, 2011). However in the recent years because of the increasing awareness in the residents, the menace of haphazard disposal has declined; this is clear from the observation. Nag River is a main river of the city which traverse through West to East in the city is now becoming a sewage disposal drain due to direct disposal of untreated city waste. The contaminants concentration is found higher in the neighbouring area of Nag River and Bhandewadi (solid waste dumping site) which is located to the East, and also has very high population density due to the presence of many slums (CGWB, 2011). Due to continuous degradation in the quality and overexploitation of groundwater in the city, assessment of groundwater vulnerable zone is essential for protecting the regional groundwater by developing efficient policies for the management and improvement of the groundwater resources.

A contamination map obtained using quality data collected from the field and processed by interpolation method is essential for determining risk in an aquifer and is more logical for groundwater planning and management (Jafari et al., 2016). It has been found that incorporation of both the hydro-chemical and hydro-geological dataset enable more realistic evaluation of groundwater vulnerability of an aquifer (Pusatli et al., 2009).

In the present study, an attempt has been made to modify standard DRASTIC obtained using hydro-geological characteristic of an area by considering field quality parameter (hydro-chemical) to evaluate groundwater vulnerability map for Nagpur City.

6.2 STUDY AREA

Nagpur is a centrally urbanized city situated in the geographical centre of India (Figure 6.1). Nagpur City is managed by Nagpur Municipal Corporation (NMC), which covers an area approximately 218 sq. km. It is situated between $21°00'–21°15'$ North latitudes and $79°00'–79°15'$ East longitudes at 310 m amsl. The city is occupied by a series of metamorphic and crystalline formation in the east, and the west is covered by Deccan trap formation (CGWB, 2009). The topographical arrangement of the city is characterized by flat to mild slope in the East to steep slopes in the West. The city comprises two major water bodies, Nag River and Pilli River, which flow within the urban setting of the city. The climatic condition of city follows a mean rainfall of 1,000 mm and high temperature in summer approx. 47°C to cold winter when temperatures drop to 8°C.

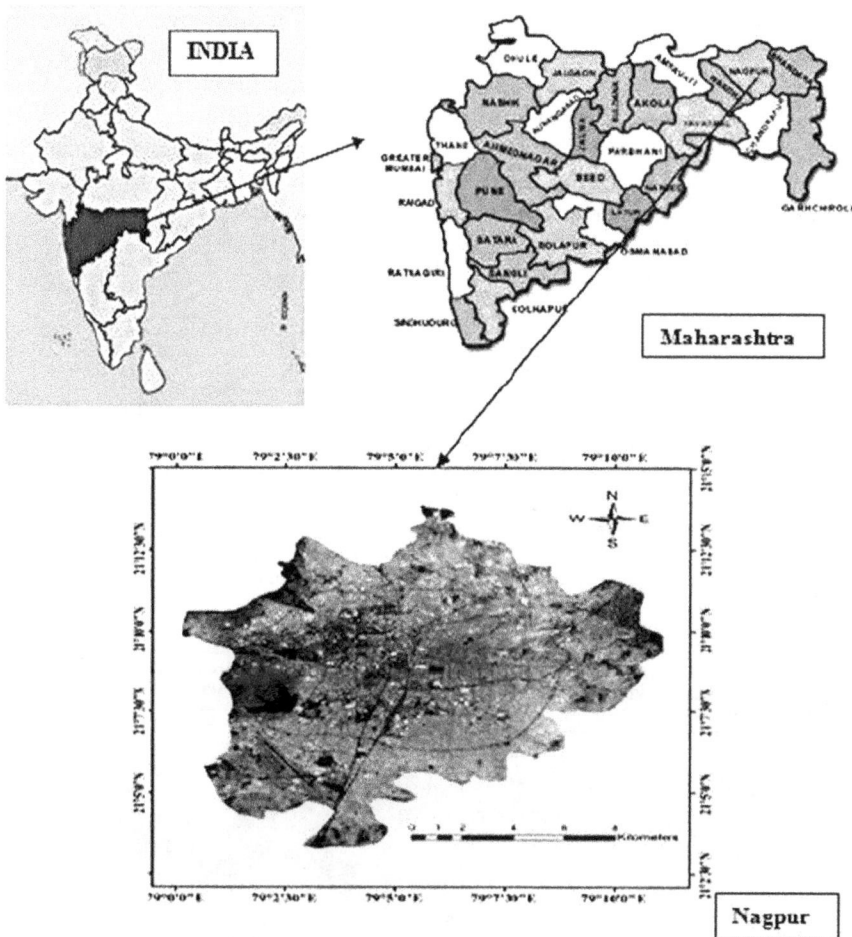

FIGURE 6.1 Map showing Nagpur City.

6.3 METHODOLOGY

For the study area, the groundwater vulnerability map is evaluated using the modified DRASTIC approach in which the field quality parameter is considered along with the intrinsic parameters of an aquifer. The methodology implemented for groundwater vulnerability assessment of Nagpur City is formulated in Figure 6.2.

The raw data used in the analysis are collected from reliable government organization and official website (Table 6.1). The collected data are in a different format like point data (groundwater level, rainfall and quality concentration), paper map (soil data and aquifer media), satellite image (DEM) and different projection system.

As collected data are not in appropriate format, they cannot be used directly for the analysis. So these data are imported into ArcGIS software to convert in a DRASTIC supportive (raster) format of a same projection system. Map showing detail of each parameter is prepared which shows different classification based on their characteristic. Thematic maps of these controlling parameters are prepared by assigning rates to the various classification of the parameters, according to their influence in groundwater quality. On the basis of each parameter impact on groundwater, weight is assigned to each thematic map and overlay analysis is performed to obtain groundwater vulnerability map. The field quality data are used to enhance the standard DRASTIC approach and to validate the resultant groundwater vulnerability maps. All the operations are performed in ArcGIS software.

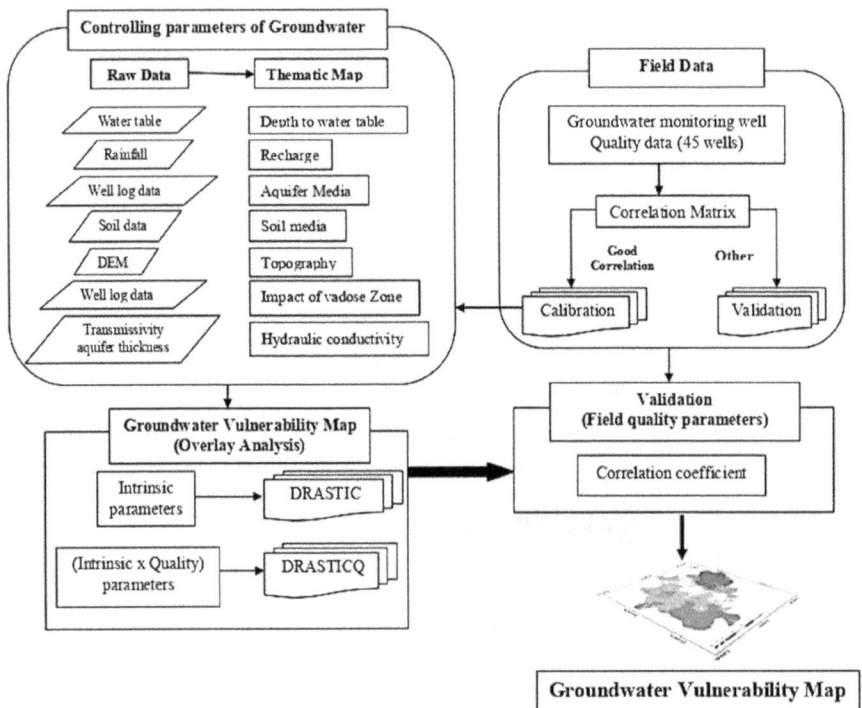

FIGURE 6.2 Methodology adopted in the present study.

TABLE 6.1

Different Sources of Data Collection

S. No.	Raw Data	Source
1	Groundwater level, saturated thickness, well log data, Pumping Data, Groundwater quality	CGWB – Central Ground Water Board
2	Soil data	NBSS – National Bureau of Soil Survey
3	Rainfall	IMD – India Meteorological Department
4	DEM	Bhuvan (official site)

6.3.1 Intrinsic Groundwater Vulnerability (Standard DRASTIC)

The DRASTIC method estimates the intrinsic vulnerability of groundwater considering the hydro-geological characteristics of aquifers following (Aller et al., 1987). DRASTIC is an overlay index method initialize by seven intrinsic parameters includes:

- D: Depth to the water table
- R: Recharge
- A: Aquifer media
- S: Soil media
- T: Topography
- I: Impact of vadose zone
- C: Hydraulic conductivity

The details classifications of each parameter are tabulated in Table 6.2 and classification of parameters in Table 6.3.

The Intrinsic Vulnerability Index (IVI) is calculated using Equation 6.1 including of all the seven standard parameters to which rates and weight are assigned using procedure documented in standard method generated by Aller et al. (1987) and tabulated in Table 6.3.

$$IVI = D_w x D_r + R_w x R_r + A_w x A_r + S_w x S_r + T_w x T_r + I_w x I_r + C_w x C_r \qquad (6.1)$$

where w is the weight, and r is the rates, assigned to seven standard parameters.

6.3.2 Sensitive Analysis

Napolitano and Fabbr (1996), developed Single Parameter Sensitivity Analysis (SPSA) is used in the present study, to identify the sensitive parameters for groundwater vulnerability of Nagpur City. The analysis was performed by comparing the calculated 'Effective' weight of the parameters with assigned 'Theoretical' weight of the respective parameter.

TABLE 6.2

Intrinsic Parameters Involve in Assessment of Groundwater Vulnerability Assessment

Parameter	Description	Preparation	Relevance
D	It shows the total media depth a pollutant has to pass through before reaching the aquifer.	The water level depths of 45 monitoring wells are collected from CGWB is interpolated using Kriging tool in ArcGIS to create a map	More the depth to groundwater less will be the chances of contamination
R	It infiltrates the contaminants present at the land surface to meet the water table.	Based on the literature (Baalousha, 2006; Gupta, 2014) 40% of rainfall is considered as recharge. From the point data, map is prepared using Kriging interpolation	Higher the rate of recharge more will be the infiltration of contaminant into the aquifer
A	It is a formation of consolidated or unconsolidated media on which the yield of groundwater depends (Aller et al., 1987).	From litho log data of CGWB, Nagpur	Higher the porosity of media or presence of fracture in the rock formation more will be the contamination
S	It is topmost layers at the land surface, which restricts the movement of contamination through recharge.	Soil map (Paper format) collected from NBSS. Geo-referencing of the map is done and processed in raster map	Larger the grain size of soil more will be the contamination as compared to finer soil
T	The water will infiltrate to media or become a part of runoff, depends upon the slope of the area.	In slope (%), extracted from DEM of the study area using ArcGIS	Flatter the slope higher will be the chance of contamination and vice versa for steeper slopes
I	It is unsaturated zone, significant for contaminant reduction because various reactions occur in this zone.	From Depth to water level and DEM using methodology expressed in (Li and Zhao, 2011)	Greater will be the Vadose zone depth lesser will be the probability of groundwater contamination
C	It is the ability of fluid to flows from the different aquifer media.	From Well log and pumping data of CGWB. By dividing saturated thickness of aquifer from transmissivity	Contaminants transportation will increase with increase in hydraulic conductivity

The parameter is said to be effective, if the evaluated effective weight is higher than assigned standard theoretical weight suggested by Aller et al. (1987) for evaluating the groundwater vulnerability. The "effective weight" is the outcome of the rating of individual parameter values and the respective assigned weight. The effective weight (W) is evaluated using Equation (6.2) of all the parameters.

$$W = \frac{P_w P_r}{IVI} \tag{6.2}$$

where P is the parameter, w is the weight, and r is the rate of respective parameter.

TABLE 6.3

DRASTIC Parameters Rates and Weights (Aller et al., 1987)

1. Depth to the water table (D) (m)

Sub parameter	R*	W
0.7–2.61	10	5
2.62–3.87	9	
3.87–4.69	8	
4.69–5.95	6	
5.95–7.86	4	
7.86–10.77	2	
10.77–15.2	1	

2. Recharge(R) (In mm)

Sub parameter	R	W
396–404	3	4
405–410	5	
411–416	7	
417–422	8	
423–433	9	

3. Aquifer Media(A)

Subparameter	R	W
Inter-trapean	1	3
Basalt (Massive)	4	
Amgaon – Gneiss complex	7	
Un-classified Gneiss–tirods	8	

4. Soil media(S)

Sub parameter	R	W
Clay loam	3	2
Clayey	7	
Alluvial	8	

5. Topography(T) (%)

Sub parameter	R	W
<2.7	10	2
2.7–5	9	
5–7.9	7	
7.9–11	5	
11–16	4	
16–23	3	
23>	1	

6. Impact of Vadose zone(I) (m)

Subparameter	R	W
0.6–3.2	8	5
3.2–3.9	7	
3.9–4.5	6	
4.5–5	5	
5–5.9	4	
5.9–7	3	
7–10.8	2	

7. Hydraulic Conductivity (C) (m/sec)

Sub parameter	R	W
$<10^{-6}$	5	3
10^{-5}–10^{-6}	6	
10^{-4}–10^{-5}	8	
10^{-3}–10^{-4}	9	

* R = rating and W = weight

6.3.3   Modified DRASTIC (DRASTICQ)

In DRASTIC, the parameters involved appraise the intrinsic vulnerability of an aquifer involve the transport process and attenuation of contaminant (Baalousha, 2006). The source of contamination, specific contaminant, and contamination scenario are not taken into consideration, as DRASTIC includes only the hydrogeological parameters. For reducing the subjectivity of DRASTIC, a quality index map is involved with the standard DRASTIC model. Quality index map takes into consideration the concentration of groundwater contaminant at a given location. Groundwater is inherently a mixture of many nutrients, chemicals, and salts, as well as contaminants, which may be due to parent rocks or by anthropogenic activities. Addressing data collection towards each contaminant and transforming the collected data to appropriate format is a herculean task. So, the preference of selecting a particular quality parameter was the availability of data in regards to that particular parameter and also its correlation with other quality parameters.

The quality parameter involved in the analysis is based on the quality classification of the contaminant. In the study, the quality data is collected from CGWB, Nagpur and the drinking water standard by Bureau of Indian Standard (BIS) is used for its classification. The Quality Index (QI) is evaluated by summation of parameters selected for DRASTIC modification using Equation 6.3 proposed by Pasatli et al. (2009).

$$QI = \Sigma (X_i)^2 \qquad (6.3)$$

where X is the determined grade of quality parameter (from 1 to 5) of contaminant 'i'.

The groundwater vulnerability Index (VI) using modified DRASTIC approach (DRASTICQ) is evaluated following Equation 6.4, to identify the area susceptible to groundwater contamination.

$$VI = IVI \times QI \qquad (6.4)$$

6.4   RESULTS AND DISCUSSION

6.4.1   Intrinsic Groundwater Vulnerability (Standard DRASTIC)

The map obtained using standard DRASTIC rate and weights (Table 6.3) showing intrinsic groundwater vulnerability of Nagpur City is shown in Figure 6.3(a).

The resultant DRASTIC map is divided into five classifications using natural break (Jenks) method (in ArcGIS), which shows area under different degree of groundwater vulnerability. From the close examination of the DRASTIC (IVI) map Figure 6.3(a) we can say that the South zone of the city is having minimum vulnerability index; where the vulnerability varies from moderate to high in the centre of city. The area lying in the North-East and North-West are having higher vulnerability index and are at higher risk of contamination.

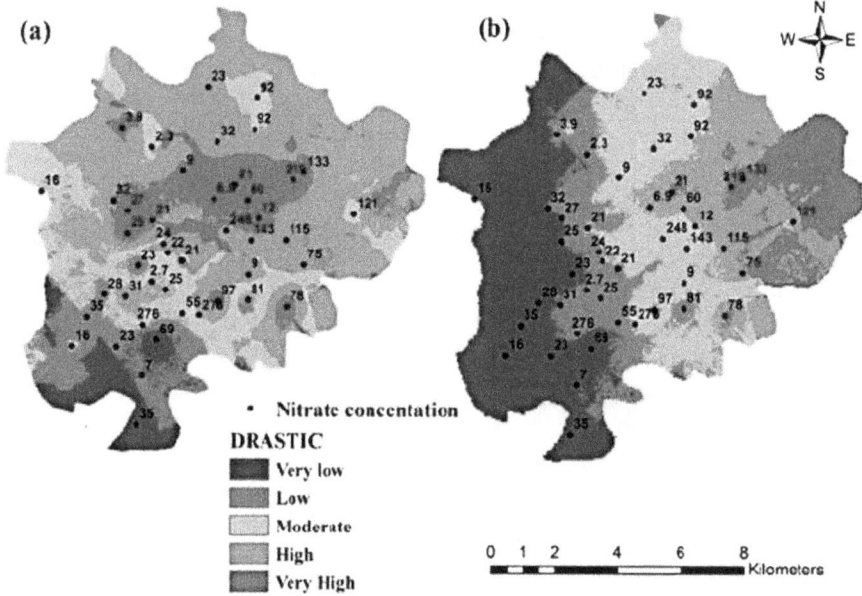

FIGURE 6.3 Map showing area under different degree of vulnerability in Nagpur City.

6.4.2 SENSITIVITY ANALYSIS

The SPSA analysis was executed in groundwater vulnerability map of the Nagpur City, to find out the most effective parameters. The result of SPSA (Table 6.4) reveals that recharge and topography are effective parameters for the study area showing effective weight 22.75% and 5.42% respectively, which is much higher than the assigned theoretical weight 17.4% and 4.35% respectively. The other parameters are also having higher effective weight and are important for assessment of groundwater vulnerability such as aquifer media, soil media, and hydraulic conductivity.

The depth to water level and impact of vadose zone are found to be less effective parameters for the study area as their calculated effective weight is smaller than assigned theoretical weight.

6.4.3 MODIFIED DRASTIC (DRASTICQ)

From correlation matrix of the quality parameters (Table 6.5), it was observed that TDS and TH show a positive correlation with all the quality parameters and are the most basic parameter evaluated for quality analysis of groundwater as per some norms of prevailing standards.

They are integrated in the inventory to account for the abandoned contaminants. The classification limit of TH and TDS as per drinking water standard is listed in Table 6.6. The monitoring data of the selected parameters (TDS and TH) are converted into integer numerical value of 1–5 (rates) considering the standard classification of I-V (Table 6.6) and the QI is evaluated considering TDS and TH using Equation 6.3.

TABLE 6.4
Result of Sensitivity Analysis

Parameter	Effective Weight (%)				Theoretical Weight	Theoretical Weight (%)
	Min	Max	Average	SD		
D	4	28	14.91	4.44	5	21.74
R	9	37	22.75	3.13	4	17.40
A	1	19	13.14	2.91	3	13.04
S	3	18	9.76	2.32	2	8.70
T	0	12	5.42	1.83	1	4.35
I	7	29	18.72	2.89	5	21.74
C	9	22	14.58	2.30	3	13.02

TABLE 6.5
Correlation Matrix between Different Quality Parameters

	TDS	TH	Na	Mg	Cl	No$_3$	So$_4$
TDS	1						
TH	0.757	1					
Na	0.826	0.461	1				
Mg	0.737	0.921	0.783	1			
Cl	0.876	0.822	0.809	0.783	1		
No3	0.766	0.739	0.525	0.768	0.708	1	
So4	0.898	0.656	0.785	0.768	0.827	0.556	1

TABLE 6.6
Classification of the Parameter Used in Quality Index Evaluation

Parameters (mg/l)	Drinking Water (IS 10500:1991, Revised 2003)]				
	Very Low	Low	Moderate	High	Very High
TDS	140–220	221–350	351–500	501–1065	1066–2510
TH	93–110	111–200	201–300	301–550	551–980

The Quality Index (QI) map is divided into five classifications varying from very high to very low same as IVI map. The obtained intrinsic vulnerability map and quality index map are normalized using Equation 6.5, so that units on a layer would be dimensionless and can be computed at the same numerical scale.

$$Q_{nor} = \frac{Q - Q_{min}}{Q_{max} - Q_{min}} \tag{6.5}$$

where Q_{nor} is the normalized map, Q is the parameter index map, and Q_{max} and Q_{min} are maximum and minimum index values of the respective map.

The map showing Groundwater Vulnerability Index (VI) of Nagpur City using modified DRASTIC approach (DRASTICQ) is evaluated considering the Intrinsic Vulnerability Index (IVI) map and Quality Index (QI) map via Equation 6.4. From the study of groundwater vulnerability map (DRASTICQ) of the city shown in Figure 6.3(b), it was observed that higher vulnerability index is found in the East zone while the Centre-North to South part lies under moderate degree of vulnerability and West to South zones are safe having least vulnerability index.

6.4.4 VALIDATION

The maps generated using standard DRASTIC and Modified DRASTIC (DRASTICQ) approach showing area under different degree of vulnerability is area specific, which need to be validated. An effort has been made to obtain the correlation between the resultant vulnerability maps and field contaminants such as Nitrate (NO_3), Sulphate (SO_4) and Chloride (Cl) using different correlation techniques shown in Table 6.7. Groundwater has no geological source of Nitrate; its presence shows the interference of anthropogenic activities on groundwater. The results reveal that DRASTICQ shows a good correlation with Nitrate as well as other field quality data in comparison to standard DRASTIC. The varying Nitrate concentration value of different monitoring wells located within the city are marked in vulnerability maps (Figure 6.4), which shows that vulnerability defined by DRASTICQ is closer to the real groundwater state in the study area.

6.5 DISCUSSION

The groundwater vulnerable zones defined in DRASTIC and DRASTICQ show areas under the different class of vulnerability. The resultant map evaluated using DRASTICQ shows better correlation with the actual field contaminant data (Table 6.7) and a good correlation is observed with Nitrate which signifies that the resultant map identifies the area susceptible to contamination due to hydrogeology of the area as well as anthropogenic activities at land surface.

Groundwater is physically protected by the earth surface; intrinsic parameters are derived from the property of protective layer and aquifer characteristics which are included as DRASTIC parameters, but they are not the only factor influencing the contamination. Quality index includes the presence of contaminant in groundwater which is combined with the fundamental aquifer characteristics which

TABLE 6.7

Correlation of DRASTICQ and DRASTIC with Different Contaminants

Correlation	DRASTIC			DRASTICQ		
	No_3	So_4	Cl	No_3	So_4	Cl
Pearson coefficient	0.206	0.321	0.229	0.702	0.806	0.845
Kendall's coefficient	0.051	0.274	0.142	0.568	0.787	0.709
Spearman coefficient	0.065	0.372	0.190	0.693	0.905	0.866

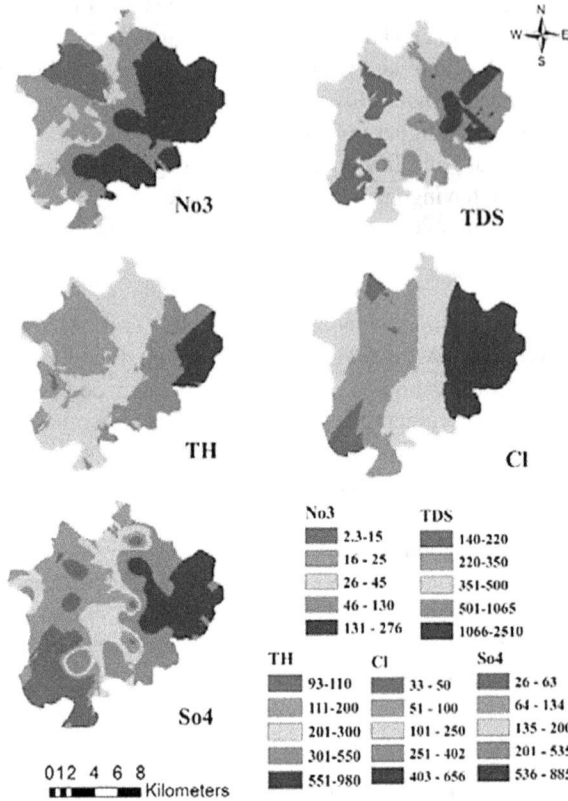

FIGURE 6.4   Spatial variations of contaminants within the city limit.

facilitate in better prediction of region which are more vulnerable to contamination and need attention.

6.6   CONCLUSION

From the study of DRASTIC and DRASTICQ groundwater vulnerability map, it is ascertained that the groundwater in the East region of the city is at higher risk of contamination. On the other hand, the groundwater at the South zone is having minimum risk of contamination. It can be a result of topography which is flat to mild slope in the East, the presence of solid waste disposable site and slum areas in the East. Also, due to drainage effect of major rivers flowing from West to East within the city. The result of single parameter sensitive analysis reveals that for groundwater vulnerability assessment of Nagpur City, recharge and topography are most effective parameters followed by aquifer media, soil media, and hydraulic conductivity. In DRASTIC only the hydro-geological parameters are considered and DRASTICQ includes both hydro-geological and hydro-chemical characteristic which shows more realistic results and resemblance with the actual field contaminant value. Therefore, the introduced method DRASTICQ shows better results than prevailing approach to evaluate the area under higher risk of groundwater contamination including hydro-geological and hydro-chemical parameters.

REFERENCES

Aller, L., Lehr, J., Petty, R., and Bennett, T. 1987. A standardized system to evaluate ground-water pollution using hydrogeologic setting. *Journal of the Geological Society of India* 29(1): 23–37.

Baalousha, H. 2006. Vulnerability assessment for the Gaza strip, Palestine using DRASTIC. *Environmental Geology* 50(3): 405–415, doi: 10.1007/s00254-006-0219-z

Barzegar, R., Moghaddam, A. A., and Baghban, H. 2015. A supervised committee machine artificial intelligent for improving DRASTIC method to assess groundwater contamination risk: A case study from Tabriz plain aquifer, Iran. *Stochastic Environmental Research and Risk Assessment* 30(1): 883–889, doi: 10.1007/s00477-015-1088-3

CGWB, 2009. *Ground Water Information Nagpur District.* Faridabad: Central Ground Water Board (CGWB).

CGWB, 2011. *Contamination of Ground Water by Sewage.* Faridabad: Central Ground Water Board.

Gupta, N. 2014 Groundwater Vulnerability Assessment using DRASTIC Method in Jabalpur District of Madhya Pradesh. *International Journal of Recent Technology and Engineering* 3(3): 36–43, doi: 10.4236/ijg.2016.74043

Huan, H., Wang, J., and Teng, Y. 2012. Assessment and validation of groundwater vulnerability to nitrate based on a modified DRASTIC model: A case study in Jilin City of northeast China. *Science of the Total Environment* 440: 14–23, doi: 10.1016/j.scitotenv.2012.08.037

IS 10500:1991, 1991. *Revised 2003 Indian Standard of Drinking Water.* Bureau of Indian Standard (BIS), New Delhi, India.

Jafari, F., Javadi, S., Golmohammadi, G., Mohammadi, K., Khodadadi, A., and Mohammadzadeh, M. 2016. Groundwater risk mapping prediction using mathematical modeling and the Monte Carlo technique. *Environmental Earth Science* 75(6): 491, doi: 10.1007/s12665-016-5335-9

Jang, C. S., Lin, C. W., Liang, C. P., Chen, J. S., 2015. Developing a reliable model for aquifer vulnerability. *Stochastic Environment Research and Risk Assessment,* 30(1), 175–187.

Javadi, S., Kavehkar, N., Mousavizadeh, M. H. and Mohammadi, K. 2011. Modification of DRASTIC Model to Map Groundwater Vulnerability to Pollution Using Nitrate Measurements in Agricultural Areas. *Journal of Agricultural Science and Technology* 13(2): 239–249.

Leone, A., Ripa, M., Uricchio, V., Deak, J., and Vargay, Z. 2009. Vulnerability and risk evaluation of agricultural nitrogen pollution for Hungary's main aquifer using DRASTIC and GLEAMS models. *Journal of Environmental Management,* 90: 2969–2978.

Li, R., and Zhao, L. 2011. Vadose zone mapping using geographic information systems and geostatistics. *IEEE,* 336: 3177–3179.

Napolitano, P., and Fabbr, A. 1996. Single-parameter sensitivity analysis for aquifer vulnerability assessment using DRASTIC and SINTACS. *HydroGIS 96: Application of Geographical Information Systems in Hydrology and Water Resources Management.* Vol. 235, pp. 559–566. Vienna: IAHS.

Neshat, A., Pradhan, B., Pirasteh, S., and Mohd Shafri, H. Z. 2013. Estimating groundwater vulnerability to pollution using a modified DRASTIC model in the Kerman agricultural area, Iran. *Environmental Earth Science* 71(7): 3119–3131, doi: 10.1007/s12665-013-2690-7

Pujari, P. R., Pardh, P., Muduli, P., Harkare, P., and Nanoti, M. V. 2007. Assessment of pollu-
    tion near landfill site in Nagpur, India by resistivity imaging and GPR. *Environmental
    Monitoring and Assessment*, 131: 489–500, doi: 10.1007/s10661-006-9494-0
Pusatli, O. T., Camur, M. Z., and Yazicigil, H. 2009. Susceptibility indexing method for irrigation
    water management planning: Applications to K. Menderes river basin, Turkey. *Journal
    of Environmental Management*, 90: 341–347, doi: 10.1016/j.jenvman.2007.10.002
Sadeghfam, S., Hassanzadeh, Y., Nadiri, A. A., and Zarghami, M. 2016. Localization of ground-
    water vulnerability assessment using catastrophe theory. *Water Resources Management*,
    30(13): 4585–4601, doi: 10.1007/s11269-016-1440-5
Samake, M., Tang, Z., Hlaing, W., Ndoh Mbue, I., Kasereka, K., and Balogun, W. O. 2011.
    Groundwater vulnerability assessment in shallow aquifer in Linfen Basin, Shanxi
    Province, China using DRASTIC model. *Journal of Sustainable Development*, 4(1):
    53–71, doi: 10.5539/jsd.v4n1p53
Sener, E., and Davraz, A. 2013. Assessment of groundwater vulnerability based on a modi-
    fied DRASTIC model, GIS and an analytic hierarchy process (AHP) method: the case
    of Egirdir Lake basin (Isparta, Turkey). *Hydrogeology Journal* 21(3): 701–714, doi:
    10.1007/s10040-012-0947-y
Singh, A. 2014. Conjunctive use of water resources for sustainable irrigated agriculture.
    Journal of Hydrology, 519: 1688–1697, doi: 10.1016/j.jhydrol.2014.09.049
Sinha, M. K., Verma, M. K., Ahmad, I., Baier, K., Jha, R., and Azzam, R. 2016. Assessment
    of groundwater vulnerability using modified DRASTIC model in Kharun Basin,
    Chhattisgarh, India. *Arabian Journal of Geosciences* 9(98): 1–22, doi: 10.1007/
    s12517-015-2180-1
Umar, R., Ahmed, I., and Alam, F. A. 2009. Mapping groundwater vulnerable zones using
    modified DRASTIC approach of an alluvial aquifer in parts of Central Ganga Plain,
    Western Uttar Pradesh. *Journal Geological Society of India* 73(2): 193–201, doi:
    10.1007/s12594-009-0075-z
Wu, H., Chen, J., and Qian, H. 2016. A modified DRASTIC model for assessing contami-
    nation risk of groundwater in the northern suburb of Yinchuan, China. *Environmental
    Earth Science* 75(6): 1–10, doi: 10.1007/s12665-015-5094-z

7 Soft Computing Techniques for Forecasting of Water Demand

Prerna Pandey, Shilpa Dongre, and Rajesh Gupta

CONTENTS

7.1 INTRODUCTION

Analyzing and forecasting urban water demand is a complex yet imperative task, as cities must meet the water demands of every citizen. Urban water demand changes hour-to-hour in a day, day-to-day in a week, season-to-season in a year, and is

DOI: 10.1201/9781003203445-8

also affected by population growth, and conservation scenarios. There exist three types of demand forecasting horizons. The first two are short-term and medium-term forecasts, which provide a forecast for hourly and monthly basis and are used for operation and management. The third is the long-term, which provide forecast for yearly basis and is necessary for planning and infrastructure design (Bougadis et al., 2005). Currently, water managers produce demand estimates using long-term climate trends. However, climate change introduces uncertainties that may limit the accuracy of this method, as historical trends will no longer be reliable for predicting future climate sensitive water demand (Gober et al., 2010).

Forecasting plays a vital role in WDNs operations for making strategic and planning decisions. However, it is worth noting that the forecasted values are just projections, and do not provide any exact value. The aim is to reduce the error between the projected values and the real values, with the help of forecasting tools and more sophisticated models. Several forecasting methods have been developed by researchers that uses the previous water consumption records. These methods include autoregressive integrated moving average (ARIMA), support vector machine (SVM), Artificial neural network (ANN), Evolutionary Artificial Neural Networks (EANNs), Long Short-Term Memory (LSTM), Convolutional Neural Network (CNN).

7.2   BACKGROUND

Water demand forecasting is the process of carrying out predictions about the future water use based on knowledge of past water demand pattern (Billings and Jones, 2011). Such forecasts will be very useful to make futuristic decisions like how much water supply or storage is required, how large the supply system needs to be designed in order to fulfil peak demand, and changes in water utility revenues in the future.

7.2.1   WATER DEMAND

There are different interpretations of the definition of the term, 'water demand' by different authors. Billings and Jones (2011) defined the term as the total volume of water necessary or required to supply to the consumer over a period of time, including leakages and all other losses. In these terms, total system water demand and total water production are directly proportional to each other. On the other hand, Rinaudo (2015) defined water demand as the quantity of water required by every consumer end for the various uses such as domestic, commercial, industrial, and agriculture. The term 'water demand' is alternatively used as 'water consumption' in the following sections.

7.2.2   FACTORS AFFECTING WATER DEMAND

There are a number of factors that have significant impact on water demand forecast. These are population, weather and climate change, migration of population to urban areas, employment, economic cycles etc. Some of these factors and their influence on the water demand is discussed.

- *Population*:

  The exponential growth of population in every country is the major trend factor in the rise in water demand. Migration of population to city and towns from rural areas for better opportunities, is another contributing factor. This however results in reduced water consumption in rural areas. Both these factors – population increase and migration of people – introduces an inherent uncertainty in the accurate prediction of water demand.

  The increase in industrialization and commercialization also impacts the rise in water demand, due to increase in the number of working professionals. The change in lifestyle also has a major impact on the water demand, such as rise in one's income, and leads to major use of available resources through many ways. Hence such abrupt and unaccountable change in consumption question the reliable estimate of future water consumption. Therefore, an accurate prediction model is needed.

- *Weather and climate change*:

  Climate behaviour prediction is one of the most challenging jobs for the meteorological department in a country. The variability of climate change affects many sectors such as, agriculture, domestic use etc. The seasonal variation, for example in summer, peak demand is very high, while at peak winter, the demand is low, however continuous flow is required to be maintained in order to avoid the freezing. Therefore, it is very necessary to determine the climate variables such as, humidity, rainfall, temperature etc., their inter-relationships and their impact on various sectors (Masri et al., 2019).

- *Lack of data in developing countries*:

  As water metering system at residential household is very uncommon, maintaining water consumption data or records causes major problems in water demand estimation in developing countries, compared to developed countries, where (almost) every house has a metered connection with proper laid pipe systems. Even though some households have metered connections, they might have separate connections from wells, bores, public taps, or may have purchased water from private vendors. Hence, it becomes very difficult for policy-makers to achieve accurate water consumption data, which ultimately makes, actual forecasting, very difficult.

7.2.3 PREPROCESSING OF DATASET

The available dataset obtained from the government offices usually lacks certainty due to improper management of data in developing countries. Hence, after selecting the model for demand forecast, the dataset requires pre-processing. There is need to train and test the dataset first, followed by the validation. In the training process, the plot is generated using the given dataset, and then it is tested using the selected model. Such validation of model requires the dataset to be divided suitably in percentages, such as 60% dataset is used for testing and 40% for training. These distributions can be changed according to the requirement.

7.2.4 Forecasting Horizons

The water demand prediction is divided into different horizons, as per the need, as shown below:

- **Short-term horizon:** Forecast horizons in the short term are commonly for hourly or monthly bases (Donkor, 2014). Such predictions are advantageous for optimizing and managing system operations and pumping. Forecast errors in the short term primarily arise from the inherent variability and unpredictability of the weather and human behaviour. These effects can be noticed and effectively analyzed using short-term forecasting methods.
- **Medium-term horizon:** This forecast horizon is for forecasting on an hourly or weekly basis. These are useful in the decision-making of budget, and revenue forecasting, and staging treatment and distribution system. These are basically focused on variability of water consumption by fixed or changing customer behaviour. Variation in demand under this horizon is basically noticed due to weather variability.
- **Long-term horizon:** These are generally useful to develop forecast horizon for decades, and sometimes even for 30 years. This helps in deciding the sizing of system capacity and raw water supply. The error in such cases arises basically due to the length of the forecast period. However, of all three stated horizons, this one possesses the highest risk: if a developed utility is larger than planned or required, it will give rise to increase in capital cost. Also, if the system's long-term capacity is inadequately addressed, it causes a shortage of water.

7.3 FORECASTING METHODS

Many advanced models have been developed in the recent past with the advent in technology, computer system, and the data availability. These models provide quite accurate forecast, if the data used is of high accuracy. These models can be grouped as: (i) Statistical Approach based models; and (ii) Soft Computing approach-based models.

7.3.1 Statistical Approaches

These approaches are in use since decades. The extrapolation of the previous year's data is one of the approach easy to opt. These are still in use and adopted by many government departments. The different types of statistical used approach are as shown in the following sections.

- *Traditional methods*:
    Traditional approaches use census data to predict future population. The water demand is then calculated by multiplying projected population with the per capita demand. Numerical methods (Arithmetic Increase (AI), Geometric Increase (GI), Incremental Increase (II), etc.) and Graphical

methods (Graphical Comparison, Logistic Curve, and Graphical Extension (GE)) are commonly adopted approaches (CPHEEO, 1999). These methods are more useful for long-term water demand forecasting, where population growth curve is usually a S-shaped curve. In S-shaped curve, initially population increase with time is observed proportional to the population, in the middle portion, rate of population increase becomes constant and finally it declines and moves towards saturation population. The accuracy of the forecast method depends on the present status of the city, i.e., whether it is in lower, middle, upper, or transitional part of growth curve. Many government authorities use average rate of increase obtained by two-three methods to reduce the error in prediction.

• *Time series models*:

   The time series models are useful for short- and medium-term forecast, wherein fluctuation in demand is observed with time. A time series is considered to be stationary, if its statistical parameters do not change with time. Further, it will be considered as linear, if any value in the series can be obtained as linear combination of past and future values.

ARIMA proposed by Al-Saati (1998) is one of the widely used linear model in time series forecasting. Arandia et al. (2016) presented a seasonal ARIMA (SARIMA) models with data assimilation. The tailoring process was adopted for identifying, estimating, and validating the models, and exploring how the length of demand history used in fitting can improve forecast performance. Oliveira et al. (2017) introduced a double seasonal ARIMA model in which Harmony Search was applied to the parameter estimation of the ARIMA model based on historical water demand data.

The Empirical mode decomposition method (EMD) as introduced by Huang et al. (1998) is found suited to both non-linear and non-stationary time series (Pandey et al., 2021). The time series is decomposed into numerous intrinsic mode function (IMF) and one residual using EMD. EMD is combined with ARIMA, ANN and feed forward back propagation NN (FFBP-NN) (Anele et al., 2017), and least square support vector machine (LSSVM) (Shabri et al., 2015) for the water demand forecast.

Wu and Huang (2009) added finite white noise to the original data, to overcome the problem of mode mixing that is inherent in EMD and suggested Ensemble empirical mode decomposition (EMDD). Wu et al. (2009) suggested random white noise having zero mean and specified standard deviation (Pandey et al., 2021), to be summed to available time series to decompose the time series by EEMD. Xu et al. (2018) proposed a dual-scale deep belief network (DSDBN) approach. In this, the original daily water demand time series was decomposed into several IMFs and one residue component with EEMD method. The DSDBN is observed to perform well as compared to EEMD-SVR, EEMD-ARIMA, and EEMD-FFNN.

Pattern sequence-based forecasting (PSF) method was proposed by Martínez–Álvarez et al. (2008) and improved by Álvarez et al. (2011). The preprocessing of data is done by means of normalization and clustering. Forecasting is carried out using the information provided by clustering and followed by denormalization. The forecasting results with the PSF method are degraded with respect to positive or negative trends. To overcome this problem, the DPSF model was proposed by

Pandey et al. (2021). Pandey et al. (2021) proposed a hybrid model by combining EEMD and DPSF. The proposed model was compared with several other hybrid models including a combination of EEMD, DPSF, and ARIMA. When compared with other statistical and soft computing approach the proposed method observed to works well.

7.3.2 Soft Computing Approaches

Soft computing methods include varieties of model which are better than statistical-based approaches when dealt with the imprecision and non-linearity of the data.

7.3.2.1 Artificial Neural Networks (ANN)

ANN belongs to machine learning family based on biological neural networks (Bennett et al., 2013). For several reasons, ANNs are an appealing tool for forecasting. To begin, they make fewer assumptions than traditional statistical models. Secondly, they can frequently generalize the results and forecast the output of previously unknown data. Further, they are capable of modeling highly non-linear relationships in data and accurately estimating any non-linear function. ANNs can forecast the future values of potentially noisy multivariate time series using historical data.

Various types of ANN models have been suggested for water demand prediction by different researchers. Feed Forward Neural Network (FNN) are the simplest one. FNN are used for water demand forecasting (Bougadis et al., 2005), and these are shown better over regression models. FNN have been improved and several other forms of FNN like wavelet ANN (WA-ANN) (Adamowski et al., 2012), radial basis function neural networks (RBNN) (Bennett et al., 2013), recurrent neural network (RNN) (Sutskever et al., 2014) etc., are developed. RNN is a type of ANN that forms a directed cycle of neurons; however, they are inefficient when learning patterns with long-term dependency (Gers et al., 2000). The LSTM can transport information across multiple time steps, preventing early signals from fading. Mu et al. (2020) were the first to use an LSTM-based model to predict water demand in on Hefei, China dataset. Later, Du et al. (2021) proposed DWT-PCA-LSTM, a hybrid LSTM model combining discrete wavelet transform (DWT) and principal component analysis (PCA).

Input data in ANN may be imprecise and can be defined using fuzzy relationships of uncertainty (Mamlook et al., 2009). Altunkaynak et al. (2005) used the fuzzy logic for water demand forecasting and shown the accuracy in prediction with less than 10% relative error. Most of the other approaches combines fuzzy with ANN resulting in neuro-fuzzy models (ANFIS). Neuro-fuzzy models combine the pattern recognition capabilities of artificial neural networks with the reasoning capabilities of fuzzy logic. Vijayalaksmi and Babu (2015) have shown the application of ANFIS on different datasets and observed them to perform better than the ANN. ANN, however, require a very large amount of data, long training time, and significant computing power for large datasets.

7.3.2.2 Support Vector Machine (SVM)

As a machine learning method for learning tasks such as classification and regression, Boser et al. (1992) developed SVM. The SVM is based on the idea that non-linear trends in the input space can be mapped to linear trends in a higher-dimensional feature space and can be used as a learning algorithm to recognize subtle patterns in complex datasets. An SVM's generalization property is not dependent on the entirety of the training data; only the support vectors are required for generalization. Unlike neural networks, which seek to minimize empirical error, SVM seeks to minimize the generalization error's upper bound (Vapnik, 2013). SVMs are classified into two types: support vector classifiers and regressions (SVR). By minimizing the generalization error bound, SVR attempts to achieve generalized performance. Many researchers have shown its application on water demand forecast by comparing it with other models (Zhang et al., 2013).

7.3.2.3 Metaheuristic Models

The metaheuristic models are solution approaches that coordinate the interaction of local optimization processes with higher-level strategies to build a process capable of escaping local optima and conducting a comprehensive search of a solution space (Glover and Kochenberger, 2006). Metaheuristics is an approach that entails encoding the tendency of a natural phenomenon towards improvement in mathematical symbols and codes and reducing problem-solving procedures to an algorithm that faithfully reproduces the dynamics of the metaphorically deployed phenomenon. Metaheuristics are typically employed in water demand forecasting models to modify the parameters of other approaches, estimate the coefficients of a function, or train an intelligent agent such as an ANN. Shirkoohi et al. (2021) opted ANN-GA, while Gutierrez-Estrada (2009) used hybrid FNN, fuzzy logic, and GA for demand forecasting.

7.4 ASSESSMENT OF FORECASTING MODELS

Determining how well a model performed is one of the most important steps of forecasting. The performance of any forecasting method can be evaluated by four commonly used indexes: root mean square error (RMSE); mean absolute error (MAE); mean absolute percentage error (MAPE). These can be mathematically expressed as below:

$$RMSE = \sqrt{\frac{1}{N} \sum_{i=1}^{N} \left| X_i - \hat{X}_i \right|^2} \qquad (7.1)$$

$$MAE = \frac{1}{N} \sum_{i=1}^{N} \left| X_i - \hat{X}_i \right| \qquad (7.2)$$

$$MAPE = \frac{1}{N} \sum_{i=1}^{N} \frac{\left| X_i - \hat{X}_i \right|}{X_i} \times 100\% \qquad (7.3)$$

where, X_i and $\hat{X}_i$ are the measured and predicted data at the time of t, and N is the number of predicted values.

RMSE represent the overall performance of the model, but are subtle with the larger errors. MAE does not assign higher importance to larger or smaller errors; it is merely an indicator for overall agreement between observed and predicted value. MAPE is the most accurate assessment as it does not depend on the units and hence easy to compare results from different studies.

7.5 SOFT COMPUTING METHODOLOGIES

Four models ANN, LSTM, fuzzy, and ANFIS are brief described in this section and applied to a case study in next section.

7.5.1 ARTIFICIAL NEURAL NETWORK (ANN)

A three-layer MLP neural network with one hidden layer is used to estimates hourly and daily water use. Hourly water demand forecasting uses data from the previous 24 hours and a binary index to indicate the day (weekday or weekend) to forecast demand for the next 24 hours. The number of neurons in the hidden layers is determined during model calibration; the goal is to use the fewest possible neurons without compromising forecasting accuracy. This layer uses a log sigmoid transfer function while the output layer uses a pure linear transfer function. The Levemberg Marquardt algorithm estimates network parameters, weights, and bias (Hagan and Menhaj, 1994). To avoid overfitting during the calibration phase, the calibration dataset is split into two subsets, each containing 80% of the data. The first subset trains the network, while the second tests it. To avoid signal saturation, the data is normalized and scaled to [0:1]. Normalization is done using the daily needs mean and standard deviation from the calibration dataset, with weekdays and weekends separated. This denormalizes and scales the network outputs.

7.5.2 LONG SHORT TERM MEMORY (LSTM)

The forecasting of water demand using LSTM is carried out using Python library. The following steps are involved:

Step 1: Visualization of the raw data for reference after importing. Find the correlation matrix for all factors influencing water demand if multivariate analysis is performed.

Step 2: To normalize the dataset, apply Min-Max Normalization so that all demands range from 0 to 1. This facilitates processing and speeds up computation.

Step 3: Use the source code to define the generator function and parameters. The parameters must be adjusted for each dataset. The gate weightages are set automatically.

Step 4: Flatten the model's layers for initialization. Apply LSTM layer next. Since demand data never goes negative, ReLU activation is used here.

Step 5: Test values for parameters like batch size, lookback, delay, min index, max index, epochs, validation split, etc. During the program's execution, the losses must decrease. Overfitting the model is also undesirable.

Step 6: Predict the values with the trained model. Plot results and validate the prediction generator function.

Step 7: Normalize the predicted demand outputs from Step 6. Remove the normalization function used in Step 2 and de-normalize them.

The LSTM network computes mapping between input and output sequence using some equations. Readers can refer Mu et al. (2020) for more details,

7.5.3 FUZZY LOGIC

The steps involved in fuzzy logic includes (i) Fuzzification of the input and output variables by using convenient language subsets (such as high, medium, low, heavy, light, hot, warm, huge, little); (ii) Construction of fuzzy IF–THEN rules based on expert knowledge and/or existing literature to represent the situation; (iii) The rules connect the input linguistic subsets to the output fuzzy sets; (iv) To defuzzify the result, the premise (antecedent) input subsets needs to be defuzzified, and then the rules must be defuzzified. For the study, Takagi-Sugeno (TS) method for defuzzification of data and triangular membership function for input data is used.

7.5.4 ADAPTIVE NEURO-FUZZY INTERFACE SYSTEM (ANFIS MODEL)

For the purpose of forecast, Artificial Neural Network with Takagi–Sugeno fuzzy inference system is used. The interface system involves following layers

Layer 1 (Input Layer): Only transmits input values to the next layer. This layer accepts m input variables.

Layer 2 (fuzzification layer): This layer finds the normalized value for each input. This layer's output is a membership value that indicates how closely an input value matches a fuzzy set.

Layer 3 (AND Operation Layer): This layer compares input values and finds the minimum. So, it considers the minimum value as an output. The output of this layer is the firing strength of the corresponding fuzzy rule.

Layer 4: This layer selects the set of fired rules that correspond to a set of input parameters. This layer selects a rule from the maximum number of rules present.

Layer 5: It computes the overall output by summing all the incoming signals.

7.6 APPLICATION OF MODELS

The application of the models is shown on two data sets, e.g., a Spanish dataset and an Indian dataset.

7.6.1 SPANISH DATASET

This case study is performed on water demand dataset measured at site located in a hydraulic sector zone in a city of south-eastern Spain (Pandey et al., 2021). The city has a population of approximately 5,000 consumers and extends over nearly 8.0 km². This dataset is considered for water demand prediction because of its appropriate seasonal variations and to compare the results obtained by similar studies (Anele et al., 2017). The basic statistical characteristics of this dataset are shown in Table 7.1.

Figure 7.1 shows the hourly water consumption pattern over a duration of 15 weeks. The first step in using any forecasting models is to train the dataset, and thereafter validating it for further time steps. Out of the total dataset, 70% is used for training, and the remaining 30% of the data is used for testing the models.

Each model is implemented on predefined training block and its accuracy is compared for the testing dataset. The performance of different approaches is compared with each other. For better understanding of each model at every time step, prediction is carried out for one step ahead. The detailed comparison of each model for short term water demand prediction (m³/hr) is done for predictions of 6, 12, and 24 hours.

TABLE 7.1
Detail Summary of Spanish Dataset

Min	Median	Mean	Max	Std. Dev.
−5.43	17.25	19.01	55.39	8.26

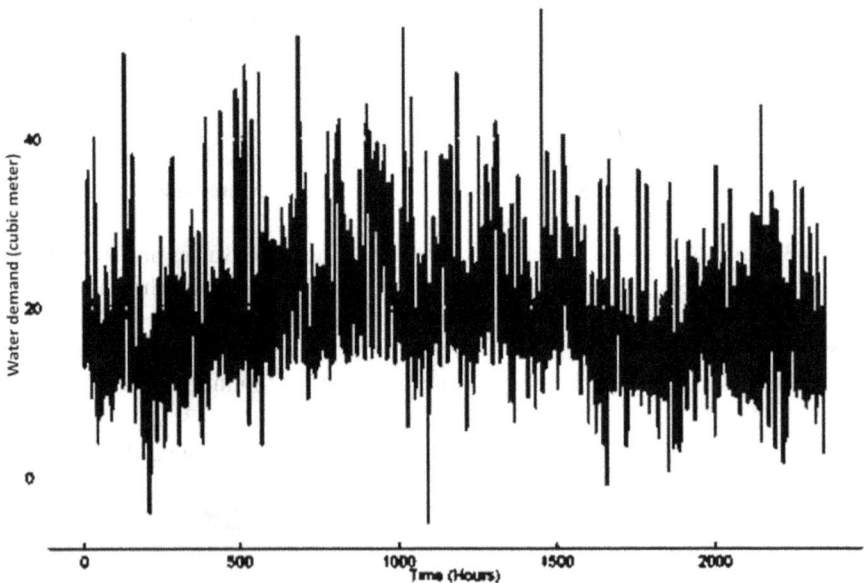

FIGURE 7.1 15 weeks' hourly water consumption dataset for of a city of Spain.

The assessments of these models in terms of different errors like RMSE, MAE and MAPE are carried out as shown in the Tables 7.2–7.4. In almost every case, the LSTM has proved to perform better among other soft computing approaches.

For the prediction with neural network, FNN model is adopted. The model parameter includes two hidden layers, which makes the information travel in one direction,

TABLE 7.2

Comparison of Soft Computing and Statistical Forecasting Models Using RMSE for Spanish Dataset (m³/hr)

	Soft Computing Approach				Statistical Approach					
Prediction Horizon	ANN	Fuzzy Logic	ANFIS	LSTM	EEMD-DPSF	EEMD-DPSF-ARIMA	EEMD-ARIMA	EEMD-PSF	EEMD-PSF-ARIMA	ARIMA
6	5.25	8.943	6.705	4.402	4.472	4.958	8.724	7.426	8.187	6.81
12	7.102	9.875	7.851	5.298	5.357	6.687	9.598	8.018	8.858	7.997
24	8.23	9.324	8.417	7.106	7.269	7.79	9.127	8.834	8.883	8.547

TABLE 7.3

Comparison of Soft Computing and Statistical Forecasting Models Using MAE for Spanish Dataset (m³/hr)

	Soft Computing Approach				Statistical Approach					
Prediction Horizon	ANN	Fuzzy Logic	ANFIS	LSTM	EEMD-DPSF	EEMD-DPSF-ARIMA	EEMD-ARIMA	EEMD-PSF	EEMD-PSF-ARIMA	ARIMA
6	3.413	7.615	5.217	2.998	3.088	3.291	6.654	6.108	7.401	5.321
12	4.315	7.861	5.896	3.568	3.685	4.186	7.607	6.681	7.277	6.093
24	5.876	8.325	6.785	4.278	4.352	5.661	8.191	6.89	7.528	6.941

TABLE 7.4

Comparison of Soft Computing and Statistical Forecasting Models Using MAPE for Spanish Dataset (%)

	Soft Computing Approach				Statistical Approach					
Prediction Horizon	ANN	Fuzzy Logic	ANFIS	LSTM	EEMD-DPSF	EEMD-DPSF-ARIMA	EEMD-ARIMA	EEMD-PSF	EEMD-PSF-ARIMA	ARIMA
6	13.945	31.781	17.105	9.215	13.604	13.723	18.637	18.112	31.168	17.223
12	14.742	31.013	18.627	10.897	13.23	14.493	20.819	19.021	30.653	18.724
24	17.408	34.251	26.845	12.943	16.155	17.293	22.735	24.032	33.909	27.099

calculating the model weights through the layers and number of neurons between 10–50. The sigmoid activation function is adopted along with Levenberg-Marquardt as training function. The number of epochs used was 100. For a LSTM model, the model parameters adopted are, two hidden layers; learning rate of 0.0085; activation function is tanh; number of epochs required at the end is set to 10; and the batch size is 230.

For the ANFIS model, the prediction parameters include number of membership functions, which is kept as 4; membership functions are assumed as triangular; the epochs are 100; FIS is generated using grid partitioning; and training of FIS is done using back propagation.

The details of another study on an Indian dataset are included in the supplemental data.

7.7   RESULTS AND DISCUSSION

The selected soft-computing approaches, ANN, fuzzy logic, ANFIS, and LSTM, for short term water demand forecasting, has shown remarkable results. The forecasting of water demand is carried out for two varied time series, one is hourly based (Spanish dataset) and the other is monthly based (Indian dataset). The calibration and validation of datasets using these models for the Spanish dataset are shown in figures 7.2, 7.3, 7.4, and 7.5. For a particular dataset, the training set is selected randomly from a block of 360 data points, and for the purpose of validation, 24 consecutive values are then opted. Each model is applied to each training block, and the accuracy of the same is examined from the test dataset.

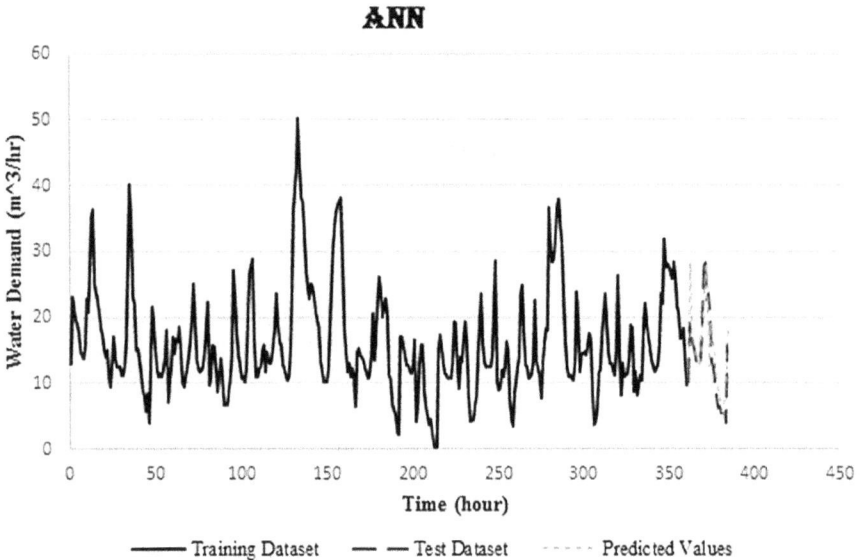

FIGURE 7.2   Calibration and validation datasets with predicted values of a city in Spain using ANN.

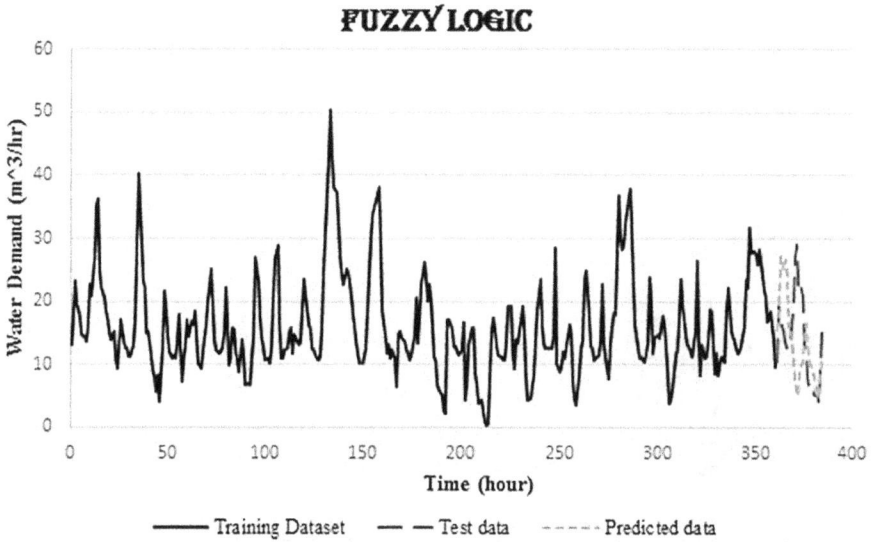

FIGURE 7.3 Calibration and validation datasets with predicted values of a city in Spain using fuzzy logic.

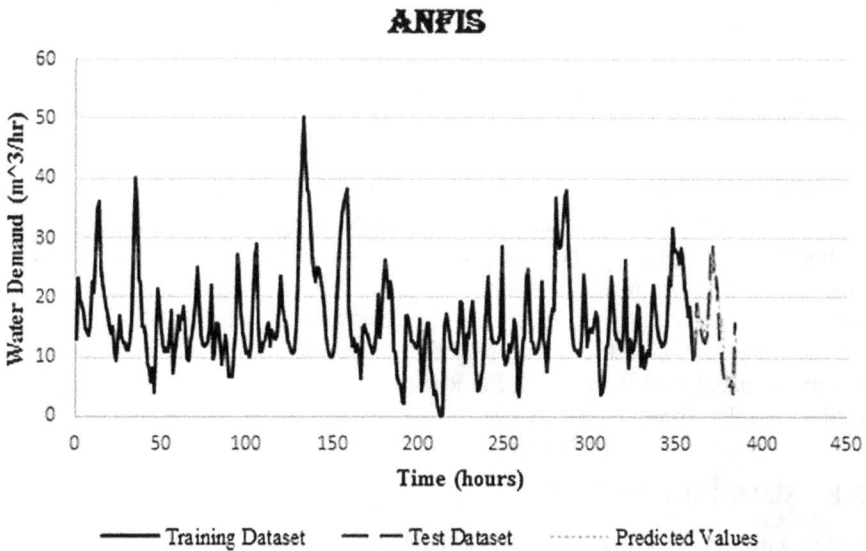

FIGURE 7.4 Calibration and validation datasets with predicted values of a city in Spain using ANFIS.

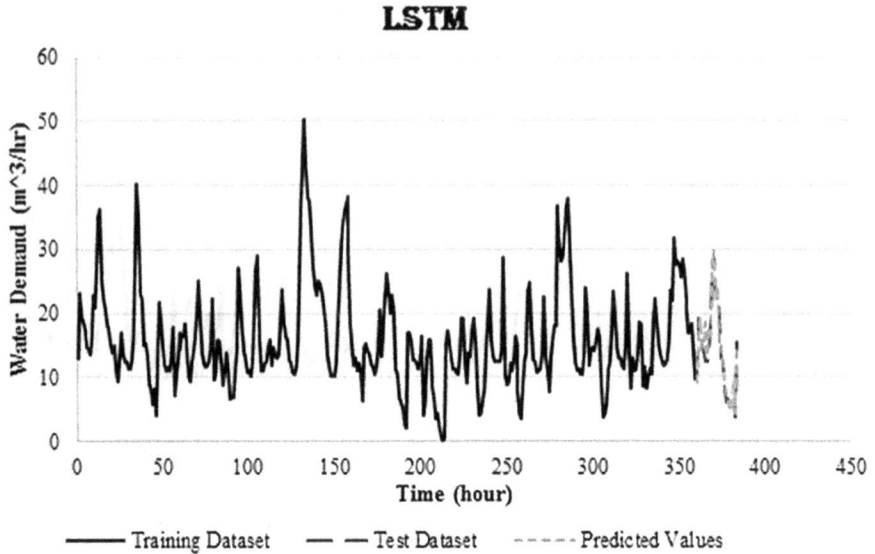

FIGURE 7.5 Calibration and validation datasets with predicted values of a city in Spain using LSTM.

From figures 7.2, 7.3, 7.4, and 7.5 it is observed that for a given dataset, LSTM model and ANFIS models validate well compared to the ANN and ANFIS models. From this, it is observed that LSTM has exhibited better performance than the other three models for the data with resolutions for 6, 12, and 24 hours. However, the LSTM's better performance can also be predicted by statistics of the prediction errors as provided in tables 7.2–7.4. As shown in table, the MAPE value for LSTM model are 9.215%, 10.897%, and 12.943%, which were lower than the values calculated by other models. The RMSE value of the LSTM models are 4.402, 5.298, and 7.106 m^3/hour, while MAE values are 2.998, 3.568, and 4.278 m^3/hour, which are again lower than other models. However, these values are quite equivalent to ARIMA based statistical model for the given time resolutions. Next, to LSTM, the hybrid ANFIS model performed well as compared to ANN and Fuzzy logic.

For the Indian dataset, where the prediction is carried out over monthly basis, the ANFIS model has shown better validation over the other two approaches. Although the non-linearity, stationarity, independence of residual etc., does not affect the fuzzy model, it still provided the worst prediction among all and is not very acceptable. When compared with the statistical models, the LSTM and ANFIS has only generated the similar results with them.

7.8   SUMMARY AND CONCLUSIONS

In this chapter, application of four soft computing-based models (LSTM, ANN, fuzzy logic, and ANFIS) to forecast the short-term water demand are discussed. These models are also compared with some of the most commonly used statistical approaches. The models are applied to water demand forecasting for Spanish and Indian datasets. However, due to lack of data points in the Indian dataset, it was not possible to forecast water consumption using LSTM. Different criteria have been used to assess the model performance, like RMSE, MAE, and MAPE. On the basis of the results, it is seen that among the soft computing approaches, LSTM has consistently outperformed every other models. The LSTM results are comparable with the statistical approach, and it is also observed to perform equally well as that of hybrid statistical approach. The following conclusions are derived.

- Although ANN is one of the most commonly adopted approaches by the practitioners for demand forecasting, LSTM model has shown significant improvement in predicting high time resolution demand with large number of datasets. This is very helpful in taking optimal decision regarding real time operation.
- The LSTM model has shown remarkable performance, which is almost equivalent to the hybrid EEMD-DPSF statistical models. The approach has shown exceptionally high performance, among others, because it filters out the worst solution at a very early stage. Hence, this leads to early convergence and output is obtained in significantly less time. The number of epochs after several runs reaches from 100 to 10.
- Also, ANFIS has shown good results and significant improvement over ANN and Fuzzy logic alone. The obtained values of MAPE are quite equivalent to statistical ARIMA approach. The only hurdle is that in deciding the appropriate number of membership functions, there is an increase in the number leads to non-linearity of the network. The study reveals that ANFIS model is one of the viable tools for forecasting.
- ANN basically lacks with the training algorithm process. Deciding which training algorithm fits particular datasets best is not an easy task. Hence the development of good training algorithms is necessary.

One of the major drawbacks of the soft computing approaches is that these are considered as black box as the level of interpretability on how the model makes the prediction remains hidden. However, interpretability of these models is important, as understanding what is going on inside the black box leads to more accuracy.

Using different algorithms on two different data sets, it is clear that no single global model can outperform the others. Depending on data availability, study area, etc., models respond differently. However, hybrid techniques like EEMD-DPSF and EEMD-DPSF-ARIMA outperform single approaches. So, developing hybrid approaches combining statistical and soft computing technologies can be a valuable study subject. For the bigger cities and communities, larger datasets may be available, and making predictions with large volumes of data results in higher accuracy, as the noise of data has less effect on it. Smaller volumes of data may be less useful for most models, with questionable capability. Hence, there is need of development of models which are able to make useful predictions with less data.

REFERENCES

Adamowski, J., Fung Chan, H., Prasher, S.O., Ozga-Zielinski, B., & Sliusarieva, A. (2012) Comparison of multiple linear and nonlinear regression, autoregressive integrated moving average, artificial neural network, and wavelet artificial neural network methods for urban water demand forecasting in Montreal, Canada. *Water Res. Research*, *48*(1): W01528.

Altunkaynak, A., Özger, M., & Çakmakci, M. (2005) Water consumption prediction of Istanbul city by using fuzzy logic approach. *Water Res. Mgmt*, *5*: 641–654.

Álvarez, F.M., Troncoso, A., Riquelme, J.C., & Ruiz, J.S.A. (2011) Energy time series forecasting based on pattern sequence similarity. *IEEE Trans. Knowl. Data Eng.*, *23*(8): 1230–1243.

Al-Saati, N. (1998) *Forecasting water consumption by Box-Jenkins ARIMA modeling. Technical Magazine*. Baghdad, Iraq: Foundation of Technical Education.

Anele, A., Y. Hamam, A. Abu-Mahfouz, & E. Todini. (2017) Overview, comparative assessment and recommendations of forecasting models for short-term water demand prediction. *Water 9*(11): 887.

Arandia, E., Ba, A., Eck, B., & McKenna, S. (2016) Tailoring seasonal time series models to forecast short-term water demand. *J. Water Res. Plan. Manag.*, *142*(3): 04015067.

Bennett, C., Stewart, R.A., & Beal, C.D. (2013) ANN-based residential water end-use demand forecasting model. *Expert Sys. App.*, *40*(4): 1014–1023.

Billings, R.B., & Jones, C.V. (2011) *Forecasting urban water demand*. American Water Works Association, Denver, USA.

Bougadis, J., Adamowski, K., & Diduch, R. (2005) Short-term municipal water demand forecasting. *Hydrological Processes: An International Journal*, *19*(1): 137–148.

Boser, B.E., Guyon, I.M., & Vapnik, V.N. (1992) A training algorithm for optimal margin classifiers. In *Proceedings of the Fifth Annual Workshop on Computational Learning Theory*, Pennsylvania USA, pp. 144–152.

CPHEEO. (1999) *Manual on water supply and treatment*. New Delhi, India: Center Public Health and Environmental Engineering Organisation. Government of India.

Donkor, E.A., Mazzuchi, T.A., Soyer, R., & Alan Roberson, J. (2014) Urban water demand forecasting: Review of methods and models. *J. Water Res. Plan. Manag.*, *140*(2): 146–159.

Du, B., Zhou, Q., Guo, J., Guo, S., & Wang, L. (2021) Deep learning with long short-term memory neural networks combining wavelet transform and principal component analysis for daily urban water demand forecasting. *Expert Systems with Applications*, *171*: 114571.

Glover, F.W., & Kochenberger, G.A. eds. (2006) *Handbook of metaheuristics* (Vol. 57). Springer Science & Business Media, Boston, UK.

Gober, P., Kirkwood, C.W., Balling, R.C., Ellis, A.W., & Deitrick, S. (2010) Water planning under climatic uncertainty in Phoenix: Why we need a new paradigm. *Annals Ass. American Geographers*, *100*(2): 356–372.

Gers, F.A., Schmidhuber, J., & Cummins, F. (2000) Learning to forget: Continual prediction with LSTM. *Neural Comput.*, *12*(10): 2451–2471.

Hagan, M.T., & Menhaj, M.B. (1994) Training feedforward networks with the Marquardt algorithm. *IEEE Trans. Neural Netw.*, *5*(6): 989–993.

Huang, N.E., Shen, Z., Long, S.R., Wu, M.C., Shih, H.H., Zheng, Q., Yen, N.-C., Tung, C.C., & Liu, H.H. (1998) The empirical mode decomposition and the Hilbert spectrum for nonlinear and non-stationary time series analysis. *Proc. R. Soc. London Ser. A*, *454*(1971): 903–995.

Mamlook, R., Badran, O., & Abdulhadi, E. (2009) A fuzzy inference model for short-term load forecasting. *Energy Policy*, *37*(4): 1239–1248.

Martínez-Álvarez, F., Troncoso, A., Riquelme, J.C., & Ruiz, J.S.A. (2008). LBF: A labeled-based forecasting algorithm and its application to electricity price time series. In *Data Mining. ICDM'08. Eighth IEEE International Conference on*, December 2008, pp. 453–461. IEEE.

Masri, B. E., Schwalm, C., Huntzinger, D. N., Mao, J., Shi, X., Peng, C., Fisher, J. B., Jain, A. K., Tian, H., Poulter, B., & Michalak, A. M. (2019) Carbon and water use efficiencies: A comparative analysis of ten terrestrial ecosystem models under changing climate, Sci. Rep., 9:14680.

Mu, L., Zheng, F., Tao, R., Zhang, Q., & Kapelan, Z. (2020) Hourly and daily urban water demand predictions using a long short-term memory-based model. *J. Water Res. Plan. Manag.*, *146*(9): 05020017.

Oliveira, P.J., Steffen, J.L., & Cheung, P. (2017) Parameter estimation of seasonal ARIMA models for water demand forecasting using the harmony search algorithm. *Proc. Eng.*, *186*: 177–185.

Pandey, P., Bokde, N.D., Dongre, S., & Gupta, R. (2021) Hybrid models for water demand forecasting. *J. Water Res. Plan. Manag.*, *147*(2): 04020106.

Rinaudo, J.D. (2015) Long-term water demand forecasting. *Understanding and managing urban water trans*, 239–268.

Shabri, A., Samsudin, R., & Teknologi, U. (2015) Empirical mode decomposition–least squares support vector machine based for water demand forecasting. *Int. J. Adv. Soft Comput. Appl.*, *7*(2): 38–53.

Shirkoohi, M.G., Doghri, M., & Duchesne, S. (2021) Short-term water demand predictions coupling an Artificial Neural Network model and a Genetic Algorithm. *Water Supply*, *21*(5): 2374–2386.

Sutskever, I., Vinyals, O., & Le, Q.V. (2014) Sequence to sequence learning with neural networks. In *NIPS'14: Proceedings of the 27th International Conference on Neural Information Processing Systems*, Vol. 2, 3104–3112.

Vapnik, V. (2013) *The nature of statistical learning theory*. Springer Science & Business Media, Boston, UK.

Vijayalaksmi, D., & Babu, K.J. (2015) Water supply system demand forecasting using adaptive neuro-fuzzy inference system. *Aquat. Proc.*, *4*: 950–956.

Wu, Z., & Huang, N.E. (2009) Ensemble empirical mode decomposition: A noise-assisted data analysis method. *Adv. Adapt. Data Anal.*, *1*(1): 1–41.

Wu, Z., Huang, N.E., & Chen, X. (2009) The multi-dimensional ensemble empirical mode decomposition method. *Adv. Adapt. Data Anal.*, *1*(3): 339–372.

Xu, Y., Zhang, J., Long, Z., & Chen, Y. (2018) A novel dual-scale deep belief network method for daily urban water demand forecasting. *Energies*, *11*(5): 1068.

Zhang, Q., Diao, Y., & Dong, J., (2013) Regional water demand prediction and analysis based on Cobb-Douglas model. *Water Res. Mgmt.*, *27*(8): 3103–3113.

8 Intervention of Computational Models for Groundwater Pollution Source Characterization

Anirban Chakraborty and Om Prakash

CONTENTS

8.1 INTRODUCTION

The demand for fresh water rises in tandem with the worldwide population. Reserves are being contaminated due to various anthropogenic activities causing rapid decline in the quality of fresh groundwater supply. Remediation of contaminated aquifers appears to be the way forward for long-term groundwater sustainability in order to meet global freshwater demand (Yadav et al., 2021). To execute an efficient groundwater clean-up, prior information of the contaminant sources' number and location, activity beginning dates, and pollutant flux release histories (magnitude of pollution in terms of pollutant mass/unit time) are required to be known with high degree of certainty (Reed and Minsker, 2004). However, the major difficulty in estimating these characteristics is that their effects are interchangeable (Guo et al., 2019), which is measured in form of pollutant concentration at monitoring well locations. Different combinations of source characteristics can result in similar scenario of groundwater pollution. Therefore, identifying unknown groundwater pollution sources' characteristics is non-unique and ill-posed problem (Neupane and Datta, 2021). Characterization of pollutant sources require forward simulation of flow and transport models to be solved backward in space and time, and therefore categorized as an inverse problem (Sun et al., 2006). The schematic diagram of groundwater pollution and remediation process is shown in Figure 8.1.

8.1.1 BACKGROUND

Groundwater pollution is a slow process and pollutants can remain undetected for long periods. Initially available spatio-temporal concentration data, suggesting pollution, are generally scanty and insufficient for accurate source characterization (Bai et al., 2021). As a result, identifying the actual source characteristics using this data becomes extremely challenging (Datta et al., 2016). In order to overcome this

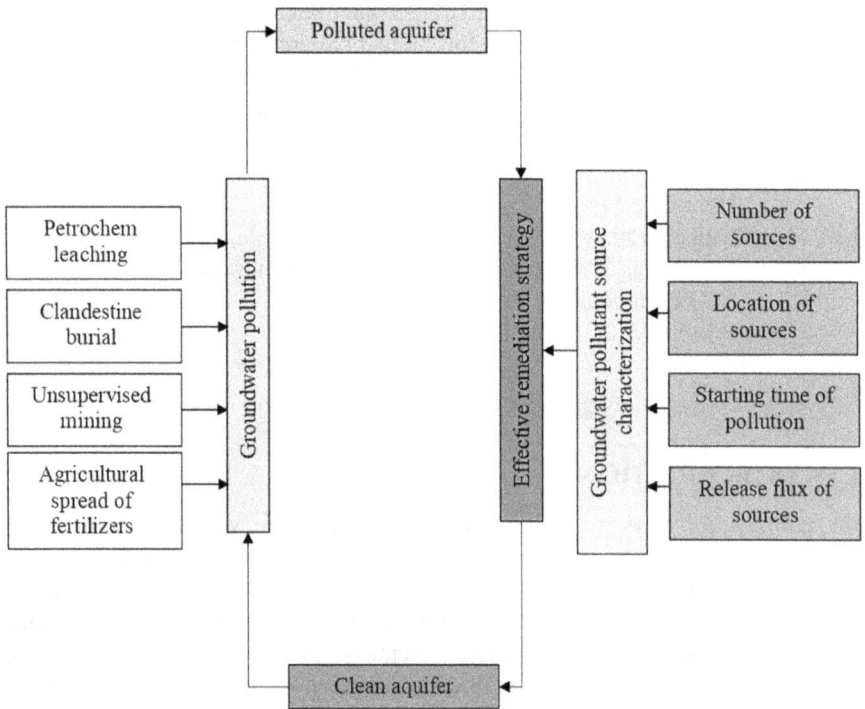

FIGURE 8.1 Schematic of pollution cycle in groundwater aquifer.

difficulty, most of the methodologies assume that the actual number of sources and their locations are well known, or at least the potential source locations are known with certainty (Mahar and Datta, 2001). However, these assumptions do not hold true, especially for clandestine sources of pollution.

8.1.2 GROUNDWATER POLLUTION SOURCES CHARACTERIZATION

All the techniques developed so far for groundwater pollution source characterization do not address a key issue of finding the number of sources (Hayford and Datta, 2021). Only when the number of sources is estimated correctly along with their spatial locations can source activity starting times and pollutant flux release histories be estimated (Barajas-solano et al., 2019). Carefully observing these techniques show that they essentially solve only for the source flux release history, and the other source characteristics are indirectly derived from the magnitude of the source flux values, such that zero or a negligible source flux magnitude represents no actual source or no source activity.

8.2 METHODS FOR CHARACTERIZATION OF GROUNDWATER POLLUTANT SOURCES

The methods of characterization of groundwater pollution sources can be broadly grouped into five major categories: 1) direct inverse methods; 2) statistical and regression methods; 3) surrogate model-based techniques; 4) optimization methods; and 5) hybrid techniques (Figure 8.2).

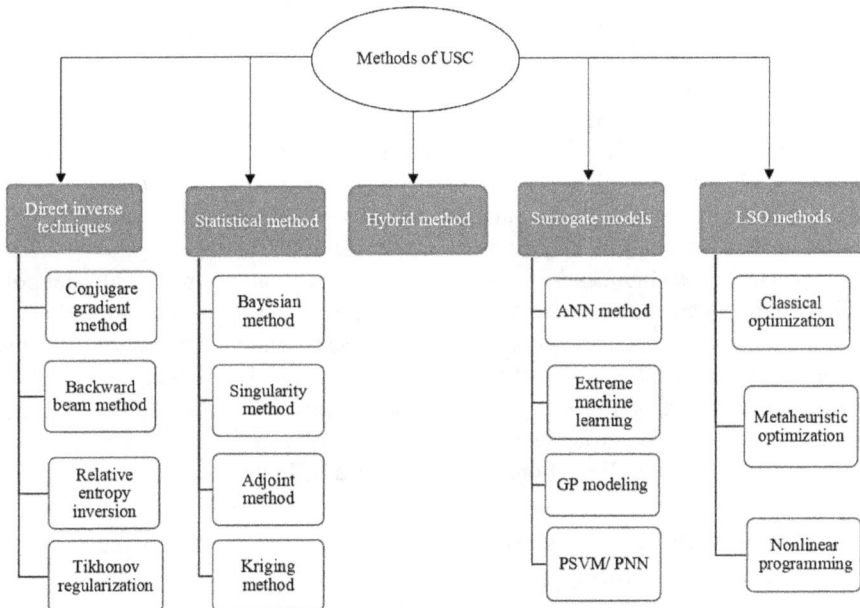

FIGURE 8.2 The summarized classification of pollutant source characterization.

8.2.1 Direct Inverse Approach for Pollutant Source Characterization

Direct inverse approach is the most classical approach for characterizing the heat flux sources and analogously used for groundwater pollution sources characterization. The solution techniques those available for back tracking the source location and starting time of the heat conduction problems can also be applied to the Unknown Source Characterization (USC) problem in groundwater.

Conjugate gradient method was used as an important tool for solving the inverse heat equation in order to find the location of heat source in purely advective medium (Su and Neto, 2001). One of the most significant tools for tackling heat conduction problem is the quasi-reversibility (QR) technique explicitly in homogeneous medium. QR method was used by Clark and Oppenheimer (1994) for ill-posed heat conduction boundary value problem.

This direct technique, though suitable for source flux recovery for homogeneous single source heat flow problems (Ababou et al., 2010), is not effective in groundwater pollution source characterization for the following reasons:

- Heat conduction medium is mostly homogeneous, whereas groundwater system is inherently heterogeneous and anisotropic.
- Generally, in the heat conduction problem there is only one source, whereas groundwater pollution sources can be multiple.
- There is no reaction involved in the case of heat transport problem, but the groundwater pollutant is mostly reactive in nature.

A study on regularization technique using Green's function for backward heat equation problem was presented by Ames and Epperson (1997). This method was used explicitly for Dirchlet's boundary condition, where the source injection is assumed to be instantaneous.

Marching Jury Backward Beam Method (MJBBM) (Atmadja and Bagtzoglou, 2001) was developed for finding the exact location of unknown groundwater pollution source. This method was further extended by Bagtzoglou and Atmadja (2003), but was limited in its applicability to only steady state groundwater flow. A state of art on comparative study of several mathematical models for characterization of pollution sources was presented by Atmadja and Bagtzoglou (2001).

Direct inverse techniques do not perform very well in unknown groundwater pollution source characterization as these are suitable to homogeneous, isotropic scenarios having simple boundary conditions as in case of heat flow problems. Also the complexities involved in terms of the unknown number of pollution sources, their locations, and starting times makes the problem unsolvable using direct inverse methods. The specific groundwater pollution source characteristics identified by these methods and their limitations are summarized in Table 8.1.

8.2.2 Statistical and Regression Methods for Pollutant Source Characterization

Data interpolation estimates the value of a variable at an unmeasured location from observed values at measured locations (Prakash and Datta, 2013). Geo-statistical data interpolation was initially developed with an emphasis on solving mining

TABLE 8.1

Summarized Description of Various Direct Inverse Methods

Method	Advantages	Limitations
Conjugate Gradient Method	Exact solution found; Well-posed;	Not robust; Limited to homogeneous idealized scenario
MJBBM	Applicable to heterogeneous flow field; Robust method;	Not applicable to transient flow field; Limited to 2-D case;
Quasi-Reversibility	Well-posed solution	Limited to homogeneous, steady state case
Relative entropy Inversion	Location can be achieved with high degree of certainty	Not applicable for missing data; limited to ideal cases;
Backward Propagation Model	Computationally intensive; exact solution achieved	Convergence error, only limited to Dirchlet's boundary

related problems but has found wide application in all major engineering fields. Geostatistical kriging has been extensively applied to various groundwater management problems.

Bayesian framework with conditional and adjoint probability are used in characterizing the groundwater pollution sources. Backward location and travel time probabilities can be used to determine the prior location of pollution in an aquifer. The forward advection-dispersion equation is used to solve the location probabilities and time travel probabilities. This method was further extended by Snodgrass and Kitanidis (1997) using Bayesian framework. This method was tested for homogeneous one-dimensional aquifer. However, this method has limited applicability in finding the USC source locations and beginning of pollution for real-life complex scenarios, comprising large scale heterogeneity and three-dimensional flows. A Bayesian framework-based technique "parameter non-negativity" was proposed by Michalak and Kitanidis (2003). The advance Bayesian framework used for identification of USC in groundwater by An et al. (2021).

A backward probability model called "adjoint" method was used by for estimating the location of the unknown groundwater pollution sources (Neupauer and Wilson, 2005). A singularity mapping technique using generalized fractal self-similarity was applied to find the location of actual sources by Datta et al. (2016). However, the degree of certainty of getting accurate potential sources decreases with increase in the number of potential sources in the domain. A kriging-based technique is used by Chakraborty and Prakash (2020) for estimating locations and release flux histories of groundwater pollutant sources. The author showed how the accuracy of results can be enhanced by using kriging based interpolation technique. The applicability and limitations of these statistical methods are summarized in Table 8.2.

8.2.3 SURROGATE MODEL BASED APPROACH

Surrogate models are used for groundwater pollution source characterization as an alternative solution of inverse problem (Jamshidi et al., 2020; Jiang et al., 2015).

TABLE 8.2

Descriptions of the Various Statistical Methods for Groundwater Pollution Source Characterization

Statistical Methods	Advantages	Limitations
Adjoint methods	Location of pollutant plume identified;	Limited to 1-D scenario; Applicable for homogeneous aquifer
Bayesian framework	Computationally intensive; Exact solution found	Pollution source injection is assumed to be instantaneous
Kriging based methods	Plume can be identified with higher degree of certainty;	Large data sets required
Fractal Singularity based approach	Applicable to heterogeneous media	Largely data sensitive and yield error in case of multiple sources

Surrogate models are generally replica of the groundwater simulation model (Asher et al., 2015).

Sampling strategy in space and time using Kalman filter was proposed by (Kollat et al., 2011). This Kalman filtering is further found in Xu and Gómez-Hernández (2017) for identifying the known groundwater pollution sources locations and estimation of spatial hydraulic conductivity.

The main limitations of these surrogate models are that they require a large volume of spatio-temporal data. However, in realistic scenarios of groundwater pollution the spatio-temporal measurement of concentration data is often sparse and limited and therefore, do not yield good results while solving source identification problem. The advantages and limitations of surrogate models-based techniques are expressed in Table 8.3.

8.2.4 Linked Simulation Optimization (LSO)

Often, USC in groundwater is formulated as an optimization problem. The flow model, MODFLOW (Harbaugh and McDonald, 1996), and solute transport model MT3D (Zheng, 1992) are considered as important binding constraint for the optimization model. LSO (Ayvaz and Karahan, 2008) works along with the groundwater flow and pollutant transport simulation models in an iterative sub-domain solution approach by minimizing the residual error (Chakraborty and Prakash, 2021) between the observed result and the simulated result. Various optimization techniques such as Simulated Annealing (SA), Particle Swarm Optimization (PSO) and Genetic Algorithm (GA) have been used independently, or in conjunction with simulation models to estimate the various USC.

GA (Sayeed and Mahinthakumar, 2005) was applied to find a single source of pollution in a 2-D homogeneous aquifer. In this study, the pollutant source characterization problem was formulated as a nonlinear optimization model considering the release history and location of pollutant sources as unknowns. Further GA technique was employed to solve USC by Chadalavada et al. (2011); Jha and Datta (2011); Rajeev Gandhi et al. (2016).

TABLE 8.3
Advantages and Limitations of Surrogate Based Models for Source Characterization

Surrogate Modes	Advantages	Limitations
Kalman filter based Approach	Robust methods; Applicable to complex 3-D heterogeneous field; Location and release flux of pollutant sources can be identified	Huge data requirement; Purely data sensitive;
Kriging clubbed Extreme Learning Machine	Can be applied for any complex boundary; Very efficient model for handling heterogeneity	Not applicable for unknown number of sources;
ANN/ GP based approach	Can be used as an alternative of groundwater simulation model	Data sensitive; Overfitting is very common feature;
PSVM approach	Applicable for complex groundwater scenarios	Data sensitive; Not tested for unknown number and starting time

SA based LSO is also used in groundwater management (Dey and Prakash, 2020). An SA based technique was used for calculating source location for three-dimensional aquifer by Jha and Datta (2012). Source flux release history and activity initiation time of pollution was recovered for known source locations by Prakash and Datta (2014).

The accuracy and the efficiency for the LSO for USC can be enhanced by using optimally chosen monitoring wells (Datta et al., 2009). A methodology integrating sequential-monitoring-network design and source identification (Prakash and Datta, 2015) was developed, where source location, source-activity initiation time, and source-flux release history are considered as explicit unknown variables. Even with sparse pollutant concentration observations, LSO methods are most successful in reconstructing source flux release histories when the number of sources and their locations are well known or the probable source locations are known with a high degree of certainty (Datta et al., 2017). The summary of various LSO methods and their specific advantages and limitations is presented in Table 8.4.

8.2.5 Hybrid Techniques

USC in groundwater is formulated using multiple techniques. By using these hybrid techniques, the accuracy of estimation is significantly enhanced. An embedded optimization technique with ANN was designed by using back propagation model (Singh et al., 2004; 2007) for finding multiple source locations of pollutant. This embedded optimization technique increases the dimensionality of the problem.

A sequential monitoring network for optimal characterization of groundwater pollution sources combining geo-statistics and optimization algorithm was presented by Prakash and Datta (2013) to find the number of sources and their locations. Large

TABLE 8.4

The Applicability and Limitations of the Various LSO Methods for Source Characterization

LSO approach	Advantages	Limitations
Classical optimization-based approach	Guaranteed convergence;	Efficiency decreases with the increase of no of variables;
Metaheuristic optimization (SA, GA, PSO)	Fast convergence; Ability to work with missing data;	Huge requirement of computational power; Not tested for unknown number of sources; May not attain global optima
Nonlinear programming	Simple problem, less computational power required;	Not very powerful tool to handle heterogeneity and complexities of groundwater management problem

number of spatio-temporal concentration data set is required for these methods. Concentration observation data from designed monitoring network is found to be more effective in identification of pollution source characteristics with higher degree of certainty.

8.3   MATHEMATICAL FRAMEWORK FOR USC

For optimal identification of USC, the first step is collection of hydrogeological parameters of the aquifer. The next step of identifying USC is the collection of pollutant concentration data. The collection of concentration data should be done from monitoring wells that it would give the best estimate of USC.

8.3.1   MONITORING WELL DESIGN FOR OBTAINING CONCENTRATION DATA

If *pob* is the maximum number of monitoring stations, where data can be measured, H is the maximum temporal steps of concentration measurement, and Q is the concentration matrix, then Q can be written as (Equation 8.1)

$$Q = \begin{bmatrix} Q_{1,1} & Q_{1,2} & \cdots & Q_{1,pob} \\ Q_{2,1} & Q_{2,2} & \cdots & Q_{2,pob} \\ \cdots & \cdots & \cdots & \cdots \\ Q_{H,1} & Q_{H,2} & \cdots & Q_{H,pob} \end{bmatrix} \tag{8.1}$$

The Eigenvalue of matrix $[Q]$ is calculated using Equation 8.2, where δ is the Eigenvalue of matrix $[Q]$ and $[I]$ is unit matrix $[Q]^T$ is the transpose of $[Q]$.

$$\left[Q\right]^T \left[Q\right] - \delta^2 \left[I\right] = 0 \tag{8.2}$$

Based on the Eigenvalues, $[\delta]$ is formed with the size of $[1 \times pob]$ given in Equation. 8.3.

$$\left[\delta\right]=\left\lfloor \delta_1,\delta_2,\delta_3,\ldots\delta_{pob}\right\rfloor \tag{8.3}$$

The elements of δ are ranked to choose the locations for installing monitoring wells from potential monitoring well locations based on their ranking. The higher value of δ signifies greater impact of the sources on the potential well locations and therefore, such potential locations which show a higher value of δ are only considered in design of the monitoring wells. It is expected that the pollutant concentration measurement from such locations is better in exhibiting the variations in source characteristics that are to be identified in LSO.

8.3.2 LSO FORMULATION

One of the most suitable techniques for identifying USC is LSO. LSO is the mathematical framework where optimization controls the simulation model. The optimal value of objective function in terms of difference between observed and simulated concentration values gives the optimum estimate of USC. The schematic diagram showing the working principle of LSO is shown in Figure 8.3.

The mathematical framework for solving USC in and LSO formulation is given by Equation 8.4.

$$Minimize \ \ F = \sum_{k=1}^{H}\sum_{iob=1}^{pob}abs\left(Cobs_{iob}^k - \delta_{iob}Cest_{iob}^k\right) \tag{8.4}$$

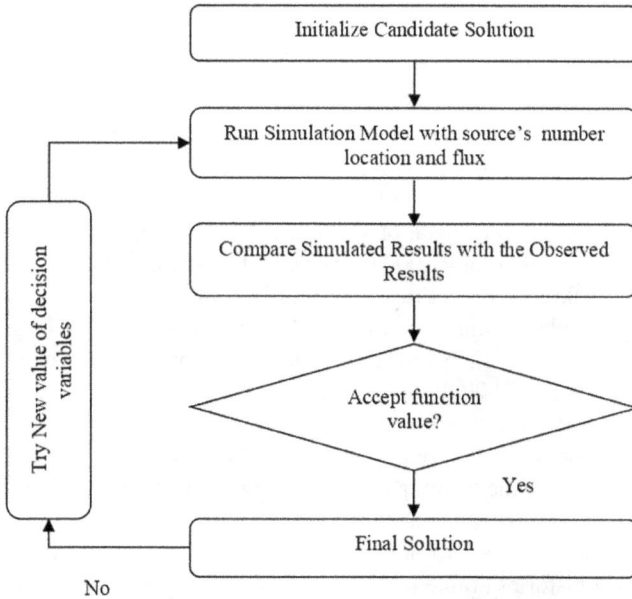

FIGURE 8.3 Schematic diagram of mathematical framework for USC.

$Cest_{iob}^k$ is the estimated concentration at k temporal time step at iob monitoring well, given by Equation 8.5. $Cobs_{iob}^k$ is observed concentration at same k time step and iob site location and δ_{iob} is the weighted factor at iob well found in Equation 8.3;

$$Cest_{iob}^k = \sum_{i=1}^n f\left(C_{si},x_i,y_i,z_i\right) \tag{8.5}$$

x_i, y_i, z_i represent the source grid ith source of contamination. The pollutant sources number is denoted by n. The basic flow chart of SA based LSO is presented in Figure 8.3.

$$C_{min} \leq C_{si} \leq C_{max} \tag{8.6a}$$

$$x_{min} \leq x_i \leq x_{max} \tag{8.6b}$$

$$y_{min} \leq y_i \leq y_{max} \tag{8.6c}$$

$$z_{min} \leq z_i \leq z_{max} \tag{8.6d}$$

x_i, y_i, z_i represent the source grid ith source of contamination; x_{max}, y_{max}, z_{max} are the limiting values of the study area. C_{si} is the ith source's flux. Equations 8.6a, 8.6b, 8.6c, and 8.6d give the constraints values of the objective function for developed LSO model.

8.3.3 RESULT OF USC USING LSO

To test the mathematical framework of proposed LSO for identifying USC a polluted site is chosen (Chakraborty and Prakash, 2021). The boundary conditions of hypothetical study area are illustrated in Figure 8.4.

The hypothetical area is made of single layer, homogeneous, anisotropic unconfined aquifer of area 1000 m × 1000 m. The detailed hydrogeological parameters are mentioned in Chakraborty and Prakash (2021). The area is divided into 400 (20 × 20) grids, having two clandestine sources of pollution denoted by red triangle. For three sampling well sites denoted by blue circles, pollutant concentration measurements at 100, 200, 300, 400, and 500 days are simulated using GMS 10.1. Two (non-erroneous and erroneous) concentration data sets are used to identify USC.

The entire activity duration of pollution is split into five stress periods of hundred days each. To validate the technique, the problem is solved in which the upper limit (n) on the number of possible sources (P) present in the research area is n = 5. SA optimization is used in LSO for identification of USC in terms of number, location, and release flux histories (Table 8.5).

The method does not know the actual number of sources location and release flux histories. It is expected that SA based LSO will detect the actual number of sources

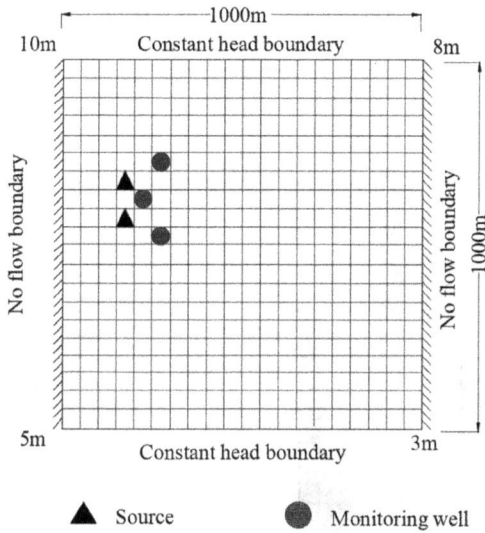

FIGURE 8.4 Plan view of the illustrative aquifer.

TABLE 8.5
Location Identification of Pollutant Sources

Actual Location of Sources	Estimated Location of Sources for Non-Erroneous Datasets	Estimated Location of Sources for Erroneous Datasets
S1 (9,4); S2(7,4)	P1 (9,4); P2 (9,4); P3 (9,4) P4 (7,4); P5 (8,4)	P1 (9,4); P2 (9,4); P3 (10,5) P4 (7,4); P5 (6,3)

present despite starting with wrong assumption about actual number of sources in the study area. In this scenario, all the four hundred grids are considered as plausible location for pollution source. It should be noted that the technique relies on the optimal value of the decision variables to estimate the number of pollution sources present in the region, their coordinates, and source flux release for each stress period given by SIJ, where I represent the source number and J represents the stress period.

The LSO results show that P1, P2, and P3 occupy the same grid location as actual source S1, for non-erroneous data, therefore, P1, P2, and P3 replicate actual source S1. Identified source locations of P4 matches with S2 and P5 is very close to actual source S2 except for one grid cell. Therefore, P4 and P5 can be said to together represent source S2. Similar results of location identification are realized while using erroneous data set, where P1, P2 and P3 combined represent S1 with a slight deviation of one grid for P3, and P4, and P5 combined represent S2 with slight deviation in P5. These deviations can be further reduced by altering SA parameter for a more rigorous search.

Figure 8.5 shows the unknown source-flux variables for every identified clandestine source for each stress period (S11, S12, S13, S14, S15, S21, S22, S23, S24, S25) marked on the x-axis. Summation of the source fluxes of potential sources P1, P2,

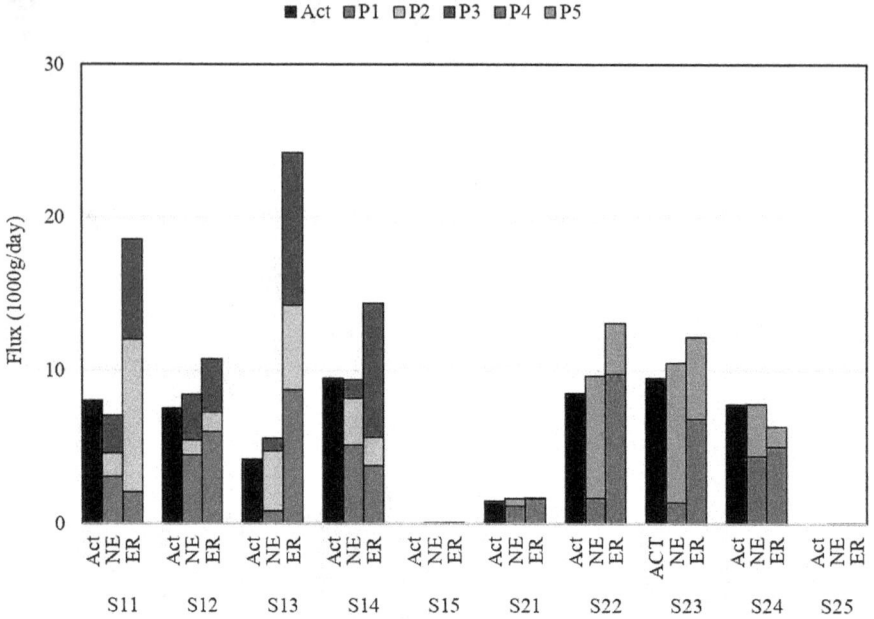

FIGURE 8.5　Identified flux histories of pollutant sources.

and P3 gives the actual source flux S1, and P4 and P5 for actual source S2 for every stress period (P1J + P2J + P3J = S1J and P4J + P5J = dS2J). The estimated source flux for both the identified sources matches closely with the actual flux value when tested with error free data. There is some deviation in identifying the release histories with erroneous data for S11, S12, and S14. The deviation of estimated flux is due to the large number of unknown variables to be identified.

The maximum number of potential sources was kept to five, which is more than the actual number of sources present in the study area. The developed methodology is able to identify the exact number of sources by superimposing multiple source locations into single coordinate location. Thus, the methodology is able to estimate number, location, and release flux histories simultaneously.

8.4　CONCLUSIONS

USC in terms of number, their locations, their activity starting time, and release flux history is the necessary first step in design of any remediation strategy. Several methodologies for identifying different source characteristics have been proposed and their specific advantages and limitations is summarized in Table 8.6. For complicated pollution situations, LSO techniques hold good in identifying USC. Though these methods are computationally expensive and may get trapped in local optima, still is the most viable alternative in limited spatio-temporal data availability scenario.

TABLE 8.6

Summarized Chart of Various Methods of Pollution Source Characterization

Method	Solver	Remarks
Direct Method	QR, Regularization, BBM, MRE, CGM	Ill-posed. Applicable for ideal scenario, all characteristics cannot be solved
Statistical method	Random walk theory, Bayesian network, Adjoint method	Data requirement is high; Heterogeneity can be solved. Release flux history can be obtained when location is implicitly known
Tracer technique	Carbon isotope is used	Financially expensive Not applicable for multiple source locations. Not robust for flux histories. Efficient for initial time
Surrogate model-based Approach	GP, ANN, DNN are used	Robust methods; Heterogeneity can be handled; Large volume of data is required; Not tested for unknown number and staring time
Linked simulation-optimization	SA, GA, PGA are being used	Applicable for heterogeneous 3-D problem. Number of potential sources is implicitly known
Hybrid Techniques	Optimization clubbed with surrogate models	Robust methods; computationally intensive; Applicable to complex USC problems

The developed LSO based technique can simultaneously identify all the unknown source characteristics, such as number, location, and release flux histories. This developed LSO overcomes one of the essential shortcomings of previous methodologies, which involved prior knowledge about the total number of sources present in the study area, or the exact possible source locations. In its present form, the approach only applies where the total number of presumed plausible sources in the study area exceeds or equals the real number of sources.

REFERENCES

Ababou, R., Bagtzoglou, A.C., Mallet, A., 2010. Anti-diffusion and source identification with the 'RAW' scheme: a particle-based censored random walk. Environmental Fluid Mechanics, 10(1): 41–76.

Ames, K.A., Epperson, J.F., 1997. A kernel-based method for the approximate solution of backward parabolic problems. SIAM Journal on Numerical Analysis, 34(4): 1357–1390.

An, Y., Yan, X., Lu, W., Qian, H., & Zhang, Z. (2021). An improved Bayesian approach linked to a surrogate model for identifying groundwater pollution sources. Hydrogeology Journal, 1–16.

Asher, M. J., Croke, B. F., Jakeman, A. J., & Peeters, L. J. (2015). A review of surrogate models and their application to groundwater modeling. Water Resources Research, 51(8), 5957–5973.

Atmadja, J., Bagtzoglou, A.C., 2001. Pollution source identification in heterogeneous porous media. Water Resources Research, 37(8): 2113–2125.

Ayvaz, M.T., Karahan, H., 2008. A simulation/optimization model for the identification of unknown groundwater well locations and pumping rates. Journal of Hydrology, 357(1): 76–92.

Bagtzoglou, A.C., Atmadja, J., 2003. Marching-jury backward beam equation and quasi-reversibility methods for hydrologic inversion: Application to contaminant plume spatial distribution recovery. Water Resources Research, 39(2).

Bai, Y., Lu, W., Li, J., Chang, Z., Wang, H., 2022. Groundwater contamination source identification using improved differential evolution Markov chain algorithm. Environmental Science and Pollution Research, 29(13):19679–19692.

Barajas-Solano, D.A., Alexander, F. J., Anghel, M., Tartakovsky, D. M., 2019. Efficient gHMC Reconstruction of Contaminant Release History. Frontiers in Environmental Science, 7:149.

Chakraborty, A., & Prakash, O., 2020. Characterization of groundwater pollution sources by kriging based linked simulation optimization. International Journal of GEOMATE, 20(81):79–85.

Chakraborty, A., & Prakash, O. (2021). Identification of clandestine groundwater pollution source locations and their release flux history. In *IOP Conference Series: Earth and Environmental Science*, Vol. 626, No. 1, p. 012003. IOP Publishing.

Chadalavada, S., Datta, B. Naidu, R., 2011. Optimisation approach for pollution source identification in groundwater: an overview. International Journal of Environment and Waste Management, 8: 40–61.

Clark, G.W., Oppenheimer, S.F., 1994. Quasireversibility methods for non-well-posed problems.

Datta, B., Amirabdollahian, M., Zuo, R., Prakash, O., 2016. Groundwater pollution plume delineation using local singularity mapping technique. International Journal, 11(25): 2435–2441.

Datta, B., Chakrabarty, D., Dhar, A., 2009. Optimal dynamic monitoring network design and identification of unknown groundwater pollution sources. Water Resources Management, 23(10): 2031–2049.

Datta, B., Petit, C., Palliser, M., Esfahani, H.K., Prakash, O., 2017. Linking a simulated annealing-based optimization model with PHT3D simulation model for chemically reactive transport processes to optimally characterize unknown contaminant sources in a former mine site in Australia. Journal of Water Resource and Protection, 9(05): 432.

Dey, S., & Prakash, O. (2020). Managing saltwater intrusion using conjugate sharp interface and density dependent models linked with pumping optimization. Groundwater for Sustainable Development, 11, 100446.

Guo, J. Y., Lu, W. X., Yang, Q. C., & Miao, T. S. (2019). The application of 0–1 mixed integer nonlinear programming optimization model based on a surrogate model to identify the groundwater pollution source. Journal of Contaminant Hydrology, 220, 18–25.

Harbaugh, A.W., McDonald, M.G., 1996. Programmer's documentation for MODFLOW-96, an update to the US Geological Survey modular finite-difference ground-water flow model. 2331-1258, US Geological Survey; Branch of Information Services [distributor].

Hayford, M. S., & Datta, B. (2021). Source characterization of multiple reactive species at an abandoned mine site using a groundwater numerical simulation model and optimization models. International Journal of Environmental Research and Public Health, 18(9), 4776.

Jamshidi, A., Samani, J. M. V., Samani, H. M. V., Zanini, A., Tanda, M. G., & Mazaheri, M. (2020). Solving inverse problems of unknown contaminant source in groundwater-river integrated systems using a surrogate transport model-based optimization. Water, 12(9), 2415.

Jha, M., Datta, B., 2012. Three-dimensional groundwater pollution source identification using adaptive simulated annealing. Journal of Hydrologic Engineering, 18(3): 307–317.

Jha, M.K., Datta, B., 2011. Simulated annealing based simulation-optimization approach for identification of unknown contaminant sources in groundwater aquifers. Desalination and Water Treatment, 32(1–3): 79–85.

Jiang, X., Lu, W., Hou, Z., Zhao, H., & Na, J. (2015). Ensemble of surrogates-based optimization for identifying an optimal surfactant-enhanced aquifer remediation strategy at heterogeneous DNAPL-contaminated sites. Computers & Geosciences, 84, 37–45.

Kollat, J.B., Reed, P.M., Maxwell, R., 2011. Many-objective groundwater monitoring network design using bias-aware ensemble Kalman filtering, evolutionary optimization, and visual analytics. Water Resources Research, 47(2).

Li, J., Lu, W., & Jiannan, L. (2021). Groundwater pollution sources identification based on the Long-Short Term Memory network. Journal of Hydrology, 126670.

Mahar, P.S., Datta, B., 2001. Optimal identification of ground-water pollution sources and parameter estimation. Journal of Water Resources Planning and Management, 127(1): 20–29.

Michalak, A.M., Kitanidis, P.K., 2003. A method for enforcing parameter nonnegativity in Bayesian inverse problems with an application to contaminant source identification. Water Resources Research, 39(2).

Neupane, R., & Datta, B. (2021). Optimal characterization of unknown multispecies reactive contamination sources in an aquifer. Journal of Hydrologic Engineering, 26(11), 04021035.

Neupauer, R. M., Wilson, J. L., 2005. Backward probability model using multiple observations of contamination to identify groundwater contamination sources at the Massachusetts Military Reservation, Water Resources Research, 41(2): 1–14.

Prakash, O., Datta, B., 2013. Multiobjective monitoring network design for efficient identification of unknown groundwater pollution sources incorporating genetic programming–based monitoring. Journal of Hydrologic Engineering, 19(11): 04014025.

Prakash, O., Datta, B., 2014. Characterization of groundwater pollution sources with unknown release time history. Journal of Water Resource and Protection, 6(4): 337–350.

Prakash, O., Datta, B., 2015. Optimal characterization of pollutant sources in contaminated aquifers by integrating sequential-monitoring-network design and source identification: methodology and an application in Australia. Hydrogeology Journal, 23(6): 1089–1107.

Rajeev Gandhi, B., Bhattacharjya, R.K., Satish, M.G., 2016. Simulation–optimization-based virus source identification model for 3D unconfined aquifer considering source locations and number as variable. Journal of Hazardous, Toxic, and Radioactive Waste: 04016019.

Reed, P.M., Minsker, B.S., 2004. Striking the balance: long-term groundwater monitoring design for conflicting objectives. Journal of Water Resources Planning and Management, 130(2): 140–149.

Sayeed, M., Mahinthakumar, G.K., 2005. Efficient parallel implementation of hybrid optimization approaches for solving groundwater inverse problems. Journal of Computing in Civil Engineering, 19(4): 329–340.

Singh, R.M., Datta, B., Jain, A., 2004. Identification of unknown groundwater pollution sources using artificial neural networks. Journal of water resources planning and management, 130(6): 506–514.

Snodgrass, M.F., Kitanidis, P.K., 1997. A geostatistical approach to contaminant source identification. Water Resources Research, 33(4): 537–546.

Su, J., Silva Neto, A.J., 2001. Heat source estimation with the conjugate gradient method in inverse linear diffusive problems. Journal of the Brazilian Society of Mechanical Sciences, 23(3): 321–334.

Sun, A.Y., Painter, S.L., Wittmeyer, G.W., 2006. A robust approach for iterative contaminant source location and release history recovery. Journal of Contaminant Hydrology, 88(3–4): 181–196.

Xu, T., Gómez-Hernández, J.J., 2017. Simultaneous identification of a contaminant source and hydraulic conductivity via the restart normal-score ensemble Kalman filter. Advances in Water Resources.

Yadav, M., Singh, G., & Jadeja, R. N. (2021). Fluoride pollution in groundwater, impacts, and their potential remediation techniques. In Groundwater Geochemistry: Pollution and Remediation Methods.

Zheng, C., 1992. MT3D: A modular three-dimensional transport model for simulation of advection, dispersion and chemical reactions of contaminants in groundwater systems. SS Papadopulos & Associates.

Part II

Air Pollution

9 Artificial Intelligence for Air Quality and Control Systems
Status and Future Trends

Divya Patel, Mridu Kulwant, Saba Shirin, Ankit Kumar, Mohammad Aurangzeb Ansari, and Akhilesh Kumar Yadav

CONTENTS

9.1   INTRODUCTION

As per the European Union, it is tough to get rid of air pollution wherever you are living. It greatly affects human health as well as the environment (EEA, 2021). Many environmental agencies such as Environmental Protection Agency (EPA), European

DOI: 10.1201/9781003203445-11

133

Union (EU) have set air quality measurement and optimum concentration parametric guidelines, as well as standards for pollutants at allowable extremes (Castelli et al., 2020). Air pollution is basically affected by chemical and physical factors and systems. Broadly, lower levels of troposphere show effects and are synchronized by the concentration of pollutants (Sahoo et al., 2016; Yadav et al., 2019). The pollutants concentration is excessively every day; pollutants like the $PM_{2.5}$, PM_{10}, SO_2, NO_2, CO, O_3 are all the way through the roof. According to US EPA, the commonly present air pollutants are widely known as criteria pollutants comprises; NO_2, CO, lead, ground-level ozone (O_3), SO_2, and PM pollutants who causes most effective health issues (Yadav, 2021).

Of the total disability-adjusted life year (DALY) attributable to air pollution in India in 2017, the largest proportions were from lower respiratory infections (29·3%), chronic obstructive pulmonary disease (29·2%), and ischaemic heart disease (23·8%), followed by stroke (7·5%), diabetes (6·9%), lung cancer (1·8%), and cataract (1·5%). The DALY rate attributable to air pollution in India in 2017 was much higher for lower respiratory infections than the rate attributable to tobacco use (Figure 9.1). For non-communicable diseases, including chronic obstructive pulmonary disease, ischaemic heart disease, stroke, diabetes, lung cancer, and cataract, the DALY rate attributable to air pollution was at least as high as the rate attributable to tobacco use (Balakrishnan et al., 2019).

Pollution free healthy air is major issues for human as well as environment as emission of air pollution have significant potential impact on growing mortality rate. For this reason, air pollution is currently drawing the attention of countries around the world (Kok et al., 2021). Development of artificial thinking systems models are effective and appropriate tools to analyse the environmental variables, its modeling complexes, and its forecasting capabilities. Additionally, these findings disclose significant correlation effect between air pollution emissions and human health like lowering down the resistance immunity to different diseases and its impact on different organs such as the kidneys, liver, and heart (Shams et al., 2021). The distorted air quality in urban areas supports premature mortalities and causes chronic diseases. Since then, the scale and quality of the parameters concerned (Cabaneros et al., 2019).

As per Shams and Jahani, ANNs are a mathematical model which can significantly help in modeling and prediction of air pollutants concentration in air with the help of its complex and nonlinear processes. ANNs can intimate the human brain's behaviour too (Shams et al., 2021). From recent years, researchers were entangled in reviewing meteorology, source emissions, and forecasting long-term pollutants like PM_{10}, $PM_{2.5}$, and oxides of nitrogen, and ozone in air. But nowadays all are keen to know the benefits and applications of AI all to control air pollution by reducing concentration of pollutants. For that reason, various studies have mainly determined air pollutant and air quality prediction models in current years (Yadav, 2015; Yadav and Jamal, 2018a, b; Cabaneros et al., 2019).

Broadly by deterministic prediction, Statistical prediction and Neural network-based prediction performed (Kok et al., 2021). Worked on pollutants (SO_2, NO_2,

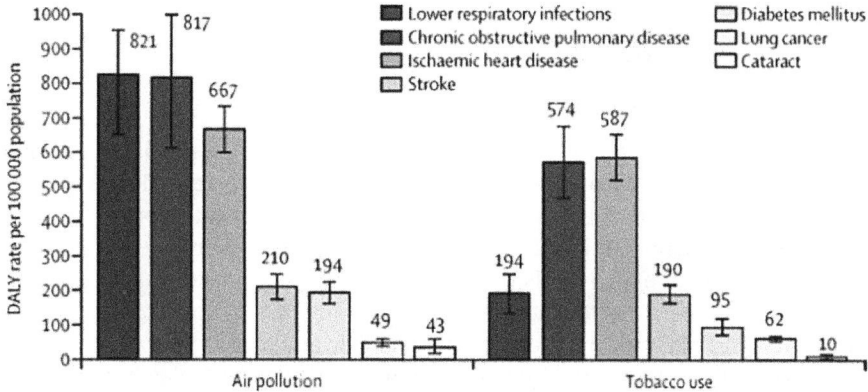

FIGURE 9.1   Daily rates attributable to air pollution and tobacco use in India, 2017.

[Reprinted from Balakrishnan (Balakrishnan et al., 2019) with permission from *The Lancet*].

CO, O_3) by using AI and statistical techniques (LSTM, GRU, etc.) to find the association of environmental pollutants, to analyse and predict the increase pattern of pollutants concentration ($PM_{2.5}$, PM_{10}, SO_2, NO_2, CO, O_3). In conclusion, air quality forecasting can support to prevent the damages caused by air pollution by forecasting its effect on ecosystem of major metropolitan cities. The command over and motion of air pollution prevention is a long-term operation (Castelli et al., 2020; Guo et al., 2020).

Broadly, the field of air quality monitoring and measuring, ML method, support vector regression (SVR), air quality forecasting via pollutants concentration and level, and AQI (air quality index) are main concern for research purposes. On the basis of concentration levels of each pollutant, the forecasting system constructs which can depict the quality of air pollution on hourly basis helps to get more flexible AQI and at the end human health. Generally, machine learning (ML) come up to SVMs, artificial neural networks (ANNs), genetic programming (GP), autoregressive integrated moving average model (ARIMA) while prediction of time series (TS) use. Deep learning (DL) helps to imitate human brain's structure and function, considered as a subdivision of ML. Widely, the autoregressive integrated moving average model (ARIMA) is a vital model to TS forecasting (Castelli et al., 2020; Kok et al., 2021).

This chapter aims to overview the analyses and predictions for upcoming scenario of air quality by using AI framework on the basis of different variables like time, season, weather forecasts along with measurements of air quality. The projected structure with different qualifications utilizes multiple ML algorithms (neural network, random forest, and logistic regression) and averages the optimum high performance to get accuracies from other specific models.

9.2 AIR QUALITY AND CONTROL SYSTEMS: CURRENT STATUS OF POLLUTION RESEARCH

9.2.1 BACKGROUND

Air pollution kills an estimated 7 million people worldwide every year. WHO data shows that 9 out of 10 people breathe air that exceeds WHO guideline limits containing high levels of pollutants with low- and middle-income countries suffering from the highest exposures (WHO, 2018). According to recent reports, nearly 3 billion people are exposed to high levels of health-damaging pollutants each and every day due to the lack of access to clean fuel and technologies for cooking, especially in rural areas. Such chronic exposure to household air pollution is the cause of nearly 4 million deaths annually from non-communicable diseases such as heart disease, stroke, chronic obstructive pulmonary disease, and cancer, as well as pneumonia. Insufficient and polluting household energy use is a health risk shared by all populations but a particularly important source of disease in women, children, and infants. Urgent action is needed to scale up access to clean cooking solutions to achieve sustainable development goals on health, gender equality, and climate, among others, and ultimately to minimize the future negative effects of the climate crisis thereby facilitating the effective enjoyment of the human right to health (WHO, 2021).

Air quality has been an important topic in the US since the creation of the Clean Air Act programme, likely all over the world too. As environmental problems and distortion events coming in light of bad air quality, many countries started working on the emerging issue. All due to bad air polluted air, almost 200,000 premature deaths happen in US each year due to combustible emissions of pollutants like particulate matter 2.5 ($PM_{2.5}$). Additionally, due to changes in ozone concentration, around 10,000 deaths happen each year in the US (Castelli et al., 2020).

As per an article of *Economic Times*, reported analyses of G20 countries comprise total share of energy related CO_2 emissions in 2018 showing in Figure 9.2, as per the graph fossil fuel contributes highest around 82 % among others like-climate mitigation, finance, and adaptation share in total emission. Among global greenhouse gas emissions of greenhouse gas (GHG), G20 countries are only responsible for highest 80% emissions. As per report, as India produces its 73% electricity via fossil fuel coal and its Nationally Determined Contributions (NDC) 2030 targets for air pollution bounded emission limits is 6–6.3 GtCO2e which is one of the highest rates (Joshi, 2019).

9.2.2 INITIATIVES FOR AIR QUALITY MANAGEMENT

- **National Ambient Air Quality Standards (NAAQS)**

  These standards set requirements for air quality that are defined in terms of an indicator, an averaging time for the measurement, a concentration, and a form. Ambient air quality refers to the condition or quality of the outdoor air. NAAQs are the standards for ambient air quality with reference to various identified pollutant notified by the Central Pollution Control Board (CPCB) under the Air (Prevention and Control of Pollution) Act, 1981.

Energy-related CO₂ emissions*

CO₂ emissions from fuel combustion (MtCO₂/year)

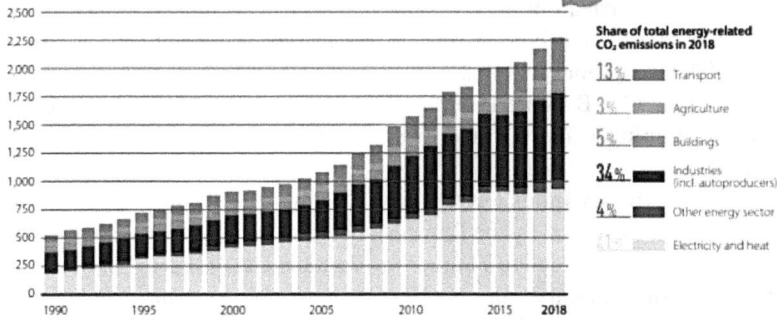

2,277 MtCO₂

Share of total energy-related CO₂ emissions in 2018

13% ■ Transport

3% ■ Agriculture

5% ■ Buildings

34% ■ Industries (incl. autoproducers)

4% ■ Other energy sector

<1% ■ Electricity and heat

The largest driver of overall GHG emissions are CO₂ emissions from fuel combustion. In India, they have steadily increased since 1990. At 41%, the electricity sector is the largest contributor, followed by industry (34%).

FIGURE 9.2 G20 nation air pollution death numbers.

[Reprinted from Joshi (Joshi, 2019) with permission from *The Economic Times*.]

Major objectives of NAAQS are: (i) to indicate necessary air quality levels and appropriate margins required to ensure the protection of vegetation, health, and property; (ii) to provide a uniform yardstick for the assessment of air quality at the national level; and (iii) to indicate the extent and need of the monitoring programme.

- **National Air Quality Monitoring Programme (NAMP)**

  The government is executing a nation-wide programme of ambient air quality monitoring known as NAMP. Under NAMP, four air pollutants have been identified for regular monitoring at all the locations: 1) SO_2; 2) NO_2; 3) suspended particulate matter (PM_{10}); and 4) fine particulate matter ($PM_{2.5}$). In addition, there are 134 real-time Continuous Ambient Air Quality Monitoring stations (CAAQMS) in 71 cities across 17 states, monitoring eight, pollutants viz. PM_{10}, $PM_{2.5}$, SO_2, NOx, ammonia (NH_3), CO, ozone (O_3), and benzene. PM_{10} are inhalable coarse particles, which are particles with a diameter between 2.5 and 10 micrometres (μm), and $PM_{2.5}$ are fine particles with a diameter of 2.5 μm or less. The smaller $PM_{2.5}$ are particularly deadly as it can penetrate deep into the lungs. The objectives of NAMP are: (i) to determine the status and trends of ambient air quality; (ii) to ascertain whether the prescribed ambient air quality standards are violated; (iii) to identify non-attainment cities; (iv) to obtain the knowledge and understanding necessary for developing preventive and corrective measures; and (v) to understand the natural cleansing process undergoing in the environment through pollution dilution, dispersion, wind-based movement, dry deposition, precipitation, and chemical transformation of the pollutants generated. The monitoring of meteorological parameters, such as wind speed and wind direction, relative humidity (RH), and temperature were also integrated with the monitoring of the air quality. The monitoring of pollutants is carried out for 24 hours (a 4-hourly sampling for gaseous pollutants and an 8-hourly sampling for particulate matter) twice a week, to have 104 observations in a year. The monitoring is being carried out with the help of the CPCB, State Pollution Control Boards (SPCB), Pollution Control Committees (PCC), and National Environmental Engineering Research Institute (NEERI).

- **National Air Quality Index (AQI)**

  The AQI was launched by the Prime Minister in April 2015, starting with 14 cities, and now extended to 71 cities in 17 states. The AQI is a tool for the effective communication of air quality status to people in terms, which are easy to understand. It transforms complex air quality data of various pollutants into a single number (index value), nomenclature and colour. There are six AQI categories: good, satisfactory, moderately polluted, poor, very poor, and severe. Each is decided based on the ambient concentration values of air pollutants and their likely health impacts (known as health breakpoints). The AQ subindex and health breakpoints are evolved for eight pollutants (PM_{10}, $PM_{2.5}$, NO_2, SO_2, CO, O_3, NH_3, and Lead (Pb)) for which short-term (up to 24h) NAAQS are prescribed.

- **Forty-Two Action Points**

  The CPCB has issued a comprehensive set of directions under Section 18 (1) (b) of Air (Prevention and Control of Pollution) Act, 1986, for the implementation of 42 measures to mitigate air pollution in the major cities, including Delhi and NCR comprising action points to counter air pollution, which include control and mitigation measures related to vehicular emissions, re-suspension of road dust, and other fugitive emissions, biomass/municipal solid waste (MSW) burning, industrial pollution, construction and demolition (C&D) activities, and other general steps.

- **Environment Pollution (Prevention and Control) Authority (EPCA)**

  EPCA was constituted with the objective of 'protecting and improving' the quality of the environment and 'controlling environmental pollution' in the National Capital Region. In the notification, jurisdiction of the EPCA has been stated as the NCR region as defined in clause (f) of Section 2 of National Capital Region Planning Board Act, 1985 (2 of 1985). The EPCA has been subsequently re-constituted from time to time, extending the tenure of the authority and/or substituting or including new members.

- **Graded Response Action Plan (GRAP)**

  The government has notified a graded response action plan for Delhi and the NCR region, which comprises the graded measures for each source framed according to the AQI categories. It also takes note of the broad health advisory for each level of the AQI that was adopted by the Government of India along with the AQI. The proposal has been framed keeping in view the key pollution sources in Delhi and the NCR region. While the major sources of pollution, including vehicles, road dust, biomass burning, construction, power plants, and industries remain continuous throughout all seasons, the episodic pollution from stubble burning, increase in biomass burning, etc., varies across seasons.

9.2.3 REGULATORY FRAMEWORK FOR AIR QUALITY MANAGEMENT AND FORECASTING

India Meteorological Department (IMD) is possibly the first institution in India to start systematic long-term environment monitoring of atmospheric aerosol properties, ozone, and precipitation chemistry. The technical coordination and overseeing of the functions of the operational air quality forecasting services in India has been entrusted to Environment Monitoring and Research Centre (EMRC), a division of IMD.

The AQ management and forecasting problems have been regulated with the help of a number of Indian Directives, which define the basic goals and quantitative criteria to be applied in order to effectively deal with air quality and air pollution problems. India recently released its much-anticipated National Clean Air Programme (NCAP), which provides a roadmap to prevent, control, and reduce air pollution. India is severely affected by air pollution; it led to 1.24 million or 12.5% of the total deaths recorded in the country during 2017 alone. It is a time-bound, national strategy to bring down levels of deadly particle air pollution ($PM_{2.5}$ and PM_{10}) by 20–30% by 2024 (compared to 2017 levels). Initially launched as a five-year action plan, the NCAP may be further extended after a mid-term review of the outcomes (Jaiswal, 2018).

The EU is considered the largest supranational body in the globe and is mandated with extensive powers in terms of policy making for environmental topics by its 28 member states. In European countries, the AQ management and forecasting problem has been regulated with the aid of a number of European Directives. EU established the current Ambient Air Quality and Clean Air for European Standards in 2008 under the thematic strategy on Air Pollution (TSAP).

The requirements for AQ forecasts include:

a) Geographical area of expected exceedance of a threshold.
b) Expected changes in pollution (improvement, stabilization or deterioration), together with the reasons for those changes.

It is interesting to note that the Directives base their requirements for AQ forecasts on the appearance of an incident or event, this being the exceedance of a threshold value concerning the con-centration level of each one of the regulated air pollutants. In addition, the same Directives state that "it is necessary to adapt procedures for data provision, assessment and reporting of air quality to enable electronic means and the Internet to be used as the main tools to make information available". This means that it is necessary to develop operational AQ management and citizen notification systems that will make use of modern ICT, and will allow for the early forecasting of the incidents of interest, on the basis of fast and efficient computational methods, as the ones based on AI algorithms (Oprea and Iliadis, 2011; Cabaneros et al., 2019).

9.3 ABBREVIATION EXPLANATION AND ERROR ASSESSMENT INDEX

For the convenience of reading, we give the abbreviations and used assessment indexes of various methods in the form of a list for easy reference (Table 9.1).

9.4 FUTURE TRENDS POTENTIAL FORECASTING METHODS

Artificial Intelligence (AI) has become influential impactor to environmental protection and climate change. Artificial intelligence is inward bound to many different areas. Machines extensively build up to resolve and minimize the after damaging effects on environment by different processes and clean the pollution (Aayush et al., 2020). Famous tech companies such as Microsoft, Google, and Tesla, despite the fact that broadening the innovation opportunities, are developing Earth friendly AI systems (Dhamija and Bag, 2020). According to a study report of Microsoft in 2018, the company was going to spend a good amount on AI technological systems, and stated that AI has the potential to minimize the air quality distortion changes by comparing and relating both data generated by AI satellite with quality monitoring data (Aayush et al., 2020). Widely, AI has great potential

TABLE 9.1
Abbreviations of Methods

S. No.	Abbreviation	Description
1	ANFIS	Fuzzy Inference Systems
2	ML	Machine Learning
3	ANN	Artificial Neural Network
4	GA	Genetic Algorithms
5	HMM	Hidden Markov Models
6	SVM	Support Vector Machines
7	AI	Artificial Intelligence
8	LSTM	Long Short-Term Memory
9	GRU	Gated Recurrent Units
10	SVR	Support Vector Regression
11	AQI	Air Quality Index
12	GP	Genetic Programming
13	AIMAM	Autoregressive Integrated Moving Average Model
14	TS	Time Series
15	NAAQS	National Ambient Air Quality Standards
16	IMD	India Meteorological Department
17	EMRC	Environment Monitoring and Research Centre
18	NCAP	National clean air programme
19	TSAP	Thematic Strategy on Air Pollution
20	ICT	
21	NMS	Network of Meteorological Stations
22	ASN	Air Stations Network
23	SILAM	System for Integrated Modeling of Atmospheric Composition
24	GFS	Global Forecast System

and application in environmental sectors, such as natural resource conservation, wildlife protection, energy management, clean energy, waste management, pollution control, and agriculture (Nishant et al., 2020).

9.4.1 AIR POLLUTION FORECASTING AND ANALYSIS

Quality of life is affected by various factors and critical environmental situations. It is important to have forecasting intelligent systems capable of alerting the population in order to prevent or to reduce the effects of an extreme environmental incident.

Figure 9.3 shows the possible effects of some environmental critical situations to other parts of the environment. For example, a severe air pollution problem would affect the soil, and the water. Also, if two problems are combined, the impact on the environment could increase significantly. For example, if there is a severe air pollution and flood in a given region, then the impact on the water and soil is greater, increasing the pollution degree.

The system is linked to a network of meteorological stations (MSN) which provides measurements and observations from specific locations. Also, the environment can be associated to specific networks of air, water, or soil quality monitoring stations capable of providing measurements and observations of various environmental parameters (e.g., concentration of air pollutants). For example, the air stations network (ASN) can provide measurements of air pollutants (SO_2, CO, NO_2, NH_3, O_3, PM_{10}, $PM_{2.5}$) in urban regions.

Forecasting and modeling of air pollution is very complex due to underlying interrelations between numerous variables of different types. That is why a number of modeling approaches have been applied in this field (Cortes et al., 2000). In recent years, several AI-based forecasting systems were presented in the literature for the prediction of different air pollutants. According to Polat, a novel feature scaling method, named neighbour-based feature scaling is combined with an ANN model and an ANFIS model for the prediction of sulphur dioxide (SO_2) concentration (Figure 9.4). The performance of the proposed method was evaluated by using four statistical indicators, including the index of agreement (IA). The results were good,

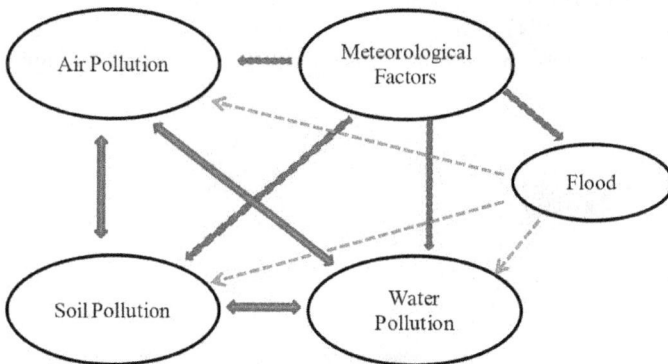

FIGURE 9.3   Interdependencies between different environmental critical situations.

showing the benefits of using the proposed method for sulphur dioxide concentration forecasting in a specific region of Turkey (the Konyia province) (Polat, 2012).

Many studies presented an overview of several soft computing approaches like ANN, SVM, and fuzzy logic as a feasible solution for urban air forecasting (Cakir and Sita, 2020). There are many that work that reported a successful use of ANN and ANFIS model for the prediction of daily CO concentration in the atmosphere of Tehran (Noori et al., 2010). Monte Carlo simulations are used to model the probability of different outcomes in a process that cannot easily be predicted due to the intervention of random variables, and the performances of the models were analyzed, with consideration to the uncertainty in forecasting. Oprea et al. presented knowledge modeling in an air pollution control DSS (Oprea et al., 2005). There are several other research project and studies that deal with air pollution forecasting in various regions worldwide, on all the continents. There are several research projects that deal with air pollution forecasting in various regions worldwide, on all the continents. Some examples of such projects and research initiatives are reported (Kolehmainen et al., 2001; Moussiopoulos et al., 2003). EUROTRAC-2 which is a subproject of SATURN, the aims of this subproject at a better understanding of urban air pollution as a prerequisite for finding effective solutions to air quality problems and for a sustainable development in the urban environment. For this purpose, the main scientific objective of SATURN is to substantially improve our ability of establishing source-receptor relationships at the urban scale. Ensuring the validity of such relationships may also facilitate assessing the impact of urban areas to regional and global scale problems of the atmospheric environment (Cortes et al., 2000; Moussiopoulos et al., 2003).

There are several dispersion models, and many universities and institutions around the world are running one or several of them for different regions (Europe, Continental US, Asia, etc.). Following is a non-exhaustive list of models being used for providing AIR Quality forecast on the 'Word Air Quality Index' project.

- **SPRINTARS East Asia**

    Forecast Model Analysis (Spectral Radiation-Transport Model for Aerosol Species). SPRINTARS has been primarily developed by Climate Change Science Section, Research Institute for Applied Mechanics, Kyushu University. SPRINTARS treats main tropospheric aerosols both from natural and anthropogenic sources (black carbon, organic matter, sulphate, soil

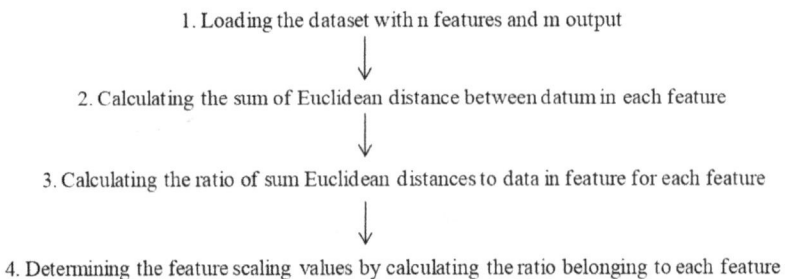

1. Loading the dataset with n features and m output

$\downarrow$

2. Calculating the sum of Euclidean distance between datum in each feature

$\downarrow$

3. Calculating the ratio of sum Euclidean distances to data in feature for each feature

$\downarrow$

4. Determining the feature scaling values by calculating the ratio belonging to each feature

FIGURE 9.4   The flowchart of the neighbour-based feature scaling method.

dust, and sea salt). They are also categorized into SPM, PM_{10}, and $PM2_{.5}$. SPRINTARS calculates transport processes of aerosols (emission, advection, diffusion, wet deposition, dry deposition, and gravitational settling). The aerosol-radiation interaction, which is scattering and absorption of solar and thermal radiation by aerosols, and the aerosol-cloud interaction, which is act of aerosols as cloud condensation nuclei and ice nuclei, are included in the calculation (Takemura, 2005).

- **SILAM Asia**

  Forecast Model Analysis (System for Integrated modeling of Atmospheric Composition) SILAM is a global-to-meso-scale dispersion model developed for atmospheric composition, air quality, and emergency decision support applications, as well as for inverse dispersion problem solution. The model incorporates both Eulerian and Lagrangian transport routines, eight chemico-physical transformation modules (basic acid chemistry and secondary aerosol formation, ozone formation in the troposphere and the stratosphere, radioactive decay, aerosol dynamics in the air, pollen transformations), 3- and 4-dimensional variation data assimilation modules (Sofiev et al., 2006).

- **Forecast Model Analysis (Global Forecast System)**

  The Global Forecast System (GFS) has been in NWS operations since 1980 and is continuously improved by the Environmental Modeling Center, whose mission is to maintain, enhance and transition-to-operations advanced numerical guidance systems for the nation's weather/water/climate enterprise and the global community for the protection of life/property and the enhancement of the economy (Baklanov and Zhang, 2020).

9.4.2 Some Implemented Systems for Air Quality Monitoring and Control

Generally, people spend most of their time at home, in the workplace, or at educational institutions (indoors). Besides, this outdoor air quality is also very harmful to environments. Hence, as well as outdoor air quality, there has been an increasing concern over indoor (IAQ) and its effects on public health. The US EPA reported that, in the US, the mean daily residential time spent indoors was 21 h, while the GerES II reported that this duration was 20 h in Germany and in India (Delhi) one study reported people spend more than 90% of their daily life in indoor environments (Datta et al., 2017). Thus, the IAQ has been recognized as a significant factor in the determination of the health and welfare of people. The Korea Ministry of Environment (KMOE) enforced the IAQ act to control five major pollutants, including PM_{10}, CO_2, CO, VOCs, and formaldehyde in indoor environments. Out of these, the IAQ standard for PM_{10} concentration is $150\,\mu g/m^3$. The IAQ is critical not only in buildings, but also in underground areas and public transportation systems. Much effort has been made for the improvement of the IAQ in subway stations.

9.4.2.1 Platform Screen Doors

Platform screen doors were installed at underground subway stations, and are being used in many subway stations in Korea to prevent the diffusion of air pollutants into the subway stations and ensure the safety of the public. Some previous studies reported that the PM concentration in subway stations significantly reduced after the installation of PSDs (Jung et al., 2010). PM measuring instruments and control system: Particulate matter with an aerodynamic diameter less than $10\,\mu m$ (PM_{10}) is one of the major pollutants in environments. As for the PM_{10} concentration, generally, measuring instruments based on β-ray absorption method are used. In order to keep the PM_{10} concentration below a healthy limit, the air quality in the underground platform and tunnels should be monitored and controlled continuously. According to some study there are many types of equipment and methods (Table 9.2) are available for measuring particulate material, which makes the choice difficult (Kleinhans et al., 2018). In addition, particulate matter can range in size from nanometres to micrometres, which makes the choice of equipment even more challenging, as reported (Yang et al., 2021).

9.4.2.2 Wireless Sensor Network for Air Quality Monitoring

Wireless sensor network (WSN) has gained worldwide attention in recent years. WSNs developed on a global scale for various activities such as environmental monitoring, habitat studying, infrastructure health monitoring, military surveillance, and traffic control. The development of intelligent sensors has been facilitated by the rise of Micro-Electro-Mechanical (MEMS) technology systems. These sensors are small, with restricted processing and computing resources, and inexpensive compared to traditional sensors. They can perceive measure and collect information from the environment and based on some local decision-making processes, they can transmit data to the user. The smart sensor nodes are low-power devices equipped with one or more sensors, a digital processor, some memory, a power supply, a radio, and an actuator. A variety of mechanical, thermal, biological, chemical, optical, and magnetic sensors can be connected to the sensor node to measure environment properties. Because the sensors have limited memory and nodes are typically developed to place them in difficult accessed locations, a radio is implemented to establish wireless communication and to transfer the data to a base station (Vinodha and Durairaj, 2021). However, WSNs have some boundaries in terms of memory, energy, computing, communication, and scalability.

Similarly, cloud computing is becoming a promising technology to provide a massive pile of computing, storage, and software services. The cloud system consists of two different layers: the core and the service layer. The first one integrates those components shared by the services including databases, e-learning algorithms, and connection mechanisms for linking and matching services. The service layer is in charge of publishing different type of services for sensor network monitoring and management (Table 9.3). Figure 9.5 shows a block diagram of the operation of the cloud system.

TABLE 9.2

Specifications of Different Instruments for Particle Measurement

Methods	Instruments for Measuring PM	Characteristics
Concentration Measurement Methods	Microbalance Methods	In this method, the particles are collected, over the surface of an oscillatory microbalance element, those microbalances use the alteration of the resonance frequency to determine the PM
	Optical Methods	In the optical detection methods, aerosol particles are lit by a light beam and irradiate this light in all directions (scattering)
	Gravimetric Method	In the gravimetric method, the particle mass concentration is determined by weighing the filters before and after the sampling period
Size Distribution Measurement Methods	Microscopy	Involves collection of particles directly from filters followed by filter preparation to improve visibility
	Impactor	Impactors are instruments for measuring size distribution in mass, which working principle is gravimetric, with multiple impact stages; in some equipment multiple orifices are found
	Diffusion Battery (EDB)	Diffusion Battery uses for particles with sizes below 0.1 µm, this is more appropriate for nanoparticles
	Mobility Analyzer	Electrical Aerosol Analyzer (EAA) and DMA are the mobility analyzers
	Centrifugal Measurement of Particle Mass	Centrifugal measurement of particle mass can be done by using a Centrifugal Particle Mass Analyzer (CPMA) or an Aerosol Particle Mass (APM)
	Differential Mobility Spectrometers (DMS)	Among the spectrometers based on particle mobility, the most known are the DMS and the Fast Mobility Particle Sizer (FMPS)
	Fast Integrated Mobility Spectrometer (FIMS)	In FIMS, the aerosol passes through a neutralizer, where the particles receive a charge distribution of bipolar equilibrium. In the sequence, the aerosol passes through a mobility analyzer, through which a gas flows (butanol-saturated). In the electrical field of the mobility analyzer, charged particles are separated in different paths, based on their electrical mobility
	Electrical Low-Pressure Impactor (ELPI)	the ELPI classifies particles according to its aerodynamic diameter, besides measuring the concentration and particle distribution in number (7 nm to 10 µm)
	Aerodynamic Sizers	Aerodynamic Particle Sizer (APS), Particle Size Distribution and Aerosol Mass Spectrometer (AMS) are the equipment used to measure aerodynamic size

9.4.2.3 Sensor-Based Wireless Air Quality Monitoring Network-SWAQMN (Polludrone)

Low cost and affordable sensors equipped device for air quality monitoring which is developed by CSIR-NEERI. Polludrone is a fully integrated air quality monitoring system that delivers reference equivalent performance. It is an air quality monitoring device, and an ample solution to monitor all the critical ambient environmental parameters related to air quality, including noise, weather, and radiation. In addition, meteorological parameters like wind speed, wind direction, rainfall, flood levels, and visibility can be monitored using external modules for ground-level weather data. Polludrone is a highly accurate yet cost-effective solution – measuring criteria pollutants to WHO air quality limits yet costing 10x less than a traditional station based on analyzer technology. It measures all major ambient parameters like $PM_{2.5}$, PM_{10},

TABLE 9.3
Types of Services Supported by Cloud System

S. No.	Services	Descriptions
1	Storage services	To this end, the following storage services have been implemented: a) create service provides the necessary actions to create new databases for data storing, b) connection service matches a sensor network to a specific database, c) save service stores a data sensor network in a concrete database, and d) retrieve service requests data from the database and returns the extracted information
2	End-user services	These types of services are focused on providing end-user data access. In this sense, the request identification service opens a user session to allow data access
3	Sensor data services	These are focused on checking the information retrieved from sensor networks. This service also adds new metadata to extend the knowledge of each measurement
4	E-learning services	These support mechanisms classify sensor data through applying e-learning mechanisms (neuronal networks). Create service builds a new neuronal network according to several setup values, while the training service allows a neuronal network to learn from an initial dataset. Finally, request service classifies new sensor network values
5	Security services	Security is one of the main problems in cloud-based systems, since services can be available from everywhere

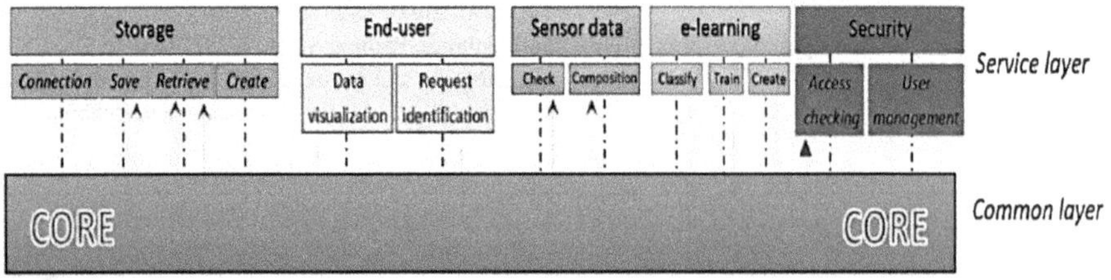

FIGURE 9.5 Cloud sensor network framework.

[Reprinted from Arroyo et al. (2019) with permission from MDPI.]

Carbon Monoxide (CO), Carbon Dioxide (CO_2), Sulphur Dioxide (SO_2), Nitrogen Dioxide (NO_2), Ozone (O_3), Hydrogen Sulfide (H_2S), Ambient Noise, Light, UV, Temperature, and Humidity. The Polludrone uses our patented e-breathing sampling technology that offers higher accuracy and near-reference data. By creating a dense network of data points, the Polludrone is powered to generate hyperlocal pollution (Gulia et al., 2020).

Application of Polludrone: Smart-city monitoring, Road-Side Monitoring, Tunnels monitoring, Campus monitoring, Airports monitoring, and Real Estate.

9.4.2.4 Some Other Applications of AI Environmental Sector

AI growing and started command over some advance technologies in a range of fields such as (Figure 9.6):

1. AI in Business: AI improves the processes of business needs by automating the commercial procedures, improves acquisition of data analysis, and customer & employee appealing techniques.
2. AI in Agriculture: AI techniques in agricultural field improve harvest capacity of crops at an advanced dimensions and quicker the pace than human labourers.
3. AI in Climate Control: AI enable enhanced and improved management techniques towards impact management of climate change and shield the environment (Aayush et al., 2020).

AI technologies on pollutant elimination range 0.64 to 1.00, lessen operational costs around 30% ability to solve practical problems (such as the wastewater treatment), provide higher accuracy, and lower error by prediction accuracy of AI technologies (Zhao et al., 2020). Similarly, there are growing implicational use has been seen in some well-known tech companies to reduce the overall efforts and side

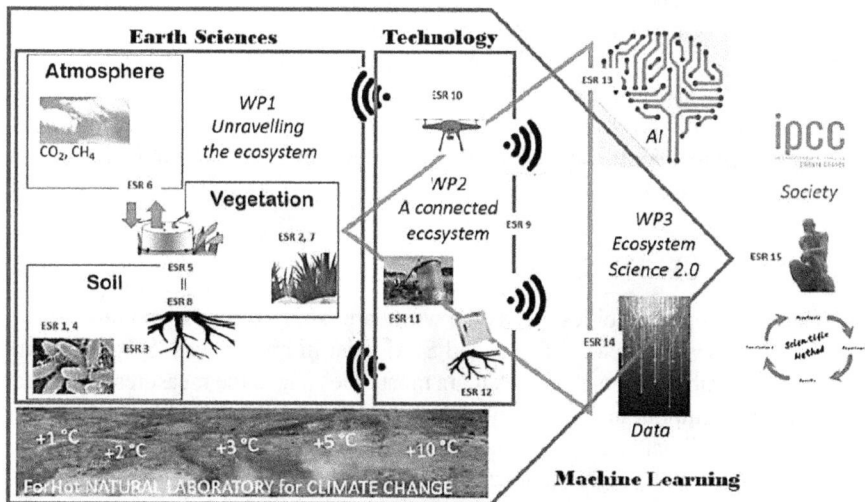

FIGURE 9.6 Ecosystem-of-things systems for key environmental variables and processes.

TABLE 9.4

List of AI Infused Earth Applications Developed by Renowned Tech Companies

S. No.	Name of AI applications	Use for
1	*iNaturalist* (AI-infused Earth applications)	Collect data from its vast circle of experts on the species encountered, which would help to keep track of their
2	*eBirds* (AI-infused Earth applications)	population, favourable eco systems, and migration patterns. These applications have also played a significant role in the better identification and protection of fresh water and marine ecosystems
3	*Plantix* (Berlin-based agricultural tech start-up PEAT has developed a DL application)	Identifies potential defects and nutrient deficiencies in soil
4	*DeepMind* (Google's very own)	AI help organization to more energy efficient and reducing overall GHG emissions by restraining their energy of data centre convention by 40%
5	*AWhere* (United States-based companies)	Uses algorithms of ML in association with satellites to forecast weather, evaluate farms for the presence of diseases and pests and analyze crop sustainability
6	*Farm Shots* (United States-based companies)	

effects of the technology, Table 9.4 comprises the list of some application name using by different companies listed (Murphy et al., 2021). Ye et al. have summarised the different applications of AI technologies to improve environmental management in detailed manner in 2020. In Figure 9.7, a number of approaches of AI techniques and research AI technological applications are represented, which clearly graphing the increasing demand and applications of AI technologies over time. AI applications help to get reliable mapping of nonlinear chemical inputs and outputs performance, biological processes in requisites prediction models to the emerging optimization and control algorithms that study the pollutants elimination progression and intelligent control systems, have been developed for environmental uncontaminated (Ye et al., 2020).

9.5 CONCLUSION

The major metropolitan cities across the world are facing severe air pollution problems. Critical levels of $PM_{2.5}$, PM_{10}, and SPM exist in many cities. SO_2, NOx, and lead levels in ambient air are decreasing in most cities due to the measures taken such as reduction of sulphur in fuel and introduction of unleaded petrol. Air pollution also responsible for climate change across the world. AI applications for the atmospheric environment verify the high efficiency of the algorithms, and also reveal their unique ability to extract knowledge from the scientific domain of interest and to help us

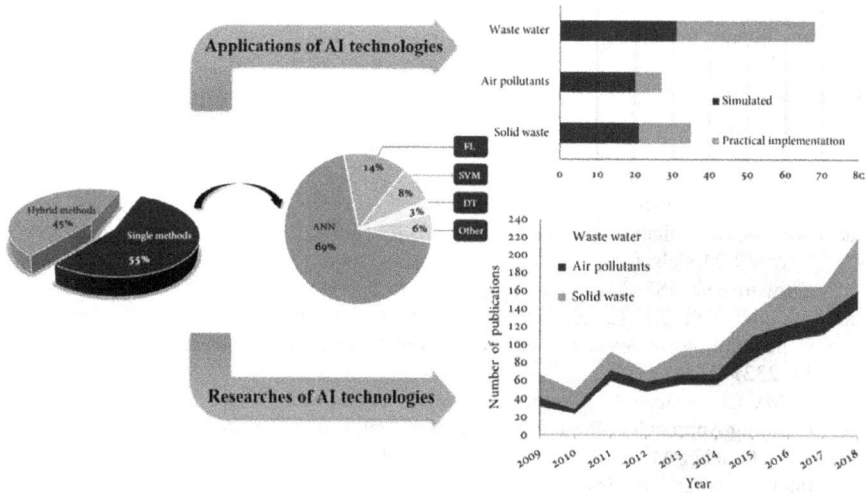

FIGURE 9.7   Applications of Artificial Intelligence Technologies in the field of Environmental management.

[Reprinted from Ye et al. (2020) with permission from Elsevier].

understand and manage environmental problems. On the basis of this literature, it is recommended that such an analysis be carried out in cases where problems related to the air quality and quality of life are concerned, in order to better inform policy and decision makers about the constrains of their decisions. AI methods may provide services related to public notification, warnings, and alerts with the necessary scientific competence, on the basis of forecasts. These systems will develop a capability of modifying their performance in order to better fit future needs, especially in terms of user-specific, quality of life information services.

ACKNOWLEDGEMENTS

We are sincerely thankful to Indian Institute of Technology (Banaras Hindu University), Varanasi, India; The Maharaja Sayajirao University of Baroda, Vadodara, India; IBM India Private Limited, Gautam Buddha Nagar, India, and Dr. Ambedkar Institute of Technology for Handicapped, Kanpur, India for this work.

CONFLICTS OF INTEREST

The authors declare no conflict of interest.

REFERENCES

Arroyo, P., Herrero, J.L., Suarez, J.I. and Lozano, J. (2019). Wireless sensor network combined with cloud computing for air quality monitoring. *Sensors Basel*, 19 (691): 1–17.

Aayush, K., Vishal, D., Hammad, N. and Manu, K.S. (2020). Application of Artificial Intelligence in Curbing Air Pollution: The Case of India. *Asian Journal of Management* 11: 285–290.

Baklanov, A. and Zhang, Y. (2020). Advances in Air Quality Modeling and Forecasting. *Global Transitions* 2: 261–270.

Balakrishnan, K., et al. (2019). The Impact of Air Pollution on Deaths, Disease Burden, and Life Expectancy across the States of India: The Global Burden of Disease Study 2017. *The Lancet Planetary Health* 3: e26–e39.

Cabaneros, S.M., Calautit, J.K. and Hughes, B.R. (2019). A Review of Artificial Neural Network Models for Ambient Air Pollution Prediction. *Environmental Modeling & Software* 119: 285–304.

Cakir, S. and Sita, M. (2020). Evaluating the Performance of Ann in Predicting the Concentrations of Ambient Air Pollutants in Nicosia. *Atmospheric Pollution Research* 11: 2327–2334.

Castelli, M., Clemente, F.M., Popovic, A., Silva, S. and Vanneschi, L. (2020). A Machine Learning Approach to Predict Air Quality in California. *Complexity* 2020: 1–23.

Cortes, U., Sanchez-Marre, M., Ceccaroni, L., R-Roda, I. and Poch, M. (2000). Artificial Intelligence and Environmental Decision Support Systems. *Applied Intelligence* 13: 77–91.

Datta, A., Suresh, R., Gupta, A., Singh, D. and Kulshrestha, P. (2017). Indoor Air Quality of Non-Residential Urban Buildings in Delhi, India. *International Journal of Sustainable Built Environment* 6: 412–420.

Dhamija, P. and Bag, S. (2020). Role of Artificial Intelligence in Operations Environment: A Review and Bibliometric Analysis. *The TQM Journal* 32: 869–896.

EEA (2021). Air Pollution Is the Biggest Environmental Health Risk in Europe. *European Environment Agency, Copenhagen, Denmark* https://www.eea.europa.eu/themes/air/air-pollution-is-the-single assessed 01/09/2021

Gulia, S., Prasad, P., Goyal, S.K. and Kumar, R. (2020). Sensor-Based Wireless Air Quality Monitoring Network (Swaqmn) - a Smart Tool for Urban Air Quality Management. *Atmospheric Pollution Research* 11: 1588–1597.

Guo, Q., He, Z., Li, S., Li, X., Meng, J., Hou, Z., Liu, J. and Chen, Y. (2020). Air Pollution Forecasting Using Artificial and Wavelet Neural Networks with Meteorological Conditions. *Aerosol and Air Quality Research* 20: 1429–1439.

Jaiswal, A. (2018). India Launches a National Clean Air Program. *Natural Resources Defense Council, New York, United States* https://www.nrdc.org/experts/anjali-jaiswal/india-launches-national-clean-air-program assessed 01/09/2021

Joshi, A. (2019). India Accounts for the Highest Number of Deaths among G20 Nations Due to Air Pollution. *The Economic Times* https://energy.economictimes.indiatimes.com/news/power/india-accounts-for-the-highest-number-of-deaths-among-g20-nations-due-to-air-pollution/72234587 assessed 01/09/2021

Jung, H.J., Kim, B., Ryu, J., Maskey, S., Kim, J.C., Sohn, J. and Ro, C.U. (2010). Source Identification of Particulate Matter Collected at Underground Subway Stations in Seoul, Korea Using Quantitative Single-Particle Analysis. *Atmospheric Environment* 44: 2287–2293.

Kleinhans, U., Wieland, C., Frandsen, F.J. and Spliethoff, H. (2018). Ash Formation and Deposition in Coal and Biomass Fired Combustion Systems: Progress and Challenges in the Field of Ash Particle Sticking and Rebound Behavior. *Progress in Energy and Combustion Science* 68: 65–168.

Kok, I., Guzel, M. and Ozdemir, S. (2021). Recent Trends in Air Quality Prediction: An Artificial Intelligence Perspective, In *Intelligent Environmental Data Monitoring for Pollution Management*, Bhattacharyya, S., Mondal, N.K., Platos, J., Snasel, V. and Kromer, P. (Eds.), Academic Press, pp. 195–221.

Kolehmainen, M., Martikainen, H. and Ruuskanen, J. (2001). Neural Networks and Periodic Components Used in Air Quality Forecasting. *Atmospheric Environment* 35: 815–825.

Moussiopoulos, N., Schlünzen, H. and Louka, P. (2003). Modeling Urban Air Pollution, In *Air Quality in Cities: Saturn Eurotrac-2 Subproject Final Report*, Moussiopoulos, N. (Ed.), Springer Berlin Heidelberg, Berlin, Heidelberg, pp. 121–154.

Murphy, K., Di Ruggiero, E., Upshur, R., Willison, D.J., Malhotra, N., Cai, J.C., Malhotra, N., Lui, V. and Gibson, J. (2021). Artificial Intelligence for Good Health: A Scoping Review of the Ethics Literature. *BMC Medical Ethics* 22: 14.

Nishant, R., Kennedy, M. and Corbett, J. (2020). Artificial Intelligence for Sustainability: Challenges, Opportunities, and a Research Agenda. *International Journal of Information Management* 53: 102104.

Noori, R., Hoshyaripour, G., Ashrafi, K. and Araabi, B.N. (2010). Uncertainty Analysis of Developed Ann and Anfis Models in Prediction of Carbon Monoxide Daily Concentration. *Atmospheric Environment* 44: 476–482.

Oprea, M. (2005). A Case Study of Knowledge Modeling in an Air Pollution Control Decision Support System. *AI Communications* 18: 293–303.

Oprea, M. and Iliadis, L. In Iliadis, L. and Jayne, C. (Eds.) Engineering Applications of Neural Networks, 2011, Springer Berlin Heidelbereg, Berlin, Heidelberg, pp. 499–508.

Polat, K. (2012). A Novel Data Preprocessing Method to Estimate the Air Pollution (SO_2): Neighbor-Based Feature Scaling (Nbfs). *Neural Computing and Applications* 21: 1987–1994.

Sahoo, S.K., Kumar, A.V., Yadav, A.K. and Tripathi, R.M. (2016). Metal Characterization of Airborne Particulate Matters in a Coastal Region. *Toxicological & Environmental Chemistry* 98: 768–777.

Shams, S.R., Jahani, A., Kalantary, S., Moeinaddini, M. and Khorasani, N. (2021). Artificial Intelligence Accuracy Assessment in No_2 Concentration Forecasting of Metropolises Air. *Scientific Reports* 11: 1805.

Sofiev, M., Siljamo, P., Valkama, I., Ilvonen, M. and Kukkonen, J. (2006). A Dispersion Modeling System Silam and Its Evaluation against Etex Data. *Atmospheric Environment* 40: 674–685.

Takemura, T. (2005). Simulation of Climate Response to Aerosol Direct and Indirect Effects with Aerosol Transport-Radiation Model. *Journal of Geophysical Research*, 110: 1-10.

Vinodha, R. and Durairaj, S. (2021). Soft Computing Approach Based Energy and Correlation Aware Cooperative Data Collection for Wireless Sensor Network. *Journal of Ambient Intelligence and Humanized Computing* 12: 5297–5308.

WHO (2018). Health, Environment and Climate Change: Report by the Director-General. World Health Assembly, World Health Organization, Geneva, Switzerland, p. 71.

WHO (2021). A Strategic Roadmap to Promote Healthier Populations through Clean and Sustainable Energy. *World Health Organization, Geneva, Switzerland* https://www.who.int/publications/m/item/a-strategic-roadmap-to-promote-healthier-populations-through-clean-and-sustainable-energy assessed 01/09/2021

Yadav, A.K. (2015). Elemental Composition and Source Apportionment of Suspended Particulate Matters and Health Risk Assessment in Mining and Non-mining Areas of Odisha, India. *Journal of Hazardous, Toxic, and Radioactive Waste* 19: 04014037.

Yadav, A.K. (2021). Human Health Risk Assessment in Opencast Coal Mines and Coal-Fired Thermal Power Plants Surrounding Area Due to Inhalation. *Environmental Challenges* 3: 100074.

Yadav, A.K. and Jamal, A. (2018a). Impact of Mining on Human Health in and around Mines. *Environmental Quality Management* 28: 83–87.

Yadav, A.K. and Jamal, A. (2018b). Technological Reviews of Particulate Matter and Their Source Identification Techniques. *Environmental Quality Management* 27: 87–95.

Yadav, A.K., Sahoo, S.K., Dubey, J.S., Kumar, A.V., Pandey, G. and Tripathi, R.M. (2019). Assessment of Particulate Matter, Metals of Toxicological Concentration, and Health Risk around a Mining Area, Odisha, India. *Air Quality, Atmosphere and Health* 12: 775–783.

Yang, W., Pudasainee, D., Gupta, R., Li, W., Wang, B. and Sun, L. (2021). An Overview of Inorganic Particulate Matter Emission from Coal/Biomass/Msw Combustion: Sampling and Measurement, Formation, Distribution, Inorganic Composition and Influencing Factors. *Fuel Processing Technology* 213: 106657.

Ye, Z., Yang, J., Zhong, N., Tu, X., Jia, J. and Wang, J. (2020). Tackling Environmental Challenges in Pollution Controls Using Artificial Intelligence: A Review. *Science of The Total Environment* 699: 134279.

Zhao, L., Dai, T., Qiao, Z., Sun, P., Hao, J. and Yang, Y. (2020). Application of Artificial Intelligence to Wastewater Treatment: A Bibliometric Analysis and Systematic Review of Technology, Economy, Management, and Wastewater Reuse. *Process Safety and Environmental Protection* 133: 169–182.

10 Fuzzy and Neural Network Model-Based Environmental Quality Monitoring System
Past, Present, and Future

Ankit Kumar, Saba Shirin, Mohammad Irfan Ansari, Govind Pandey, Shiv Nath Sharma, and Akhilesh Kumar Yadav

CONTENTS

DOI: 10.1201/9781003203445-12

10.1 INTRODUCTION

Air pollution can be defined as the presence of harmful substances such as carbon monoxide (CO), ozone (O_3), particulate matter (PM), nitric oxide (NO), and nitrogen dioxide (NO_2) in the atmosphere at higher concentrations than their normal ambient rates in such that they cause damage to human health, other living creatures, ecosystems, and the environment (Sahoo et al., 2016; Yadav and Jamal, 2018). Developing early warning systems and providing information on air pollution with forecasting is one of the most effective ways to prevent negative effects on human health and improve life quality. So far, several techniques, fundamentally creating two main categories, have been used to forecast air pollution: statistical and soft computing. The main objective of all these methods is to analyze the past behaviour of air pollutant concentrations to predict their future values under the assumption that they will behave similarly in the future.

Box Jenkins models and regression analysis lead in statistical techniques used for this purpose (Mihai and Meghea, 2011; Habermann et al., 2015). But reliability in forecasting based on statistical techniques requires that air pollution data sets satisfy some statistical assumptions such as linearity, normality, stationarity, invertibility, and large sample size. Considering that air pollution data sets generally include seasonality and consist of very different observations in terms of magnitude, satisfying these assumptions is very difficult. In recent years, soft computing techniques to forecast air pollution have increased as they do not require any statistical assumptions or constraints and provide better forecasting performance than statistical models. Artificial Neural Network (ANN) is one of the most frequently used soft computing techniques in this field (Grivas and Chaloulakou, 2006; Mishra and Goyal, 2015). However, ANN has some drawbacks, e.g., local minima problems in learning steps, poor generalization issues, difficulty in selecting appropriate network architecture, and computation complexity. Besides, ANNs cannot deal with uncertainty and subjectivity present in environmental issues and linguistic process variables such as "poor", "good" and "normal". FTS models are most appropriate for forecasting air pollution data that generally collects in the form of time series, consists of a small number of observations, is incomplete, includes uncertainty, and does not satisfy statistical assumptions. However, very few studies have been done using FTS models. Cheng et al. (2011) predicted daily O_3 concentrations in Taiwan using an FTS model based on a two-stage linguistic partition). Cagcag et al. (2013) proposed a new seasonal FTS model and applied this model to forecasting air pollution in Ankara. Rahman et al. (2015) compared the forecasting performance of Box Jenkins, ANN, and three models of FTS. FTS models used in these studies were negatively affected by outliers and abnormal observations. They use outlier-sensitive methods in the fuzzification step, the first step of the FTS building procedure, which distorts the data in favour of outliers. When considering that air pollution data generally includes outliers, this is a big disadvantage.

This study proposes a new fuzzy time series model using the Fuzzy K-Medoids clustering algorithm in the fuzzification step of FTS to minimize the negative effects of outliers and abnormal observations on the forecasting performance of fuzzy time series models. In order to evaluate the performance of the proposed method in forecasting and prediction of air pollution in comparison to that of the other fuzzy clustering-based FTS models in the literature (Cheng et al., 2008; Egrioglu et al., 2010), air pollution data consisting of weekly SO_2 concentrations measured at 65 monitoring stations in Turkey is used. According to the results of the analyses, it is observed that the proposed method provides successful forecasting results.

10.2   SCENARIO AND PROBLEMS

Air is a wonderful and precious natural resource. It is the main natural source that helps all life to sustain itself on this earth. In recent years the precious resources are getting polluted due to various human activities. As a result, pollution is becoming day-by-day more deadly for the ecosystem. According to the State of the Global Air Report 2019 (Polk, 2019), air pollution is the fifth leading risk factor of mortality worldwide. Commonly seen air pollutants include CO, SO, O_3, PM_{10}, $PM_{2.5}$, etc. They will cause environmental issues, such as soil acidification, fog, and haze but also cause health problems like heart attacks and lung diseases (Chen et al., 2019; Li et al., 2019; Zhou et al., 2019). The report emphasis on pollution-related chronic and cardiovascular diseases. Apart from these diseases, neoplasms also result from air pollution. Though the overall number of cases is fewer, it's likely to increase in the coming years if pollution is not controlled, and will add more deaths and disabilities due to air pollution. The World Health Organization (WHO) estimates that more than 90% of people live in places where air pollution is harmful (Janjua et al., 2021). The primitive pollutant affecting human health is fine particulate matter. Its size is less than or equal to 2.5 microns. The pollutant deeply penetrates the lungs causing grave trouble with health. These fine particles get into the bloodstream affecting every part of the body.

Intake of particles led to a reduction in lung functioning and an inflammatory response marked by cell influx and cytokine release (Yang et al., 2018). Some studies have focused on $PM_{2.5}$ relation with respiratory diseases (Fan et al., 2016; Khalili et al., 2018). The study carried out by Fan et al. (2016) calculated the polled risk ratio (RR) and found that asthma emergency visits added on with higher $PM_{2.5}$ (RR 1.5% per 10 µg/m³). It was found that children are more susceptible to air pollution (Khalili et al., 2018). The effects of $PM_{2.5}$ are not restricted to pulmonary diseases, but recent studies (Fu et al., 2019) have interrogated its neurological disorders. It is observed that smaller particulate matter directly crosses the blood-brain barrier. A comprehensive study was carried out by Fu et al. (2019), showing a significant association between $PM_{2.5}$ and neurodegenerative diseases and neurodevelopmental disorders. Owing to the numerous effects of air pollution, researchers tend to monitor air quality to reduce and control its severity.

Artificial intelligence methods have been observed in various fields for classification and regression (Wang and Wang, 2019; Rai et al., 2015). The accuracy of these methods has increased its applicability over typical regression methods. Many studies show good efficiency of artificial methods for predictive analysis of $PM_{2.5}$ using other pollutants and meteorological parameters (Chen, 2018). Chen carried out the predictive analysis of $PM_{2.5}$ mass concentration and found root mean square error greater than 30 for different input parameters.

Recently, much research in air pollution forecasting has been devoted to the formulation and development of models with the meteorological data-for example, statistics model (Ozel and Cakmakyapan, 2015), autoregressive integrated moving average (Samia et al., 2012), artificial neural network (ANN) (Chaudhuri and Acharya, 2012; Elangasinghe et al., 2014), community multi-scale air quality model (CMAQ) (Chen et al., 2014; Djalalova et al., 2015), weather research and forecasting model with chemistry (WRF-Chem) (Chuang et al., 2011; Saide et al., 2011), fuzzy inference system (Domańska and Wojtylak, 2012) grey model (Pai et al., 2013), and other hybrid methods (Chen et al., 2013; Russo and Soares, 2014). These methods have achieved good performances for air pollution forecasting result from their functions giving possibilities for discovering the new dependencies between data gathered in sets.

Among these methods are ANN and Fuzzy-based modeling, which have the capabilities of nonlinear mapping, self-adaption, and robustness, proved its superiority, and are widely used in forecasting fields. Recently, various ANN structures have been developed to improve the forecasting performances of air pollutants concentrations. Feng et al. (2011) applied a backpropagation neural network (BPNN) to forecast ozone concentration. Wu et al. (2011) considered dust storms when improving Elman network in predicting PM10 in Wuhan, China. Paschalidou et al. (2011) used multilayer perceptron (MLP) and radial basis function (RBF) techniques to forecast hourly PM_{10} concentrations in Cyprus. Pai et al. (2013) adopted neural network and fuzzy learning approach (ANFIS) to forecast oxidant concentration in 24-h. Antanasijević et al. (2013) employed general regression neural network (GRNN) to forecast PM_{10} concentration. The comparison with the PCR (combination of principal component analysis and multilinear regression) model has shown that the GRNN has significantly better performance than the PCR model. Ababneh et al. (2014) designed a three-layer feedforward neural network (FFNN) and recurrent Elman network to forecast PM_{10} concentrations 1 day advance in Yilan County, Taiwan.

ANFIS and ANN (artificial neural network) have many advantages in solving complex real-time problems (Jang, 1993). These methods are widely used for the prediction of nonlinear series. In ANFIS, the parameters involved are tuned using the least square and gradient descent method. The shortcoming of ANFIS in tuning the parameters is due to the drawback of the gradient descent method to get trapped in local optima. This problem results in decreasing the quality of modeling and increasing computational time. In order to optimize the computational time and achieve global optima, evolutionary techniques are introduced with ANNs (Goldberg and Holland, 1988). Ongoing research in evolutionary studies shows remarkable advantages of particle swarm optimization (PSO) and genetic algorithm (GA) (Yuan et al., 2014). The present study deals with hybridized models ANFIS-PSO and ANFIS-GA. The complexity involved in the formation of $PM_{2.5}$ results in nonlinear series with random peaks. It has been observed that it is difficult to study random fluctuations. As some random factors contribute to $PM_{2.5}$ formation leading to different features, this work introduces wavelet decomposition with proposed hybridized models to extract information through different time windows.

Although these studies have been done to improve the ANN and obtain better forecasting results, it is still required to improve the forecasting accuracy. It is a universal truth that the structures of the ANN only depend upon the input of the historical data (meteorology and pollutants) without directly taking into account the underlying physical or chemical processes and thus entail much less input and parameter data. However, the excellent forecasting performances of the ANN model generally depend on data representation because different representations can entangle and hide more or less the different explanatory factors of variation behind the data (Li et al., 2015). Therefore, feature extraction, learning, and analysis of historical data is the key to ensure forecasting accuracy.

However, ANN has some drawbacks (Local minima problems) that can occur inLearning steps, poor generalization issues, difficulty in selecting appropriate network architecture, and complexity of computation. Besides, ANNs cannot deal with uncertainty and subjectivity present in environmental issues and linguistic process variables such as "poor", "good", and "normal".

10.3   AIR POLLUTION MODELING WITH FUZZY AND NEURAL NETWORK MODEL

FTS (Fuzzy time series) and NNM (neural network Model) models are most appropriate for forecasting air pollution data that generally collects in the form of time series, consists of a small number of observations, is incomplete, includes uncertainty, and does not satisfy statistical assumptions. However, very few studies have been done using FTS models. Cheng et al. (Cheng et al., 2011) predicted daily O_3 concentrations in Taiwan using an FTS model based on a two-stage linguistic partition. Cagcag et al. (2013) proposed a new seasonal FTS model and applied this model to forecasting by Rahman et al. (2015) compared the forecasting performance of Box Jenkins, ANN, and three models of FTS. FTS models used in these studies were negatively affected by outliers and abnormal observations. They use outlier-sensitive methods in the fuzzification step, the first step of the FTS building procedure, which distorts the data in favour of outliers. When considering that air pollution data generally includes outliers, this is a big disadvantage.

Fuzzy time series was first introduced by Song and Chissom (1993), who had initiated the use of FTS to forecast scenarios with these characteristics: (i) dynamic in process; (ii) observations fulfil the characteristics of fuzzy sets; (iii) the fuzzy set's domain of discourse is a subclass of R as defined by Zadeh (1973); and (iv) the procedure is unable to be described by the normal time series models. Hence, the FTS is suitable for the decision-making process in a complex system where the circumstance of the issue is often not clear. The characteristics of membership function in the fuzzy sets have made it possible for past researchers such as Fisher (2003) to apply a fuzzy set in describing air pollution levels as air quality can be categorized into good, medium, or poor quality. Several important definitions of the FTS are outlined below:

10.4   ANALYSIS OF AVAILABLE SOFT COMPUTING MODELS (COMPARISON OF METHODOLOGY FOR AIR POLLUTION MODELING)

Artificial neural networks, support vector machines, evolutionary ANN and SVM, and fuzzy logic and neuro-fuzzy systems are widely used in AQM. Besides, the hybrid techniques combining several soft computing methods are also receiving widespread attention in AQM. The deep learning machine, ensemble model, and other soft computing techniques have been employed in a few cases for modeling purposes. This section reviews and analyses such modeling approaches thoroughly.

10.4.1 MULTIPLE LINEAR REGRESSION MODELS

Multiple regression models (MLR) or statistical methods are other approaches to forecasting air pollutant concentrations. They overcome the limitations of deterministic methods by using a large amount of observed data. Among the existing statistical methods, Autoregressive Integrated Moving Average (ARIMA), Generalized Additive Models (GAMs), Multi-layer Regression (MLR), Geographically Weighted Regression (GWR), have been widely adopted in the field of air quality prediction (Ma and Cheng, 2016). For example, Jian et al. (2012) applied the ARIMA model to submicron particle concentrations at a busy broadside in Hangzhou, China. Slini et al. (2002) developed a stochastic ARIMA model for maximum ozone concentration forecasts in Athens, Greece. Davis and Speckman (1999) adopted the GAM approach to predict one day in advance in Houston's maximum and 8h average ozone. Hu et al. (2013) used GWR to estimate ground-level $PM_{2.5}$ concentrations in the south-eastern US.

The MLR in the context of spatial interpolation is a linear methodology based on the assumption that the spatial relationship between the concentrations of a target site with one or more reference sites can be modelled using linear predictor functions. Linear regression analysis is used extensively in statistical applications and requires training to calculate the values of many coefficients, which associate the response variable (air pollutant concentrations at the target site) with the explanatory variables (air pollutant concentrations at the reference sites) according to the relationship:

$$z_i = b_0 + b_1 x_1 + b_2 x_2 + \ldots + b_i x_i \tag{10.1}$$

where z_i are the target site data, xi the reference site data and bi the regression coefficients. Higher values for the regression coefficients indicate increased importance of the reference site to the predicted concentration levels of the target site.

Zeinalnezhad et al. developed two models: (1) semi-experimental nonlinear multivariate regression model, and (2) ANFIS model, to predict the concentration of the four important pollutants and experimentally refined to obtain the least error possible. A nonlinear multivariate regression model data on pollutants containing CO, SO_2, O_3, and NO_2 are collected from a single monitoring point in Tehran. Several evaluation indices are employed to describe each model's precision and performance capability, including RMSE, R^2, and MAPE. The results of both developed models are seen to be superior to existing models. The ANFIS prediction model, related to CO, can capture more than 86% of the explained variance, and a NO_2 prediction model explains at least 76% of the total variance. The analysis demonstrates that both models showed considerable prediction accuracy and proved that the proposed ANFIS model gets higher reliability in assessing the nonlinear nature of air pollution.

However, it is notable that most of the mentioned statistical methods assumed the relationships between the variables and the target label were linear. It is inconsistent with the nonlinearity of the real world. Therefore, the prediction performance of these methods is limited.

Researchers started to adopt nonlinear machine learning (ML) models as the alternative method for air quality prediction to address the problem. Methods such as Support Vector Machine (SVM), (Ma and Cheng, 2019; Osowski and Garanty, 2007), Artificial Neural Networks (ANNs) (Alimissis et al., 2018; Yang and Wang, 2017), Backpropagation neural network (BPNN) (Feng et al., 2011), Fuzzy Logic (FL) (Lin and Cobourn, 2007) etc. have been applied in a lot of literature and discussed one by one in this study.

10.4.2 ARTIFICIAL NEURAL NETWORK MODELS

An artificial Neural Network (ANN) is a complex network with many simple mechanisms linked together, with strong nonlinearity, a system capable of complex logical operations and nonlinear relationships. It serves as a nonlinear time series forecasting method with an unspecified model term. ANN are one of the favoured techniques in predicting a complex system. They exhibit rapid information processing without any specific assumptions of the nature of the nonlinearity.

McCulloch and Pitts (1990) introduced the method of ANN, a computational model that imitates the working of a biological brain network. It is a nonlinear time series forecasting method with an unspecified model form. The ANN does not have a prior assumption on the model structure like the other modeling techniques; yet it obtains the structure by generalizing the training data. This characteristic of ANN offers itself a more flexible framework than other statistical forecasting techniques to capture nonlinear time series data (Zhang, 2003). The ANN can be trained using training data to recognize nonlinear patterns between input and output values (Abd Rahman et al., 2013). It can be trained more precisely when added with new data, and due to its generalization properties, the weights of the links can be formed by training. It can then be inserted into existing technology, helping solve difficult problems a lot quicker than the computers (Garcia et al., 2011). Based on the ANN characteristics, many researchers have applied it, particularly the multilayer perceptron (MLP) models, in predicting the environmental situation (Bai et al., 2018). The ANN is an effective tool for modeling environmental systems, especially to forecast air quality (Maier and Dandy, 2001). However, according to Heo and Kim (2004), the forecast accuracy of ANN might be affected if previously unknown data patterns are included in the training data. Also, according to Morabito and Versaci (2003), the downside of ANN is that the knowledge gained about an issue is encoded in the connections' weights which require users to further fix the strengths of the weights after the training of the data.

ANN are one of the favoured techniques in predicting a complex system. They exhibit rapid information processing without any specific assumptions of the nature of the nonlinearity (Ul-Saufie et al., 2013). According to Viotti et al. (2002), the advantage of an ANN lies in the fact that the deterministic model requires a lot of information, whereas the neural network acts as a black box, and once trained, is fast at predicting the desired value. However, the major drawback of this technique is that no deep understanding of the physical characteristics can be determined (Viotti et al.,

2002). In air pollution studies, a few types of ANN techniques have been applied. Sousa et al. (2007) used feedforward ANNs to predict hourly ozone concentration based on meteorological data. Meanwhile, Brunelli et al. (2007) used recurrent neural networks (Elman model) to predict maximum daily concentrations of several air pollutants in Palermo, Italy. Cigizoglu and Kisi (2006) used feedforward backpropagation (FFBP) and radial basis function algorithm (RBF) to estimate air pollution in Istanbul, Turkey.

Although these studies have already been done to improve the ANN and obtain better forecasting results, it is still required to improve the forecasting accuracy. It is a universal truth that the structures of the ANN only depend upon the input of the historical data (meteorology and pollutants) without directly taking into account the underlying physical or chemical processes and thus entail much less input and parameter data. However, the excellent forecasting performances of the ANN model generally depend on data representation because different representations can entangle and hide more or less the different explanatory factors of variation behind the data (Li et al., 2015). Therefore, feature extraction, learning, and analysis of historical data is the key to ensure forecasting accuracy.

10.4.3 SUPPORT VECTOR MACHINE MODELS

SVM, the solution of the universal feedforward networks, is known as the excellent tool for the classification and regression problems of good generalization ability (Vapnik, 1999). In distinction to the classical neural networks, the formulation of the learning problem of SVM leads to quadratic programming with linear constraints. The SVM is a linear machine of one output y(x), working in the high dimensional feature space formed by the nonlinear mapping of the N-dimensional input vector x into a K-dimensional feature space (K>N) through the use of the nonlinear function φ(x). The number of hidden units (K) equals the number of so-called support vectors. The learning data points closest to the separating hyperplane. The learning task is transformed to minimize the error function while keeping the network's weights at a minimum. The error function is defined through the so-called e-insensitive loss function L_ε(d, y(x)) (Vapnik, 1999).

$$L_\varepsilon\left(d,y\left(x\right)\right) = \begin{cases} \left|d-y\left(x\right)\right|-\varepsilon \ for \left|d-y\left(x\right)\right| \geq \varepsilon, \\ 0 \ for \left|d-y\left(x\right)\right| < \varepsilon, \end{cases} \tag{10.2}$$

where ε is the assumed accuracy, d is the destination, x is the input vector and y(x) the actual output of the network under excitation of x and the actual output signal of the SVM network is defined by

$$y\left(x\right) = \sum_{j=1}^{K} w_j \varphi_j\left(x\right) + b = w^T \varphi\left(x\right) + b \tag{10.3}$$

where $w = [w_1;...; w_K]^T$ is the weight vector, b the bias and $\varphi(x) = [\varphi(x_1,...\varphi_k(x)]^T$ the basis function vector.

The introduction of the Lagrangian function solves the solution of the so defined optimization problem and the Lagrange multipliers α_i, α_i' (i = 1; 2;...., p) responsible for the functional constraints defined. The minimization of the Lagrangian function has been transformed into the so-called dual problem (Vapnik, 1999).

Osowski et al. (2007) reported the method of daily air pollution forecasting using a support vector machine (SVM) and wavelet decomposition of the time series, formed based on the signals measured in the previous days. Based on the observed data of NO_2, CO, SO_2 and dust, for the past years and actual meteorological parameters, like wind, temperature, humidity, and pressure, they proposed the forecasting approach, applying the neural network of SVM type, working in the regression model.

10.4.4 BACK PROPAGATION NEURAL NETWORK MODELS

The BPNN, a multilayer feedforward network, is one of the commonly used neural networks. In the BPNN architecture, the artificial neurons are organized as layers and the informationStrictly flows forward, and the errors of the network are propagated backward. The architecture of this network consists of input layers, one or more hidden layers, and output layers. Each layer is composed of many neurons. A general structure of the three layers of BPNN is shown in Figure 10.1.

The mathematic expressions of the outputs of the hidden layer and output layer are as follows (Trigo and Palutikof, 1999)

$$h_j = f_{hidden}\left(\sum_{i=1}^{m} w_{ij}x_i\right), and\, Y_k = f_{output}\left(\sum_{j=1}^{n} w_{jk}h_j\right) \tag{10.4}$$

where $X = (x_1, x_2,..., x_i)$ (i = 1,2,..., m) represents the inputs, $h = (h_1, h_2,..., h_j)$ (j = 1,2,..., n) represents the outputs of the hidden layer, $Y = (Y_1, Y_2,..., Y_k)$ (k = 1,2,..., p) represents the outputs of the network, w represents the weight matrix between two layers, and $f_{hidden}(.)$ and $f_{output}(.)$ are transfer functions of the hidden and output layers, respectively. This paper uses the nonlinear transfer function as sigmoid in the hidden layer, and pure linear is used in the output layer as a linear transfer function. Backpropagation, meaning "error backward propagation", requires a known, desired output for each input value to calculate the loss function gradient. The gradient is fed to the optimization method, which updates the weights to minimize the loss function (Lalis et al., 2014). Generally, the least mean square error is applied to compare the network's outputs with the actual outputs (Rumelhart et al., 1986). When modeling with BPNN, the number of hidden nodes is the primary variable to be determined. Recently, the trial-and-error method and empirical formula have been applied to solve this issue (Shen et al., 2008).

$$hidden\ nodes = \sqrt{m+p} + a; a\epsilon\left[0,10\right] \tag{10.5}$$

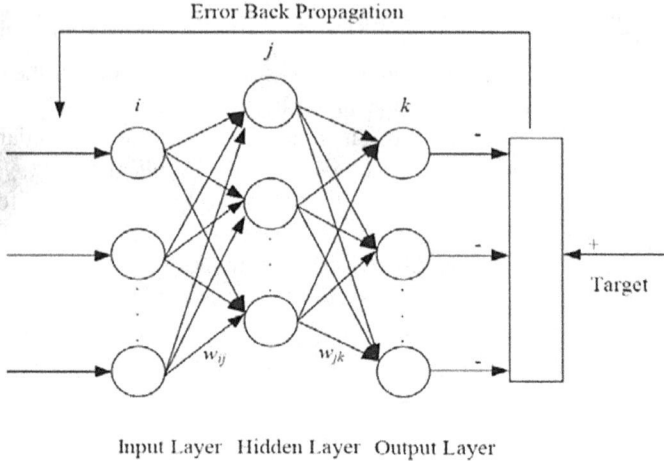

FIGURE 10.1   Structure diagrams of BPNN.

Back Propagation neural network (BPNN) is a mature algorithm of ANNs with great nonlinear regression capability. To obtain the best BPNNs for some cases, hundreds of ANNs were established with 6–30 neurons in the hidden layer. Through experiments, the best architectures were achieved with 25 neurons in the hidden layer. The sigmoid and linear functions were chosen as the activation functions for the hidden layer and the output layer, respectively. Feng et al. (2011) have reported a parallel computing method of ozone forecast with stable high forecast accuracy by using only meteorological conditions (T, H, WV and UV) in Beijing. Support Vector Machine (SVM) accurately classifies the data into its corresponding categories. BPNN was also optimized using Genetic Algorithm (GA) to achieve higher forecast stability. A contrast among BPNN, GABPNN, and SVM-GABPNN were provided. The performance evaluation showed that the predictions of SVM-GABPNN were accurate and stable. So we conclude SVM-GABPNN could be applied to the real-life ozone forecast in Beijing. A parallel computing method of ozone forecast with stable high forecast accuracy uses only meteorological conditions (T, H, WV and UV) in Beijing. Contrast among Back Propagation neural network BPNN, GABPNN, and SVM-GABPNN were provided. The performance evaluation showed that the predictions of SVM-GABPNN were accurate and stable. So we conclude SVM-GABPNN could be applied to the real-life ozone forecast in Beijing.

10.4.5   Fuzzy Logic and Neuro-Fuzzy Models

The theory of forecasting and modeling techniques based on Fuzzy Logic (Zadeh, 1965) are widely used in many fields for a great number of applications (Zhao et al., 2015) due to their advantages. One of the concepts of FL that plays a significant role in applications is the IF-THEN rules, simply known as fuzzy rules. FL has been heavily used in areas such as automation control, diagnosis, and forecasting. Besides, FL can greatly facilitate solving various problems, building expert systems and neural networks. For example, they can model complex, nonlinear and uncertain systems; include expert opinion and experiences in the modeling process; linguistic process variables; not require statistical assumptions; and, lastly, provide reasonable performance for data sets with a few observations. FTS is a modeling technique based on FL that has similar advantages. FTS models can be considered a statistical time series analysis version that applies to fuzzy sets (Dincer and Akkuş, 2018). Dincer and Akkuş reported a new FTS model for forecasting air pollution data. This model uses a robust fuzzy clustering algorithm called Fuzzy K-Medoid in the fuzzification step of the FTS model building algorithm.

10.4.6 Deep Learning Models

Deep learning is a sub-cluster of machine learning. It carries ANNs one step beyond using huge data set, solving problems without dividing, using more layers, processing simultaneously with sequential layers and providing more trustable results. All these favourable features of deep learning make it a suitable method for air pollution modeling (Ayturan et al., 2018).

Deep Learning is a subfield of machine learning concerned with algorithms inspired by the structure and function of the brain called ANNs (Copeland, 2016). It uses many nonlinear processing unit layers for feature extraction and transformation. These layers follow one after each another (Şeker et al., 2017). In the end, they create competition with high accuracy. In Figure 10.2, the development steps of AI, ML and DL according to the years are given.

The main differences between machine learning and deep learning may be explained in two matters. The first one is the size of the data, and the second one is the problem-solving methods. In deep learning, big data is used when it is compared

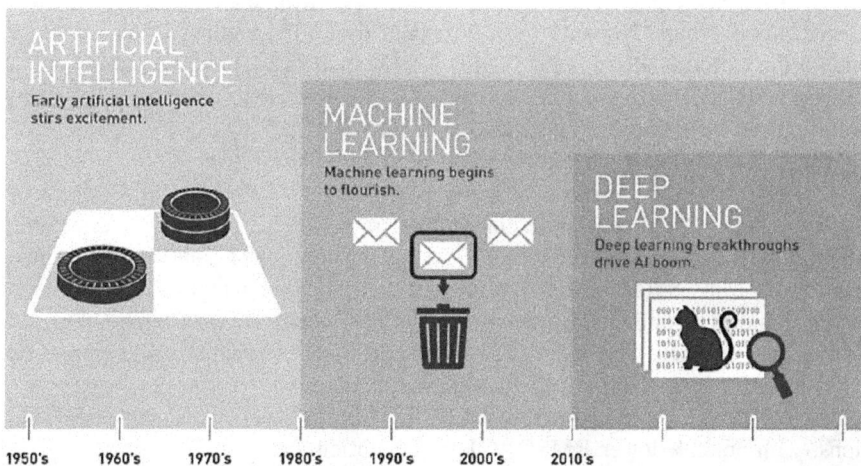

FIGURE 10.2 Development steps of AI, ML, and DL according to the years.

with machine learning. Deep learning focuses on solving end-to-end problems, while machine learning uses the divide and manage the problem method. Furthermore, deep learning makes simultaneous processes in sequential layers, while machine learning makes processes in a single layer.

The following steps develop most deep learning models: input and output vector determination, transfer function determination, network structure selection, hidden layer determination, weight specification, and learning algorithm selection. To create a prediction model with deep learning, the following steps should be implemented:

- Definition of the relevant data sets and preparation of them for analysis.
- Selection of the algorithm to be used.
- Creation of an analytical model based on the algorithm used.
- Training of the model in the test data sets and running the model to generate test scores.
- Generation of a forward prediction based on the result.

The general structure of the deep learning model consists of the input, hidden, and output layers. Each of these layers should be weighted with a numerical value (Wang, 2003). The hidden layer neurons are affected by input and output neurons, learning algorithms, network structure, and type of activation function (Alsugair and Al-Qudrah, 1998). To determine the number of neurons in the hidden layer, it is necessary to train the different networks and approach the error in the test data. Also, when the environmental problems are considered, deep learning can find the location of pollution sources, detect the areas in which the pollution levels are high and predict the future concentrations.

10.4.6.1 Air Pollution Modeling with Deep Learning

There are several modeling techniques appropriate for air pollution prediction, such as the long short-term memory (LSTM) model, deep spatiotemporal learning (STDL)-based air quality prediction method, deep air learning (DAL) and Convolutional Neural Network (CNN). Especially LSTM method is the most used one for this purpose (Reddy et al., 2018). LSTM models are part of recurrent neural networks (RNN), which use a framework for future forecasting and time series data's pollution and meteorological information (Reddy et al., 2018).

Ma et al. (2019) also reported deep learning-based techniques to predict air quality, namely transferred bi-directional long short-term memory (TL-BLSTM) model for air quality prediction. The methodology framework utilizes the bi-directional LSTM model to learn from the long-term dependencies of $PM_{2.5}$ and applies transfer learning to transfer the knowledge learned from smaller temporal resolutions to larger temporal resolutions. Especially for the predictions at larger temporal resolutions. This could help governments and researchers generate and analyze more accurate trends for long-term air quality analysis. It is found that transfer learning can effectively lower the prediction error of BLSTM for $PM_{2.5}$ at larger temporal resolutions, with 36.85% lower RMSE at daily resolution and 42.58% lower at weekly resolution. It is expected that transfer learning performs better for larger granularizes as it has less training data.

10.5   ENSEMBLE AND HYBRID MODELS

Halabi et al. proposed that the hybrid model (ANFIS merged with PSO, GA, and DE, respectively) predicts monthly global solar radiation (Halabi et al., 2018). Meanwhile, different meteorological parameters including sunshine duration, minimum and maximum ambient temperature, rainfall, and clearness index are used. The performance of the developed models is compared with other artificial intelligence (AI), hybrid AI and empirical techniques carried out by other researchers for predicting global solar radiation where the final results of this work agree with the compared studies in the highest capability of the hybrid models in predicting the solar radiation. It was observed that the proposed ANFIS-PSO model has better performance than the other models and provides a more reliable and accurate correlation throughout different input-output datasets. Where the performance evaluation parameters obtained for ANFIS-PSO are 0.3121 of RMSE, 1.8580 of RRMSE, 0.9931 of r, 0.9862 of R^2 0.2354 of MABE and 1.4159 of MAPE in training and 0.3065 of RMSE, 1.7933 of RRMSE, 0.9963 of r, 0.9921 of R^2 0.2482 of MABE and 1.4097 of MAPE in the testing process, this showed that the practical implementation of the ANFIS-PSO model could improve the accuracy and efficiency of predicting monthly global solar radiation.

Li et al. (2018) proposed a framework to support pre-monitoring for air pollutants in China. They assessed various data series, including concentrations of air contaminants (NO_2, SO_2, O_3, CO, $PM_{2.5}$, and PM_{10}) and atmospheric variables such as wind direction, wind speed, humidity, temperature, and sea level pressure hourly. The forecasting part was applied to hybrid optimization models consisting of an ensemble empirical mode decomposition and general regression neural network (EEMD-GRNN), Adaptive Neuro-Fuzzy Inference System (ANFIS), principal component regression (PCR), and linear model such as multiple linear regression (MLR) model with improved Harmony Search Algorithm with PSO strategy to forecast air contaminants concentration, while the assessment part was applied to Fuzzy Synthetic Evaluation Model with entropy weight to evaluate the air quality levels. The findings showed that the hybrid forecasting model is superior to the statistical model and other urban pollutant concentration data models.

Ausati and Amanollahi (2016) predicted the suspended $PM_{2.5}$ particles by using the precision of ANFIS comparing with EEMD (ensemble empirical mode decomposition), EEMD-GRNN, PCR, and MLR model). Cross models consisting of PCR, EEMD-GRNN, ANFIS, and MLR were used. $PM_{2.5}$ concentrations were the dependent variable in the model. Data associated with PM_{10}, SO_2, $PM_{2.5}$, CO, O_3, NO_2, average minimum temperature, average atmospheric pressure, average maximum temperature, daily total precipitation, daily wind speed, and daily relative humidity level were considered independent variables in the model.

Mirzaei et al. (2019) predicted the suspended $PM_{2.5}$ particles in Iran's urban area using a geographically and temporally weighted regression (GTWR) model. To improve the accuracy of the correlation coefficient between converted $PM_{2.5}$ from the GTWR model and $PM_{2.5}$ concentrations measured at ground monitoring station, the results of a linear model (LR), a nonlinear model (ANN), and hybrid models including GRNN and adaptive neuro-fuzzy inference system (ANFIS) were compared. The cold season's linear, nonlinear, and hybrid models displayed higher accuracy than the warm season results. Among the used models, the hybrid GRNN model showed higher accuracy compared with the other models.

10.6 POTENTIAL SOFT COMPUTING MODELS AND APPROACHES

Among many potential techniques, different variations of ANNs, evolutionary fuzzy and neuro-fuzzy models, ensemble and hybrid models, and knowledge-based models should be further explored. Besides, there is a continuous need to develop a universal model, as most of the explored models are either site-dependent or pollutant-dependent. This section discusses future research directions and potential soft computing models investigated in air quality modeling worldwide.

10.6.1 EVOLUTIONARY FUZZY AND NEURO-FUZZY MODELS

The AI methods have been observed in various fields for classification and regression (Rai et al., 2015). The accuracy of these methods has increased its applicability over typical regression methods. Many studies show good efficiency of artificial methods for predictive analysis of $PM_{2.5}$ using other pollutants and meteorological parameters. Chen et al. (2018) carried out a predictive analysis of $PM_{2.5}$ mass concentration and found root mean square error greater than 30 for different input parameters.

ANFIS (adaptive neuro-Fuzzy-Inference system) and ANN (artificial neural network) have many advantages in solving various complex real-time problems (Jang, 1993). These methods are widely used for the prediction of nonlinear series. ANFIS is a combination of ANNs and fuzzy inference that has been generally used in various areas to solve complex and nonlinear problems (Sonmez et al., 2018). This tool was developed based on the Sugeno fuzzy model, and it was first introduced by Jang (1993). In ANFIS, the parameters involved are tuned using the least square and gradient descent method. The shortcoming of ANFIS in tuning the parameters is due to the drawback of the gradient descent method to get trapped in local optima. This problem results in decreasing the quality of modeling and increasing computational time. In order to optimize the computational time and achieve global optima, evolutionary techniques are introduced with ANNs (Goldberg and Holland, 1988). Ongoing research in evolutionary studies shows remarkable advantages of particle swarm optimization (PSO) and genetic algorithm (GA) (Yuan et al., 2014). Various studies deal with hybridized models ANFIS-PSO and ANFIS-GA. The complexity involved in the formation of $PM_{2.5}$ results in nonlinear series with random peaks. It has been observed that it is difficult to study random fluctuations. As some random factors contribute to $PM_{2.5}$ formation leading to different features, this work introduces wavelet decomposition with proposed hybridized models to extract information through different time windows.

Bhardwaj and Pruthi (2020) applied the adaptive neuro-fuzzy inference system (ANFIS) to perform predictive analysis of air pollutant-fine particulate matter $PM_{2.5}$. This study covers (1) prediction analysis of $PM_{2.5}$ merely considering the lagged series (2) application of PSO and GA in tuning WANFIS for prediction of fine particulate matter; (3) using WANFIS and its improvements with evolutionary algorithms, PSO, GA for estimation of fine particulate matter at Shadipur, Delhi, which is an industrial, commercial and residential area covering all sources contributing in the formation of particulate matter; (4) comparison between ANFIS and its tuned models; and (5) finding best algorithm for tuning WANFIS. To the best of the author's knowledge, first-time evolutionary algorithms are applied for predicting air pollution, and the proposed model outperforms the existing models.

10.6.2 VARIATIONS OF ANN MODELSCASE-BASED REASONING AND KNOWLEDGE-BASED MODELS

Oprea and Liu (2016) present a knowledge-based approach applied to air pollution effects analysis in the case of $PM_{2.5}$ air pollutants in urban cities. The system provides an early warning or alerting messages to protect sensitive people (such as children) when $PM_{2.5}$ related air pollution episodes can occur next hour / next day (next 24 hours) in different zones of the pilot city (Ploiesti, in our case study experiments). The use of knowledge derived from various sources (literature, databases, questionnaires, human experts' experience, and decision tables) via manual, semi-automatic, and automatic methods are proposed for multiparameter analysis of the $PM_{2.5}$ air pollution episodes effects on vulnerable people such as children and elderly. Some measures to reduce the negative effects on human health are also proposed by this approach. The knowledge under production rules is incorporated in a knowledge base used by the ROKIDAIR intelligent decision support system (ROKIDAIR DSS). The approach applies the DSS model builder protocol, MBP-DSS, which uses two methodologies, one for selecting the knowledge-based DSS model and one for the knowledge-based DSS model development. The core module of the DSS is the knowledge base that was generated by using different knowledge discovery techniques (manual, semi-automatic, or automatic; the last two, based on inductive learning algorithms) from various sources: literature, human experts, databases, cases base, questionnaires. The experiments that were run validated the coherence of the developed knowledge base.

10.6.3 GROUP METHOD DATA HANDLING MODELS AND FUNCTIONAL NETWORK MODELS

Zhu et al. (2012) studied the monthly average concentration of SO_2, NO_2, PM_{10} and the monthly number of people in hospital because of lower respiratory disease from January 2001 to December 2005 equidistant and considered as the terms of transactions. Then based on the relational algebraic theory, they employed the optimization relation association rule to mine the association rules of the transactions. Based on the association rules revealing the effects of air pollutants on lower respiratory disease, they forecasted the number of the person who suffered from lower respiratory disease by the group method of data handling (GMDH) to reveal the risk and give a consultation to the hospital in Xigu District, the most seriously polluted district in Lanzhou.

To overcome these health issues problems and achieve high forecast precision, they described a novel approach for ranking all features according to their predictive quality using properties unique to learning algorithms based on the GMDH (Hwang, 2006). The basic building block of the GMDH consists of polynomials which are called partial descriptions. The model output was obtained by feeding the outputs of polynomials of a layer into the inputs of the next layer (i.e., the outputs of a layer become the inputs of the next layer). In each layer of GMDH, the function parameters of two variables were determined by linear regression analysis. The GMDH-type neural networks have several advantages compared with the conventional multi-layered networks. Also, this method can self-select a number of layers and a number of neurons in each layer and the ability to self-select useful input variables. The useless input variables were eliminated, and useful input variables were selected automatically (Al-Alawi et al., 2008). However, the weights of GMDH were adjusting according to the Widrow-Hoff learning rule; they did not always achieve an optimal state.

Rahman et al. (2012) investigate the primary pollutants that contribute to the increase of ozone levels, which cause negative effects on biotic health. This study investigates abductive networks based on the GMDH for ozone prediction. Abductive network models are automatically synthesized from a database of inputs and outputs. The models are developed for a location in the Empty Quarter, Saudi Arabia, using only the meteorological data and derived meteorological data. In the subsequent efforts, NO and NO_2 concentrations and their transformations were incorporated as additional inputs.

This study proposes the use of abductive networks in air quality modeling, focusing on ozone prediction. Abductive network models are self-organizing, require less intervention from the user, and can be easily implemented. To the authors' knowledge, this modeling approach is not notably used in this area of research. Training is based on a GMDH algorithm, which automatically synthesizes abductive networks from a database of inputs and outputs (Manual, 1990). To solve the limitations of prestructured models, Ivakhnenko (1968) introduced the GMDH algorithm, which provides an objective model of a high-order polynomial in the input variables to solve prediction identification, control, and other problems. The candidate models are evaluated, and the optimum model is selected automatically based on simplicity and estimated prediction performance on test data. The user has the option to set some parameters which control the complexity of the synthesized network.

10.6.4 Appropriate Input Selection Methods

There are several approaches to selecting the most significant predictors of a given model. They are divided into two categories, namely, model-free and model-based approaches (Maier et al., 2010). Model-free approaches perform input selection without relying on the performance of the developed ANN models. Model-free approaches can be further divided into two categories: ad-hoc and analytical. The selection of model predictors implemented arbitrarily or based on domain knowledge falls under the ad-hoc approach.

In contrast, the analytic approach uses a statistical measure of dependence between model predictors and target variables. This is mostly carried out through cross-correlation. However, this analytical approach can only detect linear dependence between data, leading to the omission of relevant predictors associated with the target variables in a non-linearly manner (Samarasinghe, 2016). On the other hand, model-based approaches perform input selection by determining the effect of a candidate model predictor on the overall model performance. As pointed out by Maier et al. (2000) the approach has several downsides. Firstly, the approach is time-consuming as a number of ANN models need to be developed.

Furthermore, it does not measure the impact of the utilized predictors on the model performance, as the latter is also a function of several network parameters, e.g., the number of hidden layer nodes, etc. A popular example of model-based approaches is the stepwise selection of inputs, where a network iteratively selects (e.g., forward selection), or remove (e.g., backward elimination), predictors based on the model performance. An ad hoc approach can also be done, where arbitrary combinations of model predictors are tested. A global approach can also be implemented, where a global optimization algorithm is used to select the combination of predictors that maximizes model performance. Finally, an approach based on sensitivity analysis can be undertaken, examining sensitivity plots for each predictor to the target variables.

The number of occurrences various input selection approaches have been used is shown in Figure 10.3. It can be seen that model-free approaches were implemented 99 times, compared with the 40 occasions on which model-based approaches were used. Of the model-free approaches considered, ad-hoc methods were most widely implemented with applications in 82 papers, followed by linear approaches, especially correlation analysis utilized in 13 papers. A nonlinear method was employed only 4 times. In 13 of the 40 times, a model-based approach was implemented, the process was carried out in an ad-hoc manner. A stepwise method was used 10 times, while global search approaches were implemented 7 times.

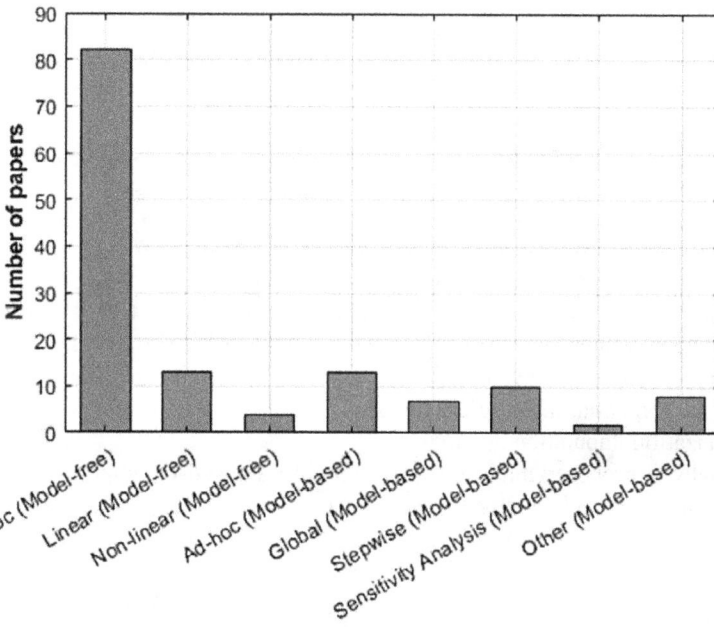

FIGURE 10.3   Number of occurrences, various input selection methods.

The results showed that the need for greater attention to predictor selection implementation. Although the selection of predictors is dependent on the external problem specifications, a systematic approach should be encouraged among modellers to reduce bias and increase the repeatability of performances of data-driven models in general. Ad-hoc approaches to predictor selection, either model-based or model-free, were used in almost 70% of the identified papers. Furthermore, linear analytical model-free approaches were widely implemented, which contradicts the rationale of using nonlinear models such as ANNs to approximate the typically nonlinear dynamics between air pollutants and predictors. This indicates the need to examine further the use of nonlinear approaches in selecting predictors. Several guidelines in determining the most suitable predictor selection technique for every problem specification can be found (Galelli et al., 2014). The proposed guidelines were evaluated by taking into account three factors, including a wide range of dataset properties that reflect the properties of real-world environmental observations, and assessment criteria selected to highlight algorithm suitability in different problem specifications, and a website for sharing data, algorithms and results (Galelli et al., 2014).

10.7 CONCLUSIONS

Soft computing models have become very popular in air quality modeling as they can efficiently model the complexity and nonlinearity associated with air quality data. This article critically reviewed and discussed existing soft computing modeling approaches. Among the many available soft computing techniques, the ANNs with variations of structures and the hybrid modeling approaches combining several techniques were widely explored in predicting air pollutant concentrations worldwide. Other approaches, including support vector machines, evolutionary ANNs, and support vector machines, fuzzy logic, and neuro-fuzzy systems, have also been used in air quality modeling for several years. Recently, deep learning and ensemble models have received huge momentum in modeling air pollutant concentrations due to their wide range of advantages over other available techniques.

Additionally, this research reviewed and listed all possible input variables for air quality modeling. It also discussed several input selection processes, including cross-correlation analysis, principal component analysis, random forest, learning vector quantization, rough set theory, and wavelet decomposition techniques. Besides, this article sheds light on several data recovery approaches for missing data, including linear interpolation, multivariate imputation by chained equations, and expectation-maximization imputation methods.

Finally, it proposed many advanced, reliable, and self-organizing soft computing models that are rarely explored and/or not explored in the field of air quality modeling. For instance, functional neural network models, variations of neural network models, evolutionary fuzzy and neuro-fuzzy systems, type-2 fuzzy logic models, group method data handling, case-based reasoning, ensemble and hybrid models, and knowledge-based systems have immense advantages potential for modeling air pollutant concentrations. Moreover, the modellers can compare the effectiveness of several input selection processes to find the most suitable one for air quality modeling. Furthermore, they can attempt to build universal models instead of developing site-specific and pollutant-specific models.

The authors believe that the findings of this review article will help researchers and decision-makers determine the suitability and appropriateness of a particular model for a specific modeling context.

ACKNOWLEDGEMENTS

We are sincerely thankful to the Indian Institute of Technology (Banaras Hindu University), Varanasi, India; Madan Mohan Malaviya University of Technology, Gorakhpur, India; Central Soil and Materials Research Station, Ministry of Jal Shakti, Department of Water Resources, River Development Ganga Rejuvenation (Government of India), Delhi, India; JMC Projects (India) Limited, Mumbai, India and Dr. Ambedkar Institute of Technology for Handicapped, Kanpur, India, for this work.

CONFLICTS OF INTEREST

The authors declare no conflict of interest.

REFERENCES

Ababneh, Mohammad F, O Ala'a, and Mohammad Hjouj Btoush. 2014. PM10 forecasting using soft computing techniques. *Research Journal of Applied Sciences, Engineering and Technology* 7 (16):3253–3265.

Abd Rahman, Nur Haizum, Muhammad Hisyam Lee, and Mohd Talib Latif. 2015. Artificial neural networks and fuzzy time series forecasting: An application to air quality. *Quality & Quantity* 49 (6):2633–2647.

Abd Rahman, Nur Haizum, Muhammad Hisyam Lee, Mohd Talib Latif, and SJJT Suhartono. 2013. Forecasting of air pollution index with artificial neural network. *Jurnal Teknologi* 63 (2).

Al-Alawi, Saleh M, Sabah A Abdul-Wahab, and Charles S Bakheit. 2008. Combining principal component regression and artificial neural networks for more accurate predictions of ground-level ozone. *Environmental Modeling & Software* 23 (4):396–403.

Alimissis, A, K Philippopoulos, CG Tzanis, and D Deligiorgi. 2018. Spatial estimation of urban air pollution with the use of artificial neural network models. *Atmospheric Environment* 191:205–213.

Alsugair, Abdullah M, and Ali A Al-Qudrah. 1998. Artificial neural network approach for pavement maintenance. *Journal of Computing in Civil Engineering* 12 (4):249–255.

Antanasijević, Davor Z, Mirjana Đ Ristić, Aleksandra A Perić-Grujić, and Viktor V Pocajt. 2013. Forecasting human exposure to PM10 at the national level using an artificial neural network approach. *Journal of Chemometrics* 27 (6):170–177.

Ausati, Shadi, and Jamil Amanollahi. 2016. Assessing the accuracy of ANFIS, EEMD-GRNN, PCR, and MLR models in predicting PM2.5. *Atmospheric Environment* 142:465–474.

Ayturan, Yasin Akın, Zeynep Cansu Ayturan, and Hüseyin Oktay Altun. 2018. Air pollution modeling with deep learning: A review. *International Journal of Environmental Pollution and Environmental Modeling* 1 (3):58–62.

Bai, Lu, Jianzhou Wang, Xuejiao Ma, and Haiyan Lu. 2018. Air pollution forecasts: An overview. *International Journal of Environmental Research and Public Health* 15 (4):780.

Bhardwaj, Rashmi, and Dimple Pruthi. 2020. Evolutionary techniques for optimizing air quality model. *Procedia Computer Science* 167:1872–1879.

Brunelli, U, V Piazza, L Pignato, F Sorbello, and S Vitabile. 2007. Two-days ahead prediction of daily maximum concentrations of SO_2, O_3, PM_{10}, NO_2, CO in the urban area of Palermo, Italy. *Atmospheric Environment* 41 (14):2967–2995.

Cagcag, Özge, Ufuk Yolcu, Erol Egrioglu, and Cagdas Hakan Aladag. 2013. A novel seasonal fuzzy time series method to the forecasting of air pollution data in Ankara. *American Journal of Intelligent Systems* 3 (1):13–19.

Chaudhuri, Sutapa, and Rajashree Acharya. 2012. Artificial neural network model to forecast the concentration of pollutants over Delhi: Skill assessment of learning rules. *Asian Journal of Water, Environment and Pollution* 9 (1):71–81.

Chen, Jianjun, Jin Lu, Jeremy C Avise, John A DaMassa, Michael J Kleeman, and Ajith P Kaduwela. 2014. Seasonal modeling of PM2. 5 in California's San Joaquin Valley. *Atmospheric Environment* 92:182–190.

Chen, Yegang. 2018. Prediction algorithm of PM2. 5 mass concentration based on adaptive BP neural network. *Computing* 100 (8):825–838.

Chen, Yuanyuan, Runhe Shi, Shijie Shu, and Wei Gao. 2013. Ensemble and enhanced PM10 concentration forecast model based on stepwise regression and wavelet analysis. *Atmospheric Environment* 74:346–359.

Chen, S., Guo, J., Song, L., Jian, L., Liu, L., Cohen, J.B., 2019. Inter-annual variation of the spring haze pollution over the North China Plain: Roles of atmospheric circulation and sea surface temperature, *International Journal of Climatetology*, 39(2): 783–798.

Cheng, Ching-Hsue, Guang-Wei Cheng, and Jia-Wen Wang. 2008. Multi-attribute fuzzy time series method based on fuzzy clustering. *Expert Systems with Applications* 34 (2):1235–1242.

Cheng, Ching-Hsue, Sue-Fen Huang, and Hia-Jong Teoh. 2011. Predicting daily ozone concentration maxima using fuzzy time series based on a two-stage linguistic partition method. *Computers & Mathematics with Applications* 62 (4):2016–2028.

Chuang, Ming-Tung, Yang Zhang, and Daiwen Kang. 2011. Application of WRF/Chem-MADRID for real-time air quality forecasting over the Southeastern United States. *Atmospheric Environment* 45 (34):6241–6250.

Cigizoglu, H Kerem, and Özgür Kisi. 2006. Methods to improve the neural network performance in suspended sediment estimation. *Journal of Hydrology* 317 (3–4):221–238.

Davis, JM, and P Speckman. 1999. A model for predicting maximum and 8 h average ozone in Houston. *Atmospheric Environment* 33 (16):2487–2500.

Dincer, Nevin Güler, and Özge Akkuş. 2018. A new fuzzy time series model based on robust clustering for forecasting of air pollution. *Ecological Informatics* 43:157–164.

Djalalova, Irina, Luca Delle Monache, and James Wilczak. 2015. PM2. 5 analog forecast and Kalman filter post-processing for the Community Multiscale Air Quality (CMAQ) model. *Atmospheric Environment* 108:76–87.

Domańska, D, and Marek Wojtylak. 2012. Application of fuzzy time series models for forecasting pollution concentrations. *Expert Systems with Applications* 39 (9):7673–7679.

Egrioglu, Erol, Cagdas Hakan Aladag, Ufuk Yolcu, Vedide R Uslu, and Murat A Basaran. 2010. Finding an optimal interval length in high order fuzzy time series. *Expert Systems with Applications* 37 (7):5052–5055.

Elangasinghe, Madhavi Anushka, Naresh Singhal, Kim N Dirks, and Jennifer A Salmond. 2014. Development of an ANN–based air pollution forecasting system with explicit knowledge through sensitivity analysis. *Atmospheric Pollution Research* 5 (4):696–708.

Fan, Jingchun, Shulan Li, Chunling Fan, Zhenggang Bai, and Kehu Yang. 2016. The impact of PM2. 5 on asthma emergency department visits: A systematic review and meta-analysis. *Environmental Science and Pollution Research* 23 (1):843–850.

Feng, Yu, Wenfang Zhang, Dezhi Sun, and Liqiu Zhang. 2011. Ozone concentration forecast method based on genetic algorithm optimized back propagation neural networks and support vector machine data classification. *Atmospheric Environment* 45 (11):1979–1985.

Fisher, Bernard. 2003. Fuzzy environmental decision-making: Applications to air pollution. *Atmospheric Environment* 37 (14):1865–1877.

Fu, Pengfei, Xinbiao Guo, Felix Man Ho Cheung, and Ken Kin Lam Yung. 2019. The association between PM2. 5 exposure and neurological disorders: A systematic review and meta-analysis. *Science of the Total Environment* 655:1240–1248.

Galelli, Stefano, Greer B Humphrey, Holger R Maier, Andrea Castelletti, Graeme C Dandy, and Matthew S Gibbs. 2014. An evaluation framework for input variable selection algorithms for environmental data-driven models. *Environmental Modeling & Software* 62:33–51.

Garcia, Ignacio, JG Rodríquez, and Yenisse M Tenorio. 2011. Artificial neural network models for prediction of ozone concentrations in Guadalajara, Mexico. *Air Quality-Models and Application*. Nicolas Mazzeo (Ed.):35–52.

Goldberg, David E, and John Henry Holland. 1988. Genetic algorithms and machine learning. Machine Learning, 3:95–99 Springer, Netherlands.

Grivas, G, and A Chaloulakou. 2006. Artificial neural network models for prediction of PM10 hourly concentrations, in the Greater Area of Athens, Greece. *Atmospheric Environment* 40 (7):1216–1229.

Habermann, Mateus, Monica Billger, and Marie Haeger-Eugensson. 2015. Land use regression as method to model air pollution. Previous results for Gothenburg/Sweden. *Procedia Engineering* 115:21–28.

Halabi, Laith M., Saad Mekhilef, and Monowar Hossain. 2018. Performance evaluation of hybrid adaptive neuro-fuzzy inference system models for predicting monthly global solar radiation. *Applied Energy* 213:247–261.

Heo, Jeong-Sook, and Dong-Sool Kim. 2004. A new method of ozone forecasting using fuzzy expert and neural network systems. *Science of the Total Environment* 325 (1–3):221–237.

Hu, Xuefei, Lance A Waller, Mohammad Z Al-Hamdan, et al. 2013. Estimating ground-level PM2. 5 concentrations in the south-eastern US using geographically weighted regression. *Environmental Research* 121:1–10.

Hwang, Heung Suk. 2006. Fuzzy GMDH-type neural network model and its application to forecasting of mobile communication. *Computers & Industrial Engineering* 50 (4):450–457.

Ivakhnenko, Alexey Grigorevich. 1968. The group method of data of handling; a rival of the method of stochastic approximation. *Soviet Automatic Control* 13:43–55.

Jang, J-SR. 1993. ANFIS: Adaptive-network-based fuzzy inference system. *IEEE Transactions on Systems, Man, and Cybernetics* 23 (3):665–685.

Janjua, Sadia, Pippa Powell, Richard Atkinson, Elizabeth Stovold, and Rebecca Fortescue. 2021. Individual-level interventions to reduce personal exposure to outdoor air pollution and their effects on people with long-term respiratory conditions. *Cochrane Database of Systematic Reviews* 8(8):1–72.

Jian, Le, Yun Zhao, Yi-Ping Zhu, Mei-Bian Zhang, and Dean Bertolatti. 2012. An application of ARIMA model to predict submicron particle concentrations from meteorological factors at a busy roadside in Hangzhou, China. *Science of the Total Environment* 426:336–345.

Khalili, Roxana, Scott M Bartell, Xuefei Hu, et al. 2018. Correction to: Early-life exposure to PM2. 5 and risk of acute asthma clinical encounters among children in Massachusetts: A case-crossover analysis. *Environmental Health* 17 (1):1–1.

Lalis, JT, BD Gerardo, and Y Byun. 2014. An adaptive stopping criterion for backpropagation learning in feedforward neural network. *International Journal of Multimedia and Ubiquitous Engineering* 9 (8):149–156.

Li, Chuan, Ming Liang, and Tianyang Wang. 2015. Criterion fusion for spectral segmentation and its application to optimal demodulation of bearing vibration signals. *Mechanical Systems and Signal Processing* 64:132–148.

Li, Ranran, and Yu Jin. 2018. The early-warning system based on hybrid optimization algorithm and fuzzy synthetic evaluation model. *Information Sciences* 435:296–319.

Lin, Y., and Cobourn, W. G. 2007. Fuzzy system models combined with nonlinear regression for daily ground-level ozone predictions. *Atmospheric Environment* 41 (16):3502–3513.

Li, Q., Li, A., Yu, X., Dai, T., Peng, Y., Yuan, D., Zhao, B., Tao, Q., Wang, C., Li, B., Gao, X., Li, Y., Wu, D., Xu, Q., 2019. Soil acidification of the soil profile across Chengdu Plain of China from the 1980s to 2010s, *Science of the Total Environment*, 698:134320.

Ma, Jun, and Jack CP Cheng. 2016. Identifying the influential features on the regional energy use intensity of residential buildings based on Random Forests. *Applied Energy* 183:193–201.

Ma, Jun. 2017. Identification of the numerical patterns behind the leading counties in the US local green building markets using data mining. *Journal of Cleaner Production* 151:406–418.

Ma, Jun, Jack CP Cheng, Changqing Lin, Yi Tan, and Jingcheng Zhang. 2019. Improving air quality prediction accuracy at larger temporal resolutions using deep learning and transfer learning techniques. *Atmospheric Environment* 214:116885.

Maier, Holger R, and Graeme C Dandy. 2000. Neural networks for the prediction and forecasting of water resources variables: A review of modeling issues and applications. *Environmental Modeling & Software* 15 (1):101–124.

Maier, Holger R, and Grame C Dandy. 2001. Neural network-based modeling of environmental variables: A systematic approach. *Mathematical and Computer Modeling* 33 (6–7):669–682.

Maier, Holger R, Ashu Jain, Graeme C Dandy, and KP Sudheer. 2010. Methods used for the development of neural networks for the prediction of water resource variables in river systems: Current status and future directions. *Environmental Modeling & Software* 25 (8):891–909.

Manual. 1990. *AIM User's*. AbTech Corporation: Charlottesville, VA, USA.

McCulloch, Warren S, and Walter Pitts. 1990. A logical calculus of the ideas immanent in nervous activity. *Bulletin of Mathematical Biology* 52 (1):99–115.

Mihai, Mihaela, and Irina Meghea. 2011. Box Jenkins methodology applied to the environmental monitoring data. *Applied Sciences* 13:74–81.

Mirzaei, Mahin, Jamil Amanollahi, and Chris G. Tzanis. 2019. Evaluation of linear, nonlinear, and hybrid models for predicting PM2.5 based on a GTWR model and MODIS AOD data. *Air Quality, Atmosphere and Health* 12 (10):1215–1224.

Mishra, Dhirendra, and Pramila Goyal. 2015. Development of artificial intelligence based NO2 forecasting models at Taj Mahal, Agra. *Atmospheric Pollution Research* 6 (1):99–106.

Morabito, Francesco Carlo, and Mario Versaci. 2003. Fuzzy neural identification and forecasting techniques to process experimental urban air pollution data. *Neural Networks* 16 (3–4):493–506.

Oprea, M., and H. Liu. 2016. A knowledge-based approach for PM2.5 air pollution effects analysis. Paper read at *2016 International Symposium on INnovations in Intelligent SysTems and Applications (INISTA)*, 2–5 Aug. 2016.

Osowski, Stanislaw, and Konrad Garanty. 2007. Forecasting of the daily meteorological pollution using wavelets and support vector machine. *Engineering Applications of Artificial Intelligence* 20 (6):745–755.

Ozel, Gamze, and Selen Cakmakyapan. 2015. A new approach to the prediction of PM10 concentrations in Central Anatolia Region, Turkey. *Atmospheric Pollution Research* 6 (5):735–741.

Pai, Tzu-Yi, Keisuke Hanaki, Han-Chang Su, and Lu-Feng Yu. 2013. A 24-h Forecast of Oxidant Concentration in Tokyo Using Neural Network and Fuzzy Learning Approach. *CLEAN–Soil, Air, Water* 41 (8):729–736.

Paschalidou, Anastasia K, Spyridon Karakitsios, Savvas Kleanthous, and Pavlos A Kassomenos. 2011. Forecasting hourly PM 10 concentration in Cyprus through artificial neural networks and multiple regression models: Implications to local environmental management. *Environmental Science and Pollution Research* 18 (2):316–327.

Polk, HS. 2019. State of global air 2019: A special report on global exposure to air pollution and its disease burden. Health Effects Institute: Boston, MA, USA.

Rahman, Syed Masiur, A. N. Khondaker, and Radwan Abdel-Aal. 2012. Self-organizing ozone model for Empty Quarter of Saudi Arabia: Group method data handling-based modeling approach. *Atmospheric Environment* 59:398–407.

Rai, A Adarsh, P Srinivasa Pai, and BR Shrinivasa Rao. 2015. Prediction models for performance and emissions of a dual fuel CI engine using ANFIS. *Sadhana* 40 (2):515–535.

Reddy, Vikram, Pavan Yedavalli, Shrestha Mohanty, and Udit Nakhat. 2018. Deep air: Forecasting air pollution in Beijing, China. *Environmental Science*.

Rumelhart, David E, Geoffrey E Hinton, and Ronald J Williams. 1986. Learning representations by back-propagating errors. *Nature* 323 (6088):533–536.

Russo, Ana, and Amílcar O Soares. 2014. Hybrid model for urban air pollution forecasting: A stochastic spatio-temporal approach. *Mathematical Geosciences* 46 (1):75–93.

Sahoo, S. K., A. V. Kumar, A. K. Yadav, and R. M. Tripathi. 2016. Metal characterization of airborne particulate matters in a coastal region. *Toxicological and Environmental Chemistry* 98 (7):768–777.

Saide, Pablo E, Gregory R Carmichael, Scott N Spak, et al. 2011. Forecasting urban PM10 and PM2. 5 pollution episodes in very stable nocturnal conditions and complex terrain using WRF-Chem CO tracer model. *Atmospheric Environment* 45 (16):2769–2780.

Samarasinghe, Sandhya. 2016. *Neural networks for applied sciences and engineering: From fundamentals to complex pattern recognition.* CRC Press, New York.

Samia, Ayari, Nouira Kaouther, and Trabelsi Abdelwahed. 2012. A hybrid ARIMA and artificial neural networks model to forecast air quality in urban areas: Case of Tunisia. Paper read at Advanced Materials Research.

Şeker, Abdulkadir, Banu Diri, and Hasan Hüseyin Balık. 2017. Derin öğrenme yöntemleri ve uygulamaları hakkında bir inceleme. *Gazi Mühendislik Bilimleri Dergisi (GMBD)* 3 (3):47–64.

Shen, Huayu, Zhaoxia Wang, Chengyao Gao, Juan QIN, Fubin YAO, and Wei XU. 2008. Determining the number of BP neural network hidden layer units. *Journal of Tianjin University of Technology* 24 (5):13.

Slini, Th, K Karatzas, and N Moussiopoulos. 2002. Statistical analysis of environmental data as the basis of forecasting: An air quality application. *Science of the Total Environment* 288 (3):227–237.

Song, Qiang, and Brad S Chissom. 1993. Fuzzy time series and its models. *Fuzzy Sets and Systems* 54 (3):269–277.

Sonmez, Adem Yavuz, Semih Kale, Rahmi Can Ozdemir, and Ali Eslem Kadak. 2018. An adaptive neuro-fuzzy inference system (ANFIS) to predict of cadmium (Cd) concentrations in the Filyos River, *Turkish Journal of Fisheries and Aquatic Sciences*, 18:1333–1343.

Sousa, SIV, Fernando Gomes Martins, MCM Alvim-Ferraz, and Maria C Pereira. 2007. Multiple linear regression and artificial neural networks based on principal components to predict ozone concentrations. *Environmental Modeling & Software* 22 (1):97–103.

Trigo, Ricardo M, and Jean P Palutikof. 1999. Simulation of daily temperatures for climate change scenarios over Portugal: A neural network model approach. *Climate Research* 13 (1):45–59.

Ul-Saufie, Ahmad Zia, Ahmad Shukri Yahaya, Nor Azam Ramli, Norrimi Rosaida, and Hazrul Abdul Hamid. 2013. Future daily PM10 concentrations prediction by combining regression models and feedforward backpropagation models with principal component analysis (PCA). *Atmospheric Environment* 77:621–630.

Vapnik, Vladimir N. 1999. An overview of statistical learning theory. *IEEE Transactions on Neural Networks* 10 (5):988–999.

Viotti, P, G Liuti, and P Di Genova. 2002. Atmospheric urban pollution: Applications of an artificial neural network (ANN) to the city of Perugia. *Ecological Modeling* 148 (1):27–46.

Wang, Sun-Chong. 2003. Artificial neural network. In *Interdisciplinary computing in java programming*. The Springer International Series in Engineering and Computer Science, vol. 743. Springer, Boston, MA.

Wang, Xianghong, and Baozhen Wang. 2019. Research on prediction of environmental aerosol and PM2. 5 based on artificial neural network. *Neural Computing and Applications* 31 (12):8217–8227.

Wu, Shengjun, Qi Feng, Yun Du, and Xiaodong Li. 2011. Artificial neural network models for daily PM10 air pollution index prediction in the urban area of Wuhan, China. *Environmental Engineering Science* 28 (5):357–363.

Yadav, A. K., and A. Jamal. 2018. Technological reviews of particulate matter and their source identification techniques. *Environmental Quality Management* 27 (4):87–95.

Yang, Biao, Jie Guo, and Chunling Xiao. 2018. Effect of PM2. 5 environmental pollution on rat lung. *Environmental Science and Pollution Research* 25 (36):36136–36146.

Yang, Zhongshan, and Jian Wang. 2017. A new air quality monitoring and early warning system: Air quality assessment and air pollutant concentration prediction. *Environmental Research* 158:105–117.

Yuan, Zhe, Lin-Na Wang, and Xu Ji. 2014. Prediction of concrete compressive strength: Research on hybrid models genetic based algorithms and ANFIS. *Advances in Engineering Software* 67:156–163.

Zadeh, Lotfi A. 1973. Outline of a new approach to the analysis of complex systems and decision processes. *IEEE Transactions on Systems, Man, and Cybernetics* (1):28–44.

Zadeh, Lotfi Aliasker. 1965. Fuzzy sets, information and control, vol. 8. *Google Scholar Digital Library Digital Library*: 338–353.

Zhang, G Peter. 2003. Time series forecasting using a hybrid ARIMA and neural network model. *Neurocomputing* 50:159–175.

Zhao, Xudong, Yunfei Yin, Ben Niu, and Xiaolong Zheng. 2015. Stabilization for a class of switched nonlinear systems with novel average dwell time switching by T–S fuzzy modeling. *IEEE Transactions on Cybernetics* 46 (8):1952–1957.

Zhu, Wenjin, Jianzhou Wang, Wenyu Zhang, and Donghuai Sun. 2012. Short-term effects of air pollution on lower respiratory diseases and forecasting by the group method of data handling. *Atmospheric Environment* 51:29–38.

Zhou, Y., Bai, Y., Yue, Y., Lü, J., Chen, S., and Xiao, H. 2019. Characteristics of the factors influencing transportation and accumulation processes during a persistent pollution event in the middle reaches of the Yangtze River, China. *Atmospheric Pollution Research*, 10:1420–1434.

Part III

Internet of Things and
Environmental Systems

11 Internet of Things (IoT) Powered Enhancements to Industrial Air Pollution Monitoring Systems

*Hemanth Kumar Bangalore Naveen,
Anand Jayachandran Jolly, Vinay
Narayanaswamy, Ajay Sudhir Bale, Baby Chithra
Ramasamy, Divyashree Neelegowda, and
Subhashish Tiwari*

CONTENTS

11.1 INTRODUCTION

As globalization increased (Sze Yuk-Hiu, 2002), the technological growth that support it has also increased. Rapid industrialization has helped make massive technological advancements. But it has also posed a huge problem to the environment as pollution has increased significantly (Lucas and Noordewier, 2016). Air pollution has seen an exponential increase in recent times, with many cities around the world becoming uninhabitable due to a severely polluted ecosystem (Nagar et al., 2021; Qadri and Faiq, 2020). Air pollution is reaching unprecedented levels and is posing a major threat to life on earth. To address these issues, pollution levels must first be monitored.

With the recent advancement in IoT technology, monitoring pollution has become much cheaper and can be controlled by remote (Miles et al., 2018, Arora et al., 2019,

Agarwal et al., 2018; Senthilkumar et al., 2020). IoT technology has helped in revolutionizing sensing and analyzing data that is procured. The need for a physical presence and monitoring the working of these sensory devices is cut short and are automated to save time and resources. In this chapter, we survey different IoT applications to monitor Air pollution specifically. Major pollutants and their effects on human life have also been discussed, along with a brief mention of popular sensors used in IoT pollution monitoring.

11.2   LITERATURE SURVEY

Given the current air pollution monitoring status in India, there are 793 stations monitoring air pollution (Kamyotra and Saha, 2011). These stations are spread across the country covering 344 cities/towns.

Comments on the accuracy of the readings from most stations (Dahiya and Rungsung, 2019) show that readings are not very accurate as they are taken manually. Appropriate air quality monitoring standards were not followed by over 80% of the country. The monitoring in these stations is done twice a week.

The methods used in the stations are manual and/or continuous (indiaenvironmentportal.org.in). The manual methods in most NAMP stations are Wet-Chemical, Gravimetric, and Chemical Analyses. There are currently 309 CAAQM (Continuous Ambient Air Quality Monitoring) stations around India which provide automated continuous monitoring round the clock, and these are found accurate ("CCR", n.d.).

India's monitoring density is 1 for every 6.8 million persons, which is well below the required rate (Brauer et al., 2019). It is estimated that we would need 1,600–4,000 extra monitors to effectively track India's overall air quality.

It is evident that India requires a lot more pollution monitoring stations, equipment, and methods to combat data inadequacy in air quality monitoring. IoT seems to be a plausible solution to this problem as it is cheaper and efficient.

11.2.1   AIR POLLUTION

Air pollution has posed the most significant threat to quality of life. With rises in greenhouse gases and various other gaseous pollutants, disease has become rampant. Understanding major contributors to air pollution is necessary before decisions can be taken to counter it.

11.2.1.1   Gaseous Pollutants

Sulphur Dioxide: This is a hydrophilic colourless gas which has adverse effects on the human body when its presence in the atmosphere is increased. Major sources of SO_2 are from the combustion of substances such as coal and oil.

Oxides of Nitrogen: This includes primarily Nitric Oxide (NO), Nitrogen Dioxide (NO_2) and Nitrous Oxide (N_2O). Its sources are mainly on-road vehicles.

Carbon Monoxide/Dioxide gases: Vehicle emissions are the primary sources of Carbon monoxide, an odourless and colourless gas which has severe effects on the human body. Carbon dioxide majorly contributes to global warming and the ongoing climate crisis.

Particulate matter: Particulate matter refers to matter present in the air with sizes measured in microns. These are found to have the most ill effects on human life. Dust, dirt, soot, and smoke are a few examples for particulate matter.

Table 11.1 explains the major sources, health effects, and permissible values for common air pollutants.

11.2.2   IoT Applications in Air Pollution Monitoring

Air pollution has been the biggest contributor to global warming and climate change. Tackling it using IoT has its own advantages and disadvantages. Monitoring it efficiently can lead to an understanding of how to approach the issue. IoT techniques to address the problem has been discussed in this section.

Given that India has some of the most air polluted cities in the world, many states have tried implementing laws to curb the pollution by reducing traffic. Since this isn't an efficient way of controlling air pollution, there have been studies of rerouting traffic into lesser polluted pathways/roads thereby reducing concentration levels in certain areas (Muthukumar et al., 2018). This requires setting up IoT modules in every road and monitoring air quality data frequently. The method uses sensors such as MQ7 for carbon monoxide, MQ135 for ammonia, MQ4 for methane gas and G37 for oxygen content. The outputs from the sensors are then converted to digital signals using an ADC. PIC16F877A microcontroller takes in these signals as inputs. An AQI algorithm then compares sensor values with threshold AQI levels. This information is then transmitted via ESP 8266 to a server from which the data can be accessed.

Communicating received data efficiently is as important as collecting it. Models that prioritize the importance of this have been explored (Yellamma et al., 2021). The sensors used are MQ7 Carbon Monoxide detector, M213 for sound, LM35 temperature sensor and SY-HS220 humidity sensor. This model also collects information on surrounding noise pollution through the M213 noise sensor. The data is collected in the Arduino and is processed and compared with an inbuilt algorithm and transmitted to cloud via ESP 8266 Wi-Fi module. Information stored in cloud can be accessed and viewed through a website or an app.

Since cities are a major source of pollution, there have been methods to monitor and track air pollution concentrations by placing IoT modules at strategic locations around the city (Alam et al., 2018). The IoT module used gas sensors MG812 for Carbon Dioxide and MQ2 (Detects LPG, $C4H_{10}$, C_3H_8, Alcohol, CH_4, H_2, CO). A sound sensor was used for measuring sound values. Dust sensor GP2Y1010AU0F was used for measuring fine particulate matter. LM35 temperature sensor measured temperature. The MCU was interfaced to a computer via USB to TTL converter. The data taken from the sensors were processed and transmitted to a pre-authorized server.

IoT's biggest advantage is its small size that makes it portable. In that context, there was a model developed to collect pollution data in the immediate vicinity of the

TABLE 11.1

Air Pollutants, Their Effects, and Values

Air Pollutants	Major Sources	Health Effects	AQI Values
Sulphur Dioxide	Combustion of substances containing Sulphur, such as Coal and Oil. (Singh, 2017) Fossil fuel combustion. Volcanic Activity	Acid rain formed from sulphuric acid (SO_2 + air + water) can have adverse effects on natural ecosystems including human life. Breathing difficulties, coughing, throat irritation. Higher concentrations can cause Asthma and similar lung diseases	Good: 0–50* (Sulphur dioxide designations – Washington State Department of Ecology, n.d.) Moderate: 51–100 Moderately Unhealthy: 101–150 Unhealthy: 151–200 Very Unhealthy: 201–300
Oxides of Nitrogen: Nitric Oxide (NO), Nitrogen Dioxide (NO_2), Nitrous Oxide (N_2O)	Combustion of gasoline in Automobiles (Libretexts, 2013) Lightning bolts	Irritation in eyes, nose, throat and lungs (Agency for Toxic Substances and Disease Registry, 2021). High concentrations can cause burns. Increase in concentration of CO_2 can cause tachycardia, cardiac arrhythmias, and impaired consciousness	AEGL (Acute Exposure Guideline Levels) For NO_2 and NO (Committee on Acute Exposure Guideline Levels et al., 2012) 10 min – 34 ppm 30 min – 25 ppm 1 h – 20 ppm 4 h – 14 ppm 8 h – 11 ppm
Carbon Monoxide/ Dioxide gases	Incomplete combustion of fuel	Increase in concentration of CO_2 can cause tachycardia, cardiac arrhythmias, higher respiratory rate, and impaired consciousness. (Langford, 2005). Chronic exposure to CO can lead to mild neurological effects (Townsend and Maynard, 2002)	Guidelines on exposure to Carbon Monoxide (Townsend and Maynard, 2002): 87.1 ppm for 15 minutes 52.3 ppm for 30 minutes 26.1 ppm for 1 hour 8.7 ppm for 8 hours.

Air Pollutants	Major Sources	Health Effects	AQI Values
Particulate Matter:PM_{10} and $PM_{2.5}$	PM consists of dust, dirt, soot, and smoke. Construction sites, Agriculture, Factories, etc. (Adams et al., 2015)	PM come in varying sizes, but they are known to cause lung diseases	PM_{10} ($\mu g/m^3$)– 24 hr (Kaku et al., 2016, Aamer et al., 2018) Good: 0–54 Moderate: 55–154 Unhealthy for sensitive: 76–185 Unhealthy: 255–354 Very Unhealthy: 355–424 Hazardous: 425–604 $PM_{2.5}$ ($\mu g/m^3$) – 24 hr (Zamora, Rice, & Koehler, 2020) Good: 0–12 Moderate:12.1–35.4 Unhealthy for sensitive: 35.5–55.4 Unhealthy: 55.5–150.4 Very Unhealthy: 150.5–250.4 Hazardous: 250.5–500.4

* An AQI value of 100 refers to 75 ppb (parts per billion) at an hour average.

IoT model and provide results that will display harmful pollutants and their corresponding AQI values (Aamer et al., 2018). Using ESP2866 MCU, multi-channel gas
sensors for CO and NO_2, temperature sensor, DHT and GPS sensor, the model was
developed. The sensors collect data and ESP8266 transmits this as an html request to
a database. The web scripts in the database perform calculations on the data using
chosen AQI formulas and then displays a wide range of AQI information.

Involving sound sensors into an IoT model designed for recording air pollution
can serve the purpose of indicating traffic near schools and hospitals (Pal et al.,
2017). Along with a sound sensor, the gas sensor used was MQ135 (Detects NH_3,
NO_x, Alcohol, C_6H_6, Smoke, CO_2). An ESP8266 was configured with an Arduino
UNO to transmit data to cloud and obtain results on a webpage. The MQ135 connected to the Arduino provided data on gas concentration, which would be displayed
in ppm (parts per million) on a webpage. The webpage would also display warnings
based on AQI standards for the gases.

Data collected from IoT monitoring setups can be readily displayed onto an
android application (Gupta et al., 2019). The sensors used were DHT11 (detects
moisture), MQ-2 and SDS021 (detects PM2.5 and PM10). An MCP3208 IC operates
as an ADC and acts as the interface between the Raspberry Pi and MQ-2, where the
MCU is used as the processing unit. The processed data from Raspberry pi is displayed on an android application which was transmitted via the integrated Wi-Fi
module.

Low Power plays an important role when an IoT network is active and recording
data. Solar energy is a very useful source of power for dispersed IoT networks in
remote locations (Mahesh and Walsange, 2020). Using an Arduino UNO R3 as the
processor unit, the sensors used were MQ-2 and a DHT11 humidity sensor. The data
from the sensors would be collected and processed by the Arduino which then transmits the processed data to cloud via ESP8266 Wi-Fi module. The data could then be
viewed on a webpage.

Cost efficiency plays a major role in setting up an IoT monitoring device (Rawal,
2019). Keeping this in mind, the gas sensors used in this model were MQ135 and
MQ7. The data was collected and processed by an Arduino UNO and transmitted to
cloud via ESP-01. The data could then be viewed on a webpage. The design is com-
paratively cost-effective. Table 11.2 summarizes the review.

11.3   DISCUSSION

From the above IoT-based work we understand that priorities are cost efficiency,
power, strategic placements, durability, number of pollutants measured, and distance
range. An ideal setup would consist of all of these, but trade-offs have to be done for
few amongst them.

Cost efficiency: Cost efficiency can be determined by overall costs on maintaining
the IoT setups and its individual cost (MCUs, Sensors, Batteries, etc.).

For setting up IoT modules around a city/locality, cost per unit matters. In that
case only specific sensors are needed depending on the most prevalent pollutant in
that location. Maintenance costs would not need emphasis as modules are set up
close by to each other.

TABLE 11.2
Summary of Literature Review

MCU	Sensors	Pollutants Measured	Highlights	Reference
Arduino UNO	MQ7, M213, LM35, SY-HS220	Carbon monoxide, noise, Humidity, Temperature	Data received from sensors were compared with AQI values before being transmitted	Ch. Harshavardhan et al. (2019)
PIC16F877A via ESP 8266	MQ7, MQ4, MQ135, G37	Carbon monoxide, Ammonia, oxygen, methane	IoT module focused on reducing road traffic in highly polluted areas,	Muthukumar et al. (2018)
ATmega328p	MG812, MQ2, GP2Y1010AU0F, LM35	Carbon dioxide, LPG, C_4H_{10}, C_3H_8, Alcohol, CH_4, H_2, CO, Dust, Temperature	Cost-effective model (45 \$) that senses a wide range of pollutants	Alam et al. (2018)
NodeMCU	Temperature sensor, DHT and GPS sensor, CO and NO_2 sensor	CO, NO_2	Data is transmitted onto a webpage and compared with AQI values	Aamer et al. (2018)
Arduino UNO	MQ135	NH_3, NOx, Alcohol, C_6H_6, Smoke, CO_2	An on-module LED gives alerts when surrounding air quality is unsuitable	Poonam et al. (2017)
Raspberry Pi	DHT11, MQ-2, SDS021	Moisture, PM2.5, PM10	Data uploaded to ThingSpeak and can be accessed through an Android app	Gupta et al. (2019)
Arduino UNO	MQ135, MQ7	NH_3, NOx, Alcohol, C_6H_6, Smoke, CO_2	Obtained data compared with AQI values and displayed graphically on a website	Ramik Rawal (2019)
Arduino UNO R3	MQ-2, DHT11 Humidity sensor	Smoke, Methane, Carbon Monoxide, humidity,	Data stored on database; hence history of values can be obtained via webpage/app	Mahesh and Walsange, (2020)

For individual monitoring units in remote locations, costs can be considered significantly higher per unit as greater emphasis is on number of pollutants monitored rather than specific pollutants. Maintenance costs will have to be given greater importance as these setups would be in remote locations. Hence, renewable sources of energy would be a viable option for powering the module.

Power: Harvesting solar energy to power an IoT seems to be the most popular option but comes with a higher cost. Batteries are equally preferred but have to be replaced over time.

Strategic placements – We know that air quality changes in accordance with height, with certain pollutants showing higher concentrations at ground level and decreasing as we go further up. Forming an IoT network would require placing the modules at appropriate heights to measure specific pollutants.

Durability: Durability serves as an important parameter when the IoT modules are setup in very remote locations. The module must be kept protected from harsh weather conditions.

Number of pollutants measured: Certain pollutants in a location will be at a higher concentration in comparison with other pollutants and will require very close monitoring. Sensors used in IoT modules must be selected based on pollutants/parameter to be measured (Example: Temperature, NO_2, CO, etc.).

Range: For IoT modules set up in remote locations, transceivers used play an important role in transmission of obtained data. Depending on the range, transceivers such as RF transmitters can be used, many surveyed have used ESP8266 WiFi module to receive and transmit data in close range data transmission.

11.4 PROPOSED FRAMEWORK

We propose a method involving ESP32, which is an open-source IoT platform for transmitting data wirelessly to a web server as shown in Figure 11.1. As ESP32 is

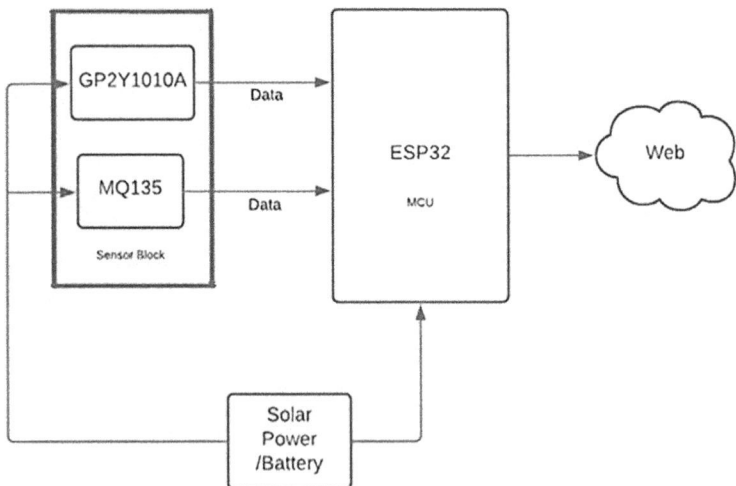

FIGURE 11.1 Proposed framework.

considerably cheap and consumes low power, it serves as the perfect option for our proposed method. It has an integrated Wi-Fi module. Given that it has 15 ADC channels, it is better suited for IoT projects involving sensors, compared to the ESP8266 that is widely used. By connecting two gas sensors to it, MQ135 (measures NH_3, NO_x, alcohol, benzene, smoke, and CO_2) and GP2Y1010AU0F (dust sensor), we are using it as a controller to obtain data from the sensors and then transmit the same. The data obtained from the sensors can be directly transmitted to an open-source web-based software or any other alternative.

11.5   FUTURE RESEARCH

The proposed model can be further implemented at various locations and can be connected to a Networked SCADA system, where data acquisition and air pollution monitoring can be performed in a wider range. This also allows for a control setup which facilitates easier maintenance as data processing can be done using previous data, moving average and can be compared with new data, and health of each module can be determined on the basis of accuracy.

11.6   CONCLUSION

The requirement of devices to monitor air quality is the need of the hour as this can provide important data which can further be used to take necessary actions to prevent pollution from getting worse. A huge environmental crisis is expected in the coming years, and it would become inevitable if the required measures are not taken well in advance. These smart IoT devices tend to work without any human intervention just by communicating with other units to capture required data for analysis. IoT is a major breakthrough in the field of technology and research, and would completely dominate the future generations with its vast application capabilities and potential in different fields. IoT's feature of being an automated process cuts down labour costs. Using an ESP32, the IoT model not only serves as a standalone IoT monitoring device, but also consumes low power and comes at a low cost. The overall size of the setup is relatively small, and provides data on certain major air pollutants. Our proposed method aims to provide a plausible solution to the problems faced in monitoring air pollution.

REFERENCES

Aamer, H., Mumtaz, R., Anwar, H., & Poslad, S. (2018). A very low cost, open, wireless, internet of things (IoT) air quality monitoring platform. *2018 15th International Conference on Smart Cities: Improving Quality of Life Using ICT & IoT (HONET-ICT)*. IEEE.

Acute Exposure Guideline Levels, C. (2012). Acute exposure guideline levels for selected airborne chemicals (Vol. 12). Washington, DC: National Academies Press.

Adams, K., Greenbaum, D. S., Shaikh, R., van Erp, A. M., & Russell, A. G. (2015). Particulate matter components, sources, and health: Systematic approaches to testing effects. Journal of the Air & Waste Management Association (1995), 65(5), 544–558.

Agarwal, A., Shukla, V., Singh, R., Gehlot, A., Garg, V., 2018. Design and development of air and water pollution quality monitoring using IoT and quadcopter, Springer, Singapore. ISBN: 978-981-10-5902-5.

Agency for Toxic Substances and Disease Registry. (2021, July 13). Retrieved July 15, 2021, from Cdc.gov website: https://www.atsdr.cdc.gov

Alam, S. S., Islam, A. J., Hasan, M. M., Rafid, M. N. M., Chakma, N., & Imtiaz, M. N. (2018). Design and development of a low-cost IoT-based environmental pollution monitoring system. *2018 4th International Conference on Electrical Engineering and Information & Communication Technology (ICEEiCT)*. IEEE.

Arora, J., Pandya, U., Shah, S., & Doshi, N. (2019). Survey – pollution monitoring using IoT. Procedia Computer Science, 155, 710–715.

Brauer, M., Guttikunda, S. K., Nishad, K. A. Dey, S., Tripathi, S. N., Weagle, C., & Martin, R. V. (2019). Examination of monitoring approaches for ambient air pollution: A case study for India. Atmospheric Environment, 216(116940), 116940.

CCR. (n.d.). Retrieved August 18, 2021, from Cpcbccr.com website: https://app.cpcbccr.com/ccr/#/caaqm-dashboard-all/caaqm-landing

Ch. Harshavardhan, K. Raghavendrakrishnasai, N. Mohanvamsi, P. Yellamma. (2019). A Smart Industrial Pollution Monitoring System using IoT. 2019 International Journal of Innovative Technology and Exploring Engineering (IJITEE), 8(7), 2278–3075.

Dahiya, S., Rungsung, P. (2019). Airpocalypse III: Assessment of air pollution in Indian cities and National Clean Air Programme (NCAP), https://storage.googleapis.com/planet4-india-stateless/2019/01/5ebbb021-airpocalypse-iii_28jan19.pdf [Accessed on August 18, 2021]

Gupta, Harsh, Bhardwaj, D., Agrawal, H., Tikkiwal, V. A., & Kumar, A. (2019). An IoT-based air pollution monitoring system for smart cities. *2019 IEEE International Conference on Sustainable Energy Technologies (ICSET)*, 173–177. IEEE

Kaku, K. C., Reid, J. S., Reid, E. A., Ross-Langerman, K., Piketh, S., Cliff, S., … Perry, K. D. (2016). Investigation of the relative fine and coarse mode aerosol loadings and properties in the Southern Arabian Gulf region. Atmospheric Research, 169, 171–182.

Langford, N. J. (2005). Carbon dioxide poisoning. Toxicological Reviews, 24(4), 229–235.

Libretexts. (2013, October 2). Sources of nitrogen oxides. Retrieved July 15, 2021, from Libretexts.org. https://chem.libretexts.org/Bookshelves/Environmental_Chemistry/Supplemental_Modules_(Environmental_Chemistry)/Acid_Rain/Sources_of_Nitrogen_Oxides

Lucas, M. T., & Noordewier, T. G. (2016). Environmental management practices and firm financial performance: The moderating effect of industry pollution-related factors. International Journal of Production Economics, 175, 24–34.

Mahesh, V., & Walsange, P. Y. V. V. (2020). Arduino and sensor-based air pollution monitoring system using IoT. International Research Journal of Engineering and Technology (IRJET), 07(07), 152–160.

Miles, A., Zaslavsky, A., & Browne, C. (2018). IoT-based decision support system for monitoring and mitigating atmospheric pollution in smart cities. Journal of Decision System, 27(sup1), 56–67

Muthukumar, S., Sherine Mary, W., Jayanthi, S., Kiruthiga, R., & Mahalakshmi, M. (2018). IoT-based air pollution monitoring and control system. *2018 International Conference on Inventive Research in Computing Applications (ICIRCA)*. IEEE.

Nagar, P. K., Gargava, P., Shukla, V. K., Sharma, M., Pathak, A. K., & Singh, D. (2021). Multi-pollutant air quality analyses and apportionment of sources in three particle size categories at Taj Mahal, Agra. Atmospheric Pollution Research, 12(1), 210–218.

Pal, P., Gupta, R., Tiwari, S., & Sharma, A. (2017). IoT-based air pollution monitoring system using Arduino. International Research Journal of Engineering and Technology, 6(3), 793–797.

Qadri, R., & Faiq, M. A. (2020). Freshwater pollution: Effects on aquatic life and human health. In Fresh Water Pollution Dynamics and Remediation (pp. 15–26). Singapore: Springer Singapore.

Rawal, R. (2019). Air Quality Monitoring System. International Journal of Computational Science and Engineering, 9(1), 1–9.

Kamyotra, J. S., Saha, D. (2011). Guidelines for the measurement of ambient air pollutants Volume-II, Central Pollution Control Board, Ministry of Environment & Forests, Govt. of India, Delhi (India) http://www.indiaenvironmentportal.org.in/files/NAAQSManualVolumeII.pdf [Accessed 18 August 2021]

Senthilkumar, R., Venkatakrishnan, P., & Balaji, N. (2020). Intelligent based novel embedded system based IoT enabled air pollution monitoring system. Microprocessors and Microsystems, 77(103172), 103172.

Singh, D. (2017, October 13). Sulphur dioxide (SO$_2$). Retrieved July 15, 2021, from Aqi.in website: https://www.aqi.in/blog/sulphur-dioxide-so2/

Sulfur dioxide designations – Washington State Department of Ecology. (n.d.). Retrieved July 15, 2021, from Ecology.wa.gov website: https://ecology.wa.gov/Regulations-Permits/Plans-policies/Areas-meeting-and-not-meeting-air-standards/Sulfur-dioxide-designations

Sze Yuk-Hiu, A. (2002). Globalization and poverty in east Asia. New Global Development, 18(1–2), 21–36.

Townsend, C. L., & Maynard, R. L. (2002). Effects on health of prolonged exposure to low concentrations of carbon monoxide. Occupational and Environmental Medicine, 59(10), 708–711.

Yellamma, P., Chandra, N., Sukhesh, P., Shrunith, P., sTeja, S.S., 2021. Arduino based vehicle accident alert system using GPS, GSM and MEMS accelerometer, *2021 5th International Conference on Computing Methodologies and Communication (ICCMC)*, IEEE, pp. 486–491.

Zamora, M. L., Rice, J., & Koehler, K. (2020). One-year evaluation of three low-cost PM2.5 monitors. Atmospheric Environment, 235(117615), 117615.

12 Impact of Temporary COVID-Related Lockdowns on Air Quality across the Globe
A Systematic Review

Rajwinder Singh, Arti Thanki, Ankita Thanki, Karanvir Singh Sohal, Anmol Kaur, and Shivani Dedakia

CONTENTS

12.1 INTRODUCTION

The usual rhythm of people's livelihood, such as the world's commerce, social, and economic compliance, national and worldwide, academic environment, and people's daily life, has been severely disturbed as a result of the lethal new

DOI: 10.1201/9781003203445-15

coronavirus. The coronavirus, which was initially discovered in Wuhan, China, in December 2019, has infected over 177 million people and killed 38 million people worldwide as of today (19th June 2021) (Worldometer, 2021). Because of the virus's rapid spread to numerous countries, the World Health Organization (WHO) announced the coronavirus as a pandemic on March 11 2020. According to previous research, Virus transmission rates were much higher in cold and dry surroundings than in warm/hot environments (Manigandan et al., 2020). The novel coronavirus is a member of the Corona viridae family and has a major impact on the human respiratory tract (Unhale et al., 2020). Coronavirus is mostly transmitted by droplet infection, and critically ill individuals must be treated in hospitals (Correia et al., 2020).

Governments have implemented lockdown procedures to control the transmission of infection and certain countermeasures such as social distancing practices. However, these practices have lowered the rate of spread of the virus but worst affects the economic position of countries worldwide. The Chinese government has imposed the social distancing strategy, and it is broadly regarded as a mandatory self-protective approach in order to cope with the coronavirus epidemic when there is no other available choice to battle this deadly virus (Huang et al., 2021). Since the end of March 2020, over half of the population of the world has been forced to live under a lockdown regime (Tosepu et al., 2020).

Although most people across the world have never experienced a long-term lockdown system, the health of the whole ecosystem, including humans, has improved considerably as a result of the closure of transport, construction, and manufacturing operations. As the vehicular and industrial activities were shut down, the maximum improvement in the quality of water bodies and air have been reported in many countries (Briz-redón et al., 2021; Yadav and Jamal, 2018a, 2018b, 2018c). The emission of air pollutants such as CO, NO_x, SO_x, PM_{10}, and $PM2_{.5}$, generated through various anthropogenic activities such as mining, burning of fossil fuels, transportation, industries were also been reduced up to an extent due to the implementation of lockdown policies as compared to the air quality before the occurrence of COVID-19 (Jiang et al., 2021; Yadav and Hopke, 2020; Yadav et al., 2014).

Many studies have been well-articulated on evaluating air characteristics standards, the state of the overall atmosphere, and the causes for variations in specific pollutant parameters in connection to COVID-19 during the lockdown period in respective countries. But a review study that provides significant information on the decrease in the number of air pollutants in worst-affected countries has not been conducted to date. Therefore, this chapter has attempted to investigate the state of air quality in different countries and focuses on the long-term environmental sustainability management strategies which would be helpful in controlling the emission of the pollutants after the lockdown. Policymakers may follow the sustainable environmental management plan recommended by this study, and this sort of scientific basic review research effort will undoubtedly aid in the development of simulation models for enhancing environmental sustainability.

12.2 PERIODS OF TEMPORARY LOCKDOWN(S) IN THE MAJOR COUNTRIES ACROSS THE GLOBE

As the first case of COVID-19 has been found in Wuhan, China, therefore, to avoid the further transmission of the virus Chinese administration has implemented the first lockdown in the world. Following the same pattern, almost all the nations have enforced partial lockdowns as the detection of COVID-19 cases were noticed to be increasing. The information about the lockdown scenarios of a few countries has been presented in Table 12.1.

Table 12.1 illustrates the time duration and length of days during which the restrictions on transportation and industrial activities in the countries mentioned above have been enforced. It can be clearly observed that China has imposed the lockdown for a shorter period than the other countries.

12.3 METHODOLOGY/FRAMEWORK ADOPTED IN PREVIOUS STUDIES

In previous studies, researchers have adopted various models to assess the dispersion of air pollutants before, during, and after the lockdown phases in their respective localities/countries. The methods of data collection, statistical tools/software and sensors used to measure the air pollutants along with their source are shown in Table 12.2.

As observed in Table 12.2, most of the researchers have used the basic comparative approach or DiD model to measure and analyze the difference in the air pollution data before, during, and after the lockdown phases. However, the implementation of

---

TABLE 12.1

COVID-19 Related Lockdowns across the Major Countries

Country	First Lockdown			Second Lockdown			Third Lockdown		
	Start Date	End Date	Length (Days)	Start Date	End Date	Length (Days)	Start Date	End Date	Length (Days)
China	23-01-2020	08-04-2020	76						
UK	23-03-2020	04-07-2020	103	05-11-2020	02-12-2020	27	05-01-2021	29-03-2021	83
France	16-03-2020	24-07-2020	133	28-10-2020	1-12-2020	34	3-04-2021	30-06-2021	88
India	25-03-2020	7-06-2020	74	24-12-2020	6-01-2021	13	15-03-2021	30-04-2021	46
Italy	21-02-2020	8-03-2020	16	09-03-2020	18-05-2020	70			
Spain	14-03-2020	09-05-2020	56	1-10-2020	9-05-2021	220			

other air pollution models such as the MERRA-2 model, GAM and PCA along with machine learning has been successful in gaining the best-fit findings (Roy et al., 2021; Sahoo et al., 2016).

Considering a new outbreak in the future having similar characteristics and patterns of dispersion or mobility as of COVID-19, the scientists or researchers can implement various available models as discussed in Table 12.2 to predict the behaviour of an outbreak on the socio-economic and environmental aspects. The implementation of such models can be helpful to prevent the escalation in the transmission of any particular outbreak.

After reviewing the above-mentioned models in Table 12.2, the authors recommend DiD models as the best approaches to know the dispersion of air pollutants during a particular period due to their easy execution and analysis procedures (Guojun et al., 2020). The approach that needs to be followed in DiD model is illustrated in Figure 12.1.

The DiD model should contain the socio-economic data such as population, area, age, GDP, food supply, consumption of natural resources while collecting the information from any region within a timeframe to support the air pollution modeling. However, the implementation of this model is limited if the comparison groups have different and unstable trends (Priva and Sanker, 2019).

12.4 VARIATION IN THE AIR QUALITY GLOBALLY DURING PRE-LOCKDOWN AND LOCKDOWN SCENARIOS IN THE MAJOR COUNTRIES ACROSS THE GLOBE

This section provides information about the change in air quality during the lockdown phases in different countries. The variations in the air quality parameters such as NO_2, SO_2, $PM_{2.5}$ and PM_{10}, CO, and O_3 have been taken into considerations.

12.4.1 UNITED KINGDOM (UK)

The UK government has imposed the first lockdown on 23 March in the year 2020 due to the increase in the total number of COVID-19 infected patients.

Oxides of Nitrogen (NO_x): The decrease in the discharge of air pollutants to the atmosphere have been decreased significantly in England during the lockdown phase. The levels of NO, NO_2, and NO_x have been observed to be reduced substantially from 67.20% to 64.86%, 36.58%- 34.92%, and 50.31% to 47.69% within the lockdown period in comparison to the pre-lockdown scenario, respectively (Munir et al., 2021). The drop in the level of NO_2 has been found in line with the recent study (Dacre et al., 2020). Even the air quality monitoring places situated in the north and centre of England also showed a reduction in the NO_2 levels during the lockdown period of 17 march to 30 April 2020. However, in the post-lockdown scenario, when the restrictions on mobility were uplifted in the UK, a minor increase in the levels of NO, NO_2 and NO_x were also observed (Tobías et al., 2020).

TABLE 12.2
Summary of Statistical Tools/Framework/Models Used in the Previous Studies

Reference	Data Collection Method/ Software	Data Analysis Method/ Model/Framework Adopted	Pollutants Measured	Sensor/Model Used	Source(s) of Data of Each Study
			Input Parameters and Analysis Methods		
(Roy et al., 2021)	Satellite-based data	GIOVANNI, MERRA-2 model	NO_2 SO_2 PM2.5 CO O_3	AURA OMI MERRA-2 Model MERRA-2 Model MERRA-2 Model AURA OMI	1. https://earthdata.nasa.gov/
(Munir et al., 2021)	LAQMS, OpenStreetMap, ArcGIS 10.5.1	GAM Along with Machine Learning	NO, NO_2, NOx PM2.5, PM10	Chemiluminescent analyzers FDMS	1. https://uk-air.defra.gov.uk/networks/network-info?view=aurn0 2. https://uk-air.defra.gov.uk/networks/monitoring-methods?view=eu-standard
(Kumari & Toshniwal, 2020b)	WAQI	Comparative approach	PM2.5, PM10, NO_2 SO_2 O		1. www.aqicn.org
(Elsaid et al., 2021)	LAQMS		PM10, PM2.5, CO, SO_2 NO_2, NOx, O_3		1. www.arpalombardia.it/Pages/Aria/Richiesta-Dati.aspx 2. http://www.inemar.eu/xwiki/bin/view/Inemar/WebHome
(Sahoo et al., 2021)	GIS, ArcGIS 10.6	Kendall rank correlation, Kendall's tau correlation matrix, EDA plot and Shapiro-Wilk's (S-W) test, Kruskal-Wallis test	PM2.5, PM10, CO, NO, NO_x, NO_2, SO_2,		1. https://app.cpcbccr.com/ccr/#/caaqm-dashboard-all/caaqm-landing 2. https://waqi.info

TABLE 12.2 (CONTINUED)

Reference	Data Collection Method/ Software	Data Analysis Method/ Model/Framework Adopted	Input Parameters and Analysis Methods		Source(s) of Data of Each Study
			Pollutants Measured	Sensor/Model Used	
(Baldasano, 2020)	LAQMS	EuropeanMonitoring and Evaluation Programme network	NO$_2$		1. http://mediambient.gencat.cat/ca/05_ambits_dactuacio/atmosfera/qualitat_de_laire/avaluacio/xarxa_de_vigilancia_i_previsio_de_la_contaminacio_atmosferica_xvpca/
(Jiang et al., 2021)	LAQMS	PDF, Significance Test	NO$_2$ SO$_2$ PM10, PM2.5 CO O3	Molybdenum converter and Chemiluminescence ultraviolet (UV) fluorescence method, β absorption method Infrared Absorption UV Absorption	1. https://quotsoft.net/air 2. https://cds.climate.copernicus.eu/cdsapp#!/dataset
(Marinello et al., 2021)	LAQMS,	Comparative Approach	NO, NO$_x$, NO$_2$, O$_3$, PM10, PM2.5		1. https://www.google.com/covid19/mobility 2. https://eur-lex.europa.eu/legalcontent/en/ALL/?uri=CELEX%3A32008L0050
(Ikhlasse et al., 2021)	LAQMS, MERA and CARA	ATMO rating index, Comparative Approach	SO$_2$, O$_3$, NO$_2$, NO$_x$, CO, PM2.5 PM10		1. French Central Air Quality Monitoring Laboratory (LCSQA) 2. https://www.lcsqa.org/fr/indices-qualite-air
(Bera et al., 2021)	LAQMS, RS & GIS, LANDSAT 8 OLI and LANDSAT 7 ETM, Arc GIS 10.3, Grapher 13 software	HCA, PCA, IBM SPSS (16.0)	CO, NO$_2$, SO$_2$, O$_3$, PM10, and PM2.5	Copernicus Sentinel-5 Precursor, TROPOMI	1. https://www.esa.int/Applications/Observing_the_Earth/Copernicus/Sentinel-5P/Air_pollution_drops_in_India_following_lockdown 2. https://earthsky.org/earth/satellite-images-air-pollution-india-covid19.

Reference	Data Source	Method	Pollutants		Source
(Adil et al., 2020)	Landsat 8 OLI Level 2	SPM algorithm	NO_2, $PM2.5$, $PM10$		1. espa.cr.usgs.gov
(Kumari and Toshniwal, 2020a)	LAQMS	Copernicus Sentinel-5 Precursor, TROPOMI	$PM10$, $PM2.5$, NO_2, SO_2 O_3	Copernicus Sentinel-5 Precursor, TROPOMI	1. https://app.cpcbccr.com/ccr/#/caaqm-dashboardall/caaqm-landing
(Junfeng et al., 2021)	LAQMS	LSDV estimation strategies, DiD model	$PM2.5$, $PM10$, SO_2, NO_2, CO		
(Liu et al., 2021)	LAQMS	Fixed-effects ordinary least squares Approach, DiD Model	$PM2.5$, $PM10$, SO_2, NO_2, CO, O_3		1. https://aqicn.org/data-platform/covid19/verify/9fbd8433-2ab9-403d-9392-43851e866615 2. http://aqicn.org/calculator/cn
(Guojun et al., 2020)	LAQMS	DiD models	CO, NO_2, $PM10$, SO_2, O_3		1. https://go.nature.com/38fFWTb
(Lovri et al., 2021)	LAQMS	PCA with Machine Learning	O_3, $PM10$, NO_2		
(Briz-redón et al., 2021)	LAQMS	R programming language, R Package Effects, GGPLOT2, RCurl, sjPlot, XML	CO, NO_2, $PM10$, O_3, SO_2		
(Higham et al., 2021)	AURN	Comparative Approach	NO_2, SO_2, $PM2.5$, $PM10$		1. UK Department for Environment Food and Rural Affairs, n.d. Site Environment Types – Defra, UK

Note – (LAQMS) Local Air Quality Monitoring Stations; (WAQI) World Air Quality Index portal; (GIOVANNI) Goddard Interactive Online Visualization and Analysis Infrastructure; (MERRA-2) Modern-Era Retrospective Analysis for Research and Applications Version 2; (GAM) Generalized Additive Models; (FDMS) Filter Dynamics Measurement System; (GIS) Geographic Information System; (MERA) Measurement and Evaluation in Rural Area; (CERA) Chemical Characterization of particles; (PDF) Probability Distribution Function; (HCA) Hierarchical Cluster Analysis; (PCA) Principal Component Analysis; (SPM) Suspended Particulate Matter; (TROPOMI) Tropospheric Monitoring Instrument; (DID) Difference in differences; (AURN) Automatic Urban and Rural Network

| Defining the parameters while considering the assumptions and constraints | ⇨ | Collection of respective data from a particular region within a defined period | ⇨ | Summarizing the data by the implementation of Statistical tools | ⇨ | Analyzing and Interpretation of the difference in the data |

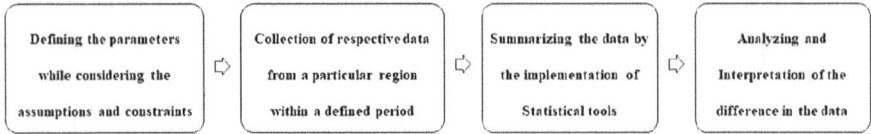

FIGURE 12.1 An approach to implement DiD model.

Sulphur Dioxide (SO₂): The substantial amount of sulphur dioxide has also been observed to increase within the UK's lockdown phase. The increase in the levels was much higher than in the last 7 years. The major source of the generation of sulphur dioxide is the industries being the point source of its origin. According to the researchers, there could be many reasons behind the increase in its level in the atmosphere (Qian et al., 2007). The increase in the rate of cremations during the lockdown has caused in causing a rapid upsurge in the concentration of sulphur oxide in China (He et al., 2014). The increase in the NO_x levels triggers the rate of conversion of SO_2 into sulphates. There are no doubts about the hazard of SO_2 to human beings. Many health-related issues such as difficulty in-breath and cough have been commonly observed (Pikhart et al., 2001). The worst effects have been noticed even by the short exposure of high levels of SO_2 in the higher age group (Conceição Martins et al., 2002).

Particulate Matter (PM): The variation in the particulate matter in pre-lockdown, lockdown, and post-lockdown cases were almost similar. Within the lockdown phase, a significant increase in PM_{10} and $PM_{2.5}$ were noticed compared to the pre-lockdown scenarios, which further observed a slight decrease in its concentration in the post-lockdown period. Several other researchers have also found an increase in the amount of PM_{10} and $PM_{2.5}$ (Ropkins and Tate, 2021; Shi et al., 2021). The reason behind the rise in the level of particulate matter could be due to the excessive use of indoor heating systems due to the lockdown conditions, people spend their time using indoor facilities which emit particulate matter (Donzelli et al., 2020; Marinello et al., 2021).

12.4.2 INDIA

The nationwide lockdown was imposed in India on 25 March 2020 which helped the atmosphere to regain its original quality. The pollutants such as $PM_{2.5}$, PM_{10}, NO_x, NO_2, CO showed a noteworthy drop in their concentrations within the lockdown in comparison to their concentration beforehand the implementation of lockdown (Bera et al., 2021).

Particulate Matter (PM): The concentration of PM_{10} has been reported to be reduced up to 51% and 80% in the lockdown and post-lockdown periods when compared to the pre-lockdown time, respectively. In addition to that, the concentration of $PM_{2.5}$ has been lowered up to 46% and 78% within the lockdown period and post-lockdown time, respectively. The level of $PM_{2.5}$ was found to be much below the permissible limit that is 60 µg/m³ (Sahoo et al., 2021).

Oxides of Nitrogen (NO_x): Similar pattern has been noticed for the NO_x i.e., NO and NO_2; as their concentration was dropped up to 62% throughout the lockdown and

50% afterwards the lockdown i.e., the post-lockdown phase in comparison to the pre-lockdown phase. The maximum reduction in the NO_2 amount was found to be 12.98% in April of the year 2020, whereas the maximum amount of NO_2 emission in the year 2019 was 40.19% in March (Bera et al., 2021).

Sulphur dioxide (SO_2): The reduction in the SO_2 levels was also lowered than the lockdown and post-lockdown phases. The average accumulation of the SO_2 has been noticed around 9.47 $\mu g/m^3$ in April of the year 2018, which showed a significant drop in its concentration with 5.36 $\mu g/m^3$ in March of the year 2020 (Bera et al., 2021).

Ozone (O_3): There has been a considerable decrease in the amount of ozone during and after the lockdown. The noticed reduction in the ozone concentration during and post-lockdown phases were around 18% and 66%, respectively. Even after the significant drop in the concentration in the post-lockdown phase, the ozone concentration was found within prescribed limits i.e., 100 $\mu g/m^3$ (Sahoo et al., 2021).

12.4.3 ITALY

In Italy, the first case of coronavirus was spotted on the 20 February 2020 in Codogno, and on the next day, the total number of cases was 34 in Lombardy, which provoked the administration to announce the region of the outbreak as red zone on 23 February 2020 (IMH, 2020; Lombardy Region, 2020).

Oxides of Nitrogen (NO_x): There has been a drastic decrease in the level of NO_2 around Italy when compared to the data of the previous three years of month February and March. The mean level of NO_2 in the second and third months of the past three years was about 50–60 $\mu g\ m^{-3}$ and 45–50 $\mu g\ m^{-3}$, respectively; whereas the concentration of NO during pre-lockdown and the total lockdown was measured to be nearby 31.9 ± 1.9 $\mu g\ m^{-3}$ and 22.1±1.2 $\mu g\ m^{-3}$, respectively. Generally, the major sources of the emission of NO_2 are traffic which contributes 68%-72%, household activities 10.4–18.7% and industrial process 7.6–14.9% to the total NO_2 emission (Cristina et al., 2020).

Particulate Matter (PM): The levels of PM during pre-lockdown and total lockdown has been significantly reduced. During the pre-lockdown scenario, the concentration of particulate matter has been decreased by 32.7% to 40.5% in the sub-areas of Italy (Cristina et al., 2020). The reduction in the pollutant has been reported to be directly linked to the restrictions on vehicular activities, as the transportation sector of Italy is responsible for the 40%–45% of total particulate matter emissions. The overall reduction in the percentage of pollutant emission during pre-lockdown and total lockdown is around 57.5% and 71% when compared to the air quality before the implementation of lockdown, respectively (Cristina et al., 2020). In Milan, the major decrease in the concentration of PM is attributed to traffic, combustion sources, industrial activities and household heating appliance (ARPA Lombardia, 2020).

Sulphur dioxide (SO_2): In the case of SO_2 emissions, a remarkable reduction in the concentrations were observed when comparison of before lockdown and pre-lockdown (19.9%) and a small decrease during the period of pre-lockdown and total lockdown periods (6.8%). There has been an overall drop in the SO_2 concentrations around 25.4% when compared to the before lockdown and total lockdown periods. The emissions from the power plants and some industrial processes are the major contributor of SO_2 in the atmosphere (INEMAR, 2017).

Ozone (O_3): The concentration of ozone possesses a remarkable increase in Milan city of Italy. The factors dependent on ozone concentrations are the density of solar radiation and the time duration of daylight. The crucial factor of growth in the daytime periods from 9.5 ± 0.2 h before the lockdown period to the 12 ± 0.2 h during the total lockdown scenarios has affected the overall ozone production (Cristina et al., 2020).

Carbon Monoxide (CO): The largest reduction in the levels of CO were measured at around 57.6% across Milan city during the total lockdown phase in comparison to the concentration of this contaminant before lockdown. The major origins of CO emissions are the incomplete combustion processes during vehicular and household activities (INEMAR, 2017). Before implementing lockdowns, the Milan city of Italy contributes 78% of total CO emissions which further drops by 74.6 and 55.5% in some parts of the city. An overall decrease in the CO emission during pre-lockdown and total lockdown has been observed around 1.4–3.5% (Cristina et al., 2020).

12.4.4 SPAIN

The nationwide lockdown in Spain was implemented on 14 March 2020 by restricting vehicle movements and industrial activities across the country.

Oxides of Nitrogen (NO_x): The analysis of the pollutant was observed during the month of March of the year 2020. In Madrid, the level of NO_2 fell in all the 23 stations by 53% in the first week of the lockdown compared to the second week; whereas Barcelona and its other metropolitan areas were examined for the decrease in the level of NO_2 and it was found to be reduced on average by 34% throughout the first week then the next week. The reduction in traffic had also affected the overall and maximum hourly values of NO_2. The reduction in these two cities could be due to the different dispersive capacities of both zones and meteorological conditions (Gonçalves et al., 2009). The area of Madrid is affected by inland surroundings; whereas Barcelona is situated near the coastal area which also affects the emissions pattern of NO_2 (Baldasano, 2020). NO_2 is generally generated from the oxidation of the NO by the O_3 and peroxyl radicals, where NO is produced from the emission from ignition processes. It has been reported that NO_x concentrations tend to fall much faster than the other pollutants due to their participation in the other reactions being carried out in the atmosphere (Dadashi et al., 2020).

Particulate Matter (PM): The lockdown period has successfully reduced particulate matter across Spain and all over the world (Briz-redón et al., 2021). Considerable reduction in the concentration of particulate matter has been observed in three major cities of Spain (Barcelona, Valencia, and Sevilla). The larger the cities, the higher the man-made actions due to the combustions of fossil fuel, transportation and industrial activities (Pengfei et al., 2020). Madrid is a large city, but the PM reduction was negligible compared to the other metropolitan cities. The reason behind this could be the longer settling period of the particulate matter which have been emitted in the larger amount before the lockdown (Xiao et al., 2018).

Carbon Monoxide (CO): The noteworthy reduction in the concentration of CO across Spain in the major cities such as Barcelona, Sevilla, and Santiago de Compostela are related to the already mentioned anthropogenic activities. However, there were a few cities in which the levels of the CO didn't change (Briz-redón et al., 2021).

Sulphur Dioxide (SO_2): The concentrations of SO_2 and ozone were also affected by the COVID-related lockdown across the major cities of Spain. The concentration of SO_2 was reduced due to the fewer emissions from the combustion processes. The reduction in SO_x and NO_x is beneficial for human health and the environment because these compounds have the characteristics to be transformed into H_2SO_4 and HNO_3 when comes into contact with rain (Gerhardsson et al., 1997).

Ozone (O_3): The increase in ozone production was associated with the reduction in the NO_2 and CO concentrations which were found in line with the previous studies (Tobías et al., 2020).

12.4.5 FRANCE

The first case in France was confirmed on 24 January 2020. The government of France announced the first lockdown on 12 March 2020, which resulted in the closure of crowded places.

Oxides of Nitrogen and Particulate Matter (NO_x & PM): During the containment phases in France, the concentrations of all the pollutants were decreased considerably besides ozone, which increases up to 27.19% during the lockdown phase and kept on increasing by 21.35% after the lockdown periods. The prime reason for the increase in the ozone concentration has been noticed due to the drop of the concentration of NO_x which ultimately lowers the O_3 titrations through the reactions with NO (Sicard et al., 2020); whereas the concentration of other pollutants was not kept constant afterwards the lockdown period for the pollutants such as NO_2, and PM_{10}. The levels of the pollutants such as NO_2 and PM_{10} got raised to 42.32% and 38.15% after the lockdown period, respectively (Ikhlasse et al., 2021).

Other Pollutants: The pollutants such as SO_2, NO_2 O_3, CO, $PM_{2.5}$ and PM_{10} measured throughout the lockdown periods represented that the lockdown has helped the atmosphere to improve its air quality. The significant improvement in the average value of pollutants during pre-lockdown, during lockdown and post-lockdown for SO_2;152.18 ($\mu g/m^3$), 124.52 ($\mu g/m^3$), 120.25 ($\mu g/m^3$), NO_2; 143.03 ($\mu g/m^3$), 89.9 ($\mu g/m^3$), 127. 96 ($\mu g/m^3$); O_3, 100.6 ($\mu g/m^3$), 127.95 ($\mu g/m^3$), 155.27 ($\mu g/m^3$), CO; 1.26, (mg/m^3), 1.00(mg/m^3), 0.79 (mg/m^3), $PM_{2.5}$; 101.17 ($\mu g/m^3$), 96.00 ($\mu g/m^3$), 59.59 ($\mu g/m^3$), PM_{10}; 237.34 ($\mu g/m^3$), 145.80 ($\mu g/m^3$), 201.43 ($\mu g/m^3$), respectively (Ikhlasse et al., 2021).

12.4.6 CHINA

The first lockdown in China has been enforced in Wuhan city on 23 January 2020, and other regions in Hubei considering the coronavirus outbreak.

Oxides of Nitrogen (NO_x): In China, with the implementation of COVID-related lockdowns, the concentration of NO_2 in 2020 was considerably less than its level in 2019. China had adopted extreme precautions and imposed the lockdowns which lead to shut down the works places and traffic in response to the spread of COVID-19 (Guojun et al., 2020).

Sulphur Dioxide (SO_2): The reduction in the SO_2 concentration in the atmosphere before and after the lockdown is linked to the continuous emissions from the power plants and water boiling facilities. From pre-lockdown to lockdown period, the densities of SO_2 and NO_2 declined, and it showed contrary trends from lockdown to post-lockdown, which implied that SO_2 and NO_2 were significantly reduced over the lockdown period together but then got increase after the lockdown time frame (Jiang et al., 2021).

Other Pollutants: The concentrations of SO_2, NO_2, CO, and PM_{10} have continued to fall since 2014 and have reduced at approximate rates of 13%, 6.3%, 5%, 5.6%, respectively (Wang et al., 2021). This is most likely to be linked with China's stringent national environmental protection policies and initiatives. In other words, before the exitance of lockdowns, the environmental policies help to reduce harmful emissions to the atmosphere. The decrease in the number of emissions from vehicles and industrial activities NO_2, CO, PM_{10}, and $PM2_{.5}$ are considerably greater amid lockdown and pre-lockdown periods (Chu et al., 2021). However, massive decreases have been observed during the shutting down period in 2020 than in 2019 and previous years which means that the lockdown procedures have caused a major impact on air quality (Jiang et al., 2021).

12.5 DISCUSSION AND RECOMMENDATIONS

The local, state, and central governments of most of the countries have prioritized the health of their citizens to prevent them from the infectious COVID-19 through providing door to door testing facilities, vaccination programmes, and other facilities. But the air condition during the post-lockdown scenario has got even worse compared to the air quality before the employment of the first lockdown due to excessive transportation and industrial activities for the production of products. Therefore, a few recommendations have been provided for the concerned authorities to take the measure according to the environmental conditions:

- Air quality around the most populated cities should be monitored from time to time.
- Transportation activities should be limited nearby the industries to avoid the greater emission of air pollutants. Products in vehicles should be filled up to their full capacities as it will lessen the requirement of the number of rounds required for the transportation of goods.
- Random checking of exhaust from vehicles and industries.
- Implementation of partial lockdowns to major parts of cities where the air quality is about to cross the permissible limits.
- Application of newspaper articles and social media for the awareness about the improvement in the air quality during the lockdowns should be published in order to coordinate and enhance the atmosphere health.

• Policymakers to build the pollution control measure according to the need of time through the anticipation of degradation in the air quality.

However, the lockdowns have brought risk to most countries' national economies, but the programmes for the refurbishment of the atmosphere should also progress hand in hand.

12.6 CONCLUSION

Quarantine actions to restrict the transmission of SARS-CoV-2 illness and the locking up of social, commercial, and business activities have been enacted in numerous countries. Because of the outbreak, the global energy pattern has shifted dramatically. Due to the shutdown of industry and transportation facilities, the interdiction of air transport, quarantine procedures and so on, energy consumption by various sectors has been decreased dramatically. The actions mentioned above have benefitted the environment by significantly improving air quality worldwide. This review found that it is evident that due to several efforts by the governments of many nations. Furthermore, thanks to COVID-19 related lockdowns, the emissions of PM_{10}, $PM_{2.5}$, CO, NO_x, and SO_2 were significantly reduced. The closure of industries and other energy sectors has caused a positive impact on the O_3 levels significantly. This improvement in air quality can be attributed partially to a decrease in heating systems as a result of the closure of industries. However, additional data are needed for a more precise assessment. WHO stated that this pandemic of COVID-19 would last for a long period, therefore, all the countries are working on preventative programmes, research, and development of vaccines for every age group. The pandemic conditions will also influence environmental sustainability on a long-term basis. The author anticipates that the data provided in the present chapter will trigger the researchers, policymakers, people to think for other alternate policies to handle the risks of deterioration of atmospheric characteristics.

REFERENCES

Baldasano, J. M. (2020). COVID-19 lockdown effects on air quality by NO 2 in the cities of Barcelona and Madrid (Spain). *Science of the Total Environment*, 741(2). https://doi.org/10.1016/j.scitotenv.2020.140353

Bera, B., Bhattacharjee, S., & Kumar, P. (2021). Significant impacts of COVID-19 lockdown on urban air pollution in Kolkata (India) and amelioration of environmental health. *Environment, Development and Sustainability*, 23(5), 6913–6940. https://doi.org/10.1007/s10668-020-00898-5

Briz-Redón, Á., Belenguer-Sapiña, C., & Serrano-Aroca, Á. (2021). Changes in air pollution during COVID-19 lockdown in Spain: A multi-city study. *Journal of Environmental Sciences*, 101, 16–26. https://doi.org/10.1016/j.jes.2020.07.029

Chu, B., Zhang, S., Liu, J., Ma, Q., & He, H. (2021). Significant concurrent decrease in PM2.5 and NO$_2$ concentrations in China during COVID-19 epidemic. *Journal of Environmental Sciences (China)*, 99(2), 346–353. https://doi.org/10.1016/j.jes.2020.06.031

Concejão Martins, L., De Oliveira Latorre, M. do R. D., Do Nascimento Saldiva, P. H., & Ferreira Braga, A. L. (2002). Air pollution and emergency room visits due to chronic lower respiratory diseases in the elderly: An ecological time-series study in São Paulo, Brazil. *Journal of Occupational and Environmental Medicine*, 44(7), 622–627. https://doi.org/10.1097/00043764-200207000-00006

Correia, G., Rodrigues, L., Gameiro da Silva, M., & Gonçalves, T. (2020). Airborne route and bad use of ventilation systems as non-negligible factors in SARS-CoV-2 transmission. *Medical Hypotheses, 141*(April), 109781. https://doi.org/10.1016/j.mehy.2020.109781

Cristina, M., Abbà, A., Bertanza, G., Pedrazzani, R., Ricciardi, P., & Carnevale, M. (2020). Lockdown for CoViD-2019 in Milan: What are the effects on air quality? *Science of the Total Environment, 732*(February), 139280. https://doi.org/10.1016/j. scitotenv.2020.139280

Dacre, H. F., Mortimer, A. H., & Neal, L. S. (2020). How have surface NO2concentrations changed as a result of the UK's COVID-19 travel restrictions? *Environmental Research Letters, 15*(10). https://doi.org/10.1088/1748-9326/abb6a2

Dadashi, M., Pages Farre, D., Hernandez, I., & Tauler, R. (2020). Chemometrics modeling of temporal changes of ozone half hourly concentrations in different monitoring stations. *Chemometrics and Intelligent Laboratory Systems, 201*(October 2019). https://doi.org/10.1016/j.chemolab.2020.104015

Donzelli, G., Cioni, L., Cancellieri, M., Morales, A. L., & Suárez-Varela, M. M. M. (2020). The effect of the covid-19 lockdown on air quality in three Italian medium-sized cities. *Atmosphere, 11*(10). https://doi.org/10.3390/atmos11101118

Gerhardsson, L., Skerfving, S., & Oskarsson, A. (1997). Effects of acid precipitation on the environment and on human health. *Advances in Environmental Control Technology: Health and Toxicology*, 355–364. https://doi.org/10.1016/b978-088415386-3/50017-5

Gonçalves, M., Jiménez-Guerrero, P., & Baldasano, J. M. (2009). Contribution of atmospheric processes affecting the dynamics of air pollution in Southwestern Europe during a typical summertime photochemical episode. *Atmospheric Chemistry and Physics, 9*(3), 849–864. https://doi.org/10.5194/acp-9-849-2009

Guojun, H., Pan, Y., & Tanaka, T. (2020). The short-term impacts of COVID-19 lockdown on urban air pollution in China. *Nature Sustainability, 3*(December). https://doi.org/10.1038/s41893-020-0581-y

He, H., Wang, Y., Ma, Q., Ma, J., Chu, B., Ji, D., Tang, G., Liu, C., Zhang, H., & Hao, J. (2014). Mineral dust and NOx promote the conversion of SO 2 to sulfate in heavy pollution days. *Scientific Reports, 4*(2), 1–6. https://doi.org/10.1038/srep04172

Huang, X., Ding, A., Gao, J., Zheng, B., Zhou, D., Qi, X., Tang, R., Wang, J., Ren, C., Nie, W., Chi, X., Xu, Z., Chen, L., Li, Y., Che, F., Pang, N., Wang, H., Tong, D., Qin, W., ... He, K. (2021). Enhanced secondary pollution offset reduction of primary emissions during COVID-19 lockdown in China. *National Science Review, 8*(2). https://doi.org/10.1093/nsr/nwaa137

Ikhlasse, H., Benjamin, D., & Vincent, C. (2021). Environmental impacts of pre / during and post – lockdown periods on prominent air pollutants in France. *Environment, Development and Sustainability, 0123456789*. https://doi.org/10.1007/s10668-021-01241-2

IMH. (2020). *Ordinance February, 23 2020 – Urgent Measures for the Containment and Management of the Epidemiological Emergency from COVID-19. Lombardy Region (in Italian)*. Italian Ministry of Health, Milan and Rome. https://www.gazzettaufficiale.it/eli/id/2020/02/25/20A01273/sg

INEMAR. (2017). *INEMAR (AIR Emissions Inventory) (in Italian)*. ARPA Lombardy.

Jiang, S., Zhao, C., & Hao, F. (2021). Toward understanding the variation of air quality based on a comprehensive analysis in Hebei province under the influence. *Atmosphere, 12*, 267. https://doi.org/10.3390/atmos12020267

ARPA Lombardia. (2020). *Preliminary Analysis of Air Quality in Lombardy during the COVID-19 Emergency (in Italian)*. Milan.

Lombardy Region. (2020). *Coronavirus – updated epidemiological data*. Lomb. Reg. https://www.regione.lombardia.it/wps/portal/istituzionale/HP/%0ADettaglioRedazionale/servizi-e-informazioni/cittadini/salute-e-prevenzione/coronavirus/coronavirus-piattaforma-dati-lombardia/coronavirus-piattaforma-dati-lombardia

Manigandan, S., Wu, M. T., Ponnusamy, V. K., Raghavendra, V. B., Pugazhendhi, A., & Brindhadevi, K. (2020). A systematic review on recent trends in transmission, diagnosis, prevention and imaging features of COVID-19. *Process Biochemistry, 98*(May), 233–240. https://doi.org/10.1016/j.procbio.2020.08.016

Marinello, S., Lolli, F., & Gamberini, R. (2021). The impact of the COVID-19 emergency on local vehicular traffic and its consequences for the environment: The case of the city of Reggio Emilia (Italy). *Sustainability (Switzerland), 13*(1), 1–22. https://doi.org/10.3390/su13010118

Munir, S., Gulnur, C., Jassim, M. S., Aina, Y. A., Ali, A., & Mayfiel, M. (2021). Changes in air quality associated with mobility trends and meteorological conditions during COVID-19 lockdown in. *Atmosphere Article, 12*, 1–26.

Pengfei, W., Chen, K., Zhu, S., Wang, P., & Zhang, H. (2020). Severe air pollution events not avoided by reduced anthropogenic activities during COVID-19 outbreak. *Resources, Conservation and Recycling, 158*(February), 104814. https://doi.org/10.1016/j.resconrec.2020.104814

Pikhart, H., Bobak, M., Gorynski, P., Wojtyniak, B., Danova, J., Celko, M. A., Kriz, B., Briggs, D., & Elliott, P. (2001). Outdoor sulphur dioxide and respiratory symptoms in Czech and Polish school children: A small-area study (SAVIAH). *International Archives of Occupational and Environmental Health, 74*(8), 574–578. https://doi.org/10.1007/s004200100266

Priva, U. C., & Sanker, C. (2019). Limitations of difference-in-difference for measuring convergence. *Laboratory Phonology, 10*(1), 1–29. https://doi.org/10.5334/labphon.200

Qian, Z., He, Q., Lin, H. M., Kong, L., Liao, D., Yang, N., Bentley, C. M., & Xu, S. (2007). Short-term effects of gaseous pollutants on cause-specific mortality in Wuhan, China. *Journal of the Air and Waste Management Association, 57*(7), 785–793. https://doi.org/10.3155/1047-3289.57.7.785

Ropkins, K., & Tate, J. E. (2021). Early observations on the impact of the COVID-19 lockdown on air quality trends across the UK. *Science of the Total Environment, 754*(January 2020), 142374. https://doi.org/10.1016/j.scitotenv.2020.142374

Roy, S., Saha, M., Dhar, B., Pandit, S., & Nasrin, R. (2021). Geospatial analysis of COVID-19 lockdown effects on air quality in the South and Southeast Asian region. *Science of the Total Environment, 756*, 144009. https://doi.org/10.1016/j.scitotenv.2020.144009

Sahoo, P. K., Mangla, S., Pathak, A. K., Salāmao, G. N., & Sarkar, D. (2021). Pre-to-post-lockdown impact on air quality and the role of environmental factors in spreading the COVID-19 cases – a study from a worst-hit state of India. *International Journal of Biometeorology*, 205–222.

Sahoo, S.K., Kumar, A. V., Yadav, A.K., & Tripathi, R.M. (2016). Metal characterization of airborne particulate matters in a coastal region. Toxicological & Environmental Chemistry, 98:7, 768–777, https://doi.org/10.1080/02772248.2016.1145222

Shi, Z., Song, C., Liu, B., Lu, G., Xu, J., Van Vu, T., Elliott, R. J. R., Li, W., Bloss, W. J., & Harrison, R. M. (2021). Abrupt but smaller than expected changes in surface air quality attributable to COVID-19 lockdowns. *Science Advances, 7*(3). https://doi.org/10.1126/sciadv.abd6696

Sicard, P., De Marco, A., Agathokleous, E., Feng, Z., Xu, X., Paoletti, E., Rodriguez, J. J. D., & Calatayud, V. (2020). Amplified ozone pollution in cities during the COVID-19 lockdown. *Science of the Total Environment, 735*, 139542. https://doi.org/10.1016/j.scitotenv.2020.139542

Tobías, A., Carnerero, C., Reche, C., Massagué, J., Via, M., Minguillón, M. C., Alastuey, A., & Querol, X. (2020). Changes in air quality during the lockdown in Barcelona (Spain) one month into the SARS-CoV-2 epidemic. *Science of the Total Environment, 726*, 138540. https://doi.org/10.1016/j.scitotenv.2020.138540

Tosepu, R., Gunawan, J., Effendy, D. S., Ahmad, L. O. A. I., Lestari, H., Bahar, H., & Asfian, P. (2020). Correlation between weather and Covid-19 pandemic in Jakarta, Indonesia. *Science of the Total Environment, 725*, 138436. https://doi.org/10.1016/j.scitotenv.2020.138436

Unhale, S. S., Ansar, Q. B., Sanap, S., Thakhre, S., & Wadatkar, S. (2020). A review on corona virus (Covid-19). *World Journal of Pharmaceutical*, 6(4), 109–115.

Wang, Q., Juan, J., Xiao, T., Zhang, J., Chen, H., Song, X., & Chen, M. (2021). The physical structure of compost and C and N utilization during composting and mushroom growth in Agaricus bisporus cultivation with rice, wheat, and reed straw-based composts. *Environmental Biotechnology*, (Nbsprc 2019).

Worldometer. (2021). *COVID-19 Cases*. https://www.worldometers.info/coronavirus/?fbclid= IwAR35ZFiRZJ8tyBCwazX2N-k7yJjZOLDQiZSA_MsJAfdK74s8f2a_Dgx4iVk

Xiao, Y., Murray, J., & Lenzen, M. (2018). International trade linked with disease burden from airborne particulate pollution. *Resources, Conservation and Recycling*, *129*(May 2017), 1–11. https://doi.org/10.1016/j.resconrec.2017.10.002

Yadav, A. K., & Hopke, P. K. (2020). Characterization of radionuclide activity concentrations and lifetime cancer risk due to particulate matter in the Singrauli Coalfield, India. *Environmental Monitoring and Assessment*, *192*(11). https://doi.org/10.1007/ s10661-020-08619-1

Yadav, A. K., & Jamal, A. (2018a). Impact of mining on human health in and around mines. *Environmental Quality Management*, *28*(1), 83–87. https://doi.org/10.1002/tqem.21568

Yadav, A. K., & Jamal, A. (2018b). Suspended particulate matter and its management system surrounding opencast coal mines. *Environmental Quality Management*, *28*(2), 123–128. https://doi.org/10.1002/tqem.21592

Yadav, A. K., & Jamal, A. (2018c). Technological reviews of particulate matter and their source identification techniques. *Environmental Quality Management*, *27*(4), 87–95. https://doi.org/10.1002/tqem.21554

Yadav, A. K., Sahoo, S. K., Patra, A. C., Dubey, J. S., Lenka, P., Sagar, D. V., Kumar, A. V., & Tripathi, R. M. (2014). Source identification of particulate matter and associated intake of elements through inhalation in an industrial area of Odisha, India. *Toxicological and Environmental Chemistry*, *96*(3), 410–425. https://doi.org/10.1080/02772248.2014.94 3225

13 Impact of Lockdown on Air Quality during COVID-19 Outbreak
A Global Scenario

Anant Patel, Neha Keriwala, Prutha Patel, and Arohi Singh

CONTENTS

13.1 INTRODUCTION

The WHO estimates that poor air quality kills 7 million people annually (WHO 2014; Agarwal et al., 2021). It is caused by pollutants like PM_{10}, $PM_{2.5}$, NO_2, SO_2, CO, and O_3 from mining, manufacturing, transportation, home cooking, biomass, and garbage burning (Ahmad and Ahmad, 2020; Yadav and Jamal, 2018). Barnes et al. (2019) claim that persons who reside near major roadways are more susceptible to pollution. Natural and man-made disasters negatively impact human and environmental wellbeing.

Air pollution is the discharge of pollutants into the atmosphere that harms human and environmental health. Nearly nine out of ten people share the same air, with poorer nations suffering from pollution levels above WHO safety requirements (Yadav et al., 2013). Air is necessary for Earth's survival and development. It has a huge influence on the country's health and economic prosperity. People are growing more worried about declining air quality due to increased pollution from industrialization, private automobiles, and fossil fuel consumption. As a result, pollutants such sulphur dioxide, nitrogen dioxide, carbon dioxide, nitric oxide, carbon monoxide, and $PM_{2.5}$ are prevalent in the atmosphere (Pandya et al., 2017; Yadav et al., 2019). Air pollution is the presence of toxic compounds in the atmosphere that impair human health, the environment, or materials. Some contaminants are gases, particles, and biological. Air pollution may harm people's health, plants and animals, and even the built environment.

Poor air quality affects respiratory and cardiovascular health. Both urban and interior air pollution are major global pollution issues (Patel, 2020). Around 90% of the world's population is affected by pollution. A chemical species that has greater quantities or qualities than the naturally occurring components of air is an air pollution. However, a pollutant is a substance that has the potential to harm human, animal, or plant health, or ecological systems.

13.1.1 GLOBAL OVERVIEW OF COVID-19 PANDEMIC

The first COVID-19 pandemic, SARS-CoV-2, was harmful to human health and negatively impacted many aspects of life (Brodawka et al., 2021). WHO was the first worldwide public health organization to investigate the virus's origins. The virus started spreading a few months before it was found in December 2019 (Cole et al., 2020). The US, UK, and South Korea have criticized the report's lateness, and say that the WHO did not have access to full or original Chinese data and samples (Garg et al., 2021).

As of 13 June 2021, the Centers for Disease Control and Prevention estimated global COVID-19 cases and fatalities at 175 million and 3,792,777, respectively. Initially, the pandemic hit Wuhan, but now, India, US and Europe have the most instances (Othman and Talib, 2021). To counteract the pandemic, several governments have imposed restrictions. By the end of March 2020, over 90% of the population was predicted to be homebound. Many states cancelled public events, closed schools, restaurants, and other entertainment venues. Decreased case counts led to a substantial spike in new coronavirus infections globally in late October/ early November 2020 (Patel and Chitins, 2021). This may be due to a state-wide vaccination drive that saw a significant quantity of COVID-19 vaccines supplied and given. The incidence of new coronavirus infections has grown dramatically since March 2021 (Figure 13.1).

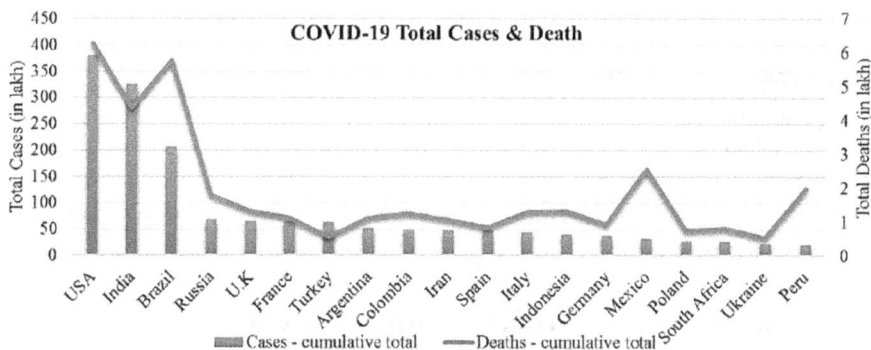

FIGURE 13.1   COVID-19 cumulative cases and death by country up to 29 Aug 2021. Lakh = 100,000.

13.2 AIR POLLUTION MODELING

It is possible to predict the environmental effects of contaminants. For most applications, an ideal model should be able to reliably predict pollutant concentration variations in both space and time. Air quality modeling is crucial in developing air pollution control and management strategies, since it may provide suggestions for better air quality planning. Measuring advancements must constantly accompany model development. Models are useful when a government needs to determine who is responsible for what proportion of a receptor's concentration. Regulators may also utilize models to forecast future concentration changes.

13.2.1 DIFFERENT MODELS USED FOR ANALYSIS USING COMPUTATIONAL APPROACH

For each species, the independent variables are related to the concentration. Modeling assesses current and future air quality, allowing policymakers to make "informed" decisions. Scale, time frame (hours, days, months, or years), and contaminant of concern are all equally important. While air pollution models are classified based on their structure and technique. There are deterministic mathematical (analytical and numerical), statistical, and physical models. These models also need a lot of data and computational resources. Among the various model groupings are: I. Dispersion models; II. Receptor models; III. Stochastic models; and IV. Box models.

13.2.2 RECENT MODELING TECHNIQUES AND TRENDS IN STATISTICAL MODELING TOOLS

Regression, multiple regression, and time series analysis are all mathematical methods used to anticipate urban air pollution levels. After establishing the essential correlations between these components, the pollutant concentrations may be determined. There are several limitations to statistical modeling, including the lack of long-term historical data, physical interpretation issues, and site-specific data. As cities struggle with air pollution, more experts are turning to ANN and FLT to forecast how long it will take to clear up the mess. In contrast to traditional modeling approaches, ANNs offer several advantages due to their absence of assumptions about the input data set. Computational models are appealing in applications where little or no understanding of the problem exists, but massive volumes of training data are accessible. The neural network creates a model and predicts using it. Based on the research, the structure of multi-perceptron neural networks seemed to be the most suitable for usage in atmospheric sciences.

13.3 METHODOLOGY FOR AIR QUALITY INDEX

Governments utilize AQI to inform the public about current and potential air pollution levels. As the AQI rises, so do the risks to public health. While different countries use different Air Quality Indices to measure air quality. Various countries have devised their own air quality indices to help the public understand pollution levels.

Each country's air quality index corresponds to its specific standards. In 1968, the newly formed National Air Pollution Control Administration produced an air quality index to assist monitor air quality in cities. For the technique's development and collection of air quality and emissions data necessary to test and calibrate the indices. The initial version of the air quality index adjusted pollutant concentrations in the air to yield various pollutant indices. This overall air quality index was weighted and summed. The whole strategy might employ concentrations predicted by a diffusion model or received from ambient monitoring data. The concentrations were converted into a standard statistical distribution with a mean and standard deviation. Individual pollutant indices may be computed using values other than unity. It may also contain any number of contaminants. Figure 13.2 shows a flowchart for analyzing changes in air quality and the various models employed by different nations.

13.3.1 AIR QUALITY INDICES AND AQI MODEL FOR INDIA

The AQI is a pollution control board-set indicator that measures current air quality. It measures CO, NO_X, PM_{10}, SO_2, and $PM_{2.5}$ pollutants in the atmosphere. The AQI is an integer from 0 to 500, with 500 being bad air quality. As with the daily air pollution levels, the citywide mean concentrations of AQI and five other contaminants were obtained by averaging values across all locations. Finally, the AQI and five air contaminants were averaged daily and annually. The AQI and five air pollutants, including daily mean CO, NO_X, PM_{10}, SO_2, and $PM_{2.5}$ values from 2020 to 2021. According to the 'linear segmentation concept', the sub-index (Ip) for a given pollutant concentration (Cp) is computed as follows:

$$Ip = \left\{ \frac{(IHI - ILO)}{(BHI - BLO)} \times (Cp - BLO) \right\} + ILO \qquad (13.6)$$

Where: BHI indicates Breakpoint concentration that is higher than or equal to the specified concentration, BLO indicates a breakpoint concentration that is less than

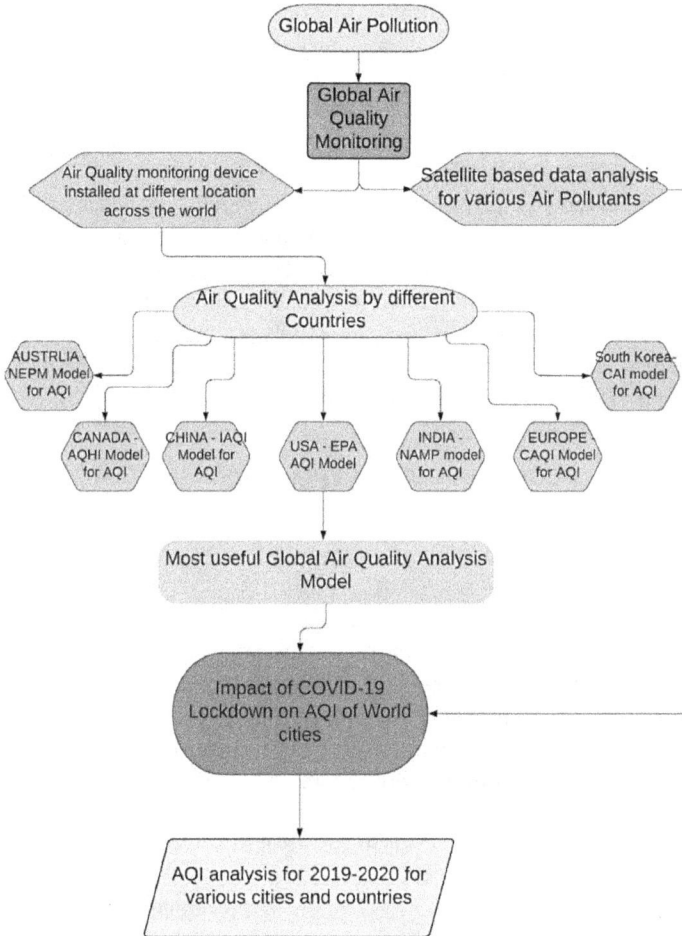

FIGURE 13.2   Methodology for the air quality analysis during COVID-19 Global lockdown.

or equal to the specified concentration, IHI indicates AQI value that corresponds to BHI, ILO indicates AQI w.r.t BLO; if ILO is higher than 50, deduct one from ILO, AQI indicates Maximum (Ip) (where p = 1, 2, 3, 4, 5, ..., n indicates the number of contaminants).

13.4   RESULTS: IMPACT OF LOCKDOWN ON GLOBAL AIR QUALITY

The world is still learning about the dangers of COVID-19 infection, a hidden threat to our respiratory and cardiovascular health. Gleichzeitig, the pandemic has brought attention to another global airborne public health issue, air pollution, which is currently being ignored in many parts of the world (Ambade et al., 2021). While COVID-19's effects may be seen in weeks, the health effects of air pollution may take years to manifest as chronic disease. The dramatic changes caused by economic downturns have highlighted the human contribution to air pollution. However, the COVID-19 problem has provided an unexpected opportunity to learn more about how to properly combat air pollution. Diverse ongoing initiatives will be needed to promote public awareness of the health hazards of air pollution, identify necessary legal changes, and monitor progress. Some pollutants like nitrogen dioxide (NO_2) have decreased significantly while others like $PM_{2.5}$ have decreased somewhat (Claeys et al., 2020; El-Sayed et al., 2021). At the same time, ozone levels seem to have increased owing to reduced NO_2 levels and climate changes (Balasubramaniam et al., 2020). As evidence from numerous countries shows, these changes are temporary. Loosening limitations boosted emissions, erasing any gain in air quality. COVID-19 has only offered a temporary respite from air pollution since the most important health impacts of air pollution occur over time.

13.4.1 IMPACT OF LOCKDOWN ON AIR QUALITY OF ASIAN COUNTRIES

The COVID-19 pandemic originated in Asia, in the city of Wuhan in the province of Hubei, China, and has spread throughout the continent. Mostly every nation in Asia has at least one positive case of COVID-19 as of June 2021. India, China, Pakistan Turkey, Iran, and Indonesia are the Asian nations with the most confirmed corona virus infections. Despite being the first region of the globe to be affected by the pandemic, several Asian countries, including China, India, Bhutan, Singapore, Taiwan, and Vietnam, have fared rather well because of their quick and widespread reaction. China has been chastised for downplaying the seriousness of the pandemic at first, but its massive response has effectively suppressed the sickness since March 2020.

13.4.1.1 Impact of Lockdown on Air Quality of India

According to the 2019 WHO study, India has 21 of the world's 30 most polluted cities. According to statistics from a world-renowned air quality website, this moved India up to sixth overall. The US AQI score was 152, and the $PM_{2.5}$ level was 58.08g/m^3 (Kabiraj and Vyankat, 2020). This is up from 72.54g/m^3 last year. This indicates that the majority of the country is ill. Cars contribute for 27% of pollution, agriculture consumption for 17%, and household cooking for 7%. More than 2 million Indians die prematurely due to air pollution (Sahoo et al., 2021).

Mumbai, Ahmedabad, Chennai, Patna, Kolkata, Bengaluru, Indore, and Kanpur all improved. In 2020, all cities except Kolkata had lower SO_2 levels than in 2019. It is important to note that the annual average population weighted $PM_{2.5}$ concentrations do not represent daily or seasonal variations in concentrations near cities or big pollution sources. As the GBD study focuses on long-term exposures, rather than short-term exposure spikes, which may have health effects. Cities in the Delhi-NCR area had similar declines, perhaps due to sharing an airshed. Except for NO_2 levels in Gurugram during lockdown Phase I and SO_2 levels during lockdown Phases I and II, all pollutants (PM_{10}, $PM_{2.5}$, NO_2, SO_2, Benzene) were lower in Delhi and four NCR cities in 2020 than in 2019.

These sources contributed to regional $PM_{2.5}$ and maybe reached downwind urban areas through vaporization. Except for Chennai and Patna, most cities had considerable NO_2 decreases in 2020 compared to 2019. Except for Mumbai, other cities had lower $PM_{2.5}$ readings (Singh and Chauhan, 2020). The analysis of satellite derived $PM_{2.5}$ revealed high local and public consumption of fuel and agricultural products, resulting in elevated $PM_{2.5}$ levels in certain areas, which may be transported to downwind urban zones through optional airborne development (Sreekanth et al., 2021). Figure 13.3 shows the air pollutants and AQI for Ahmedabad and Delhi cities from March 2020 to May 2021.

13.4.1.2   Impact of Lockdown on Air Quality of China

Several pneumonia cases were detected in late December in Wuhan. Most of these instances involved the Wuhan Seafood Market. On 23 January 2020, the COVID-19 lockdown in Wuhan began, with strict home quarantining. On 12 February 2020, thousands more instances were confirmed in Wuhan owing to an improvement in the

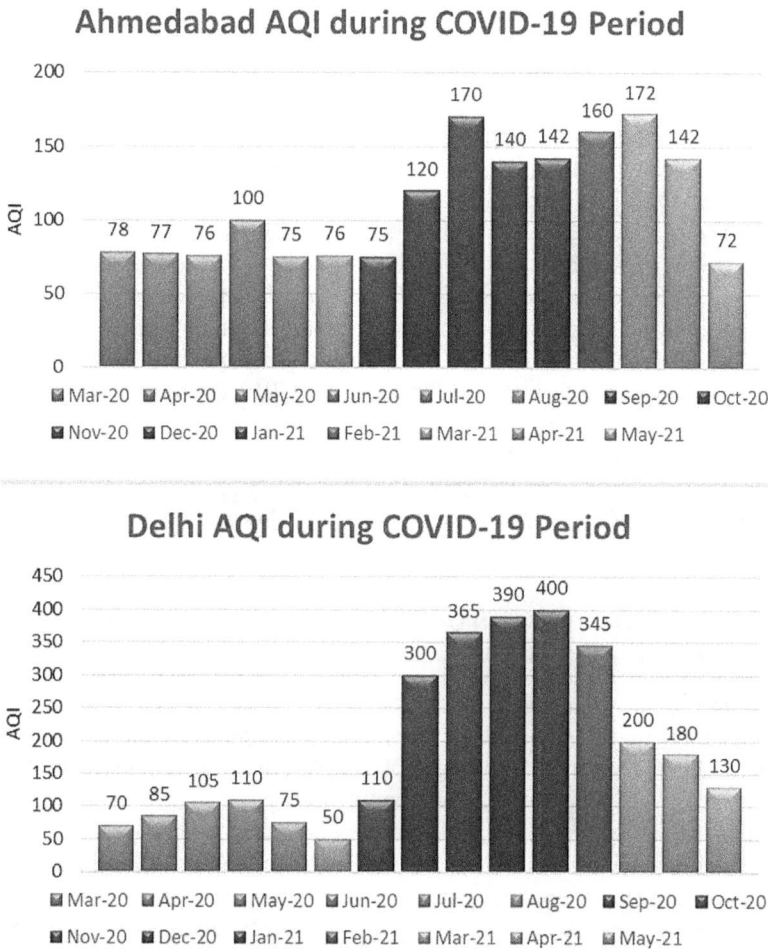

FIGURE 13.3   Average monthly value of air quality parameters from pre lockdown to post lockdown phase for Ahmedabad and Delhi.

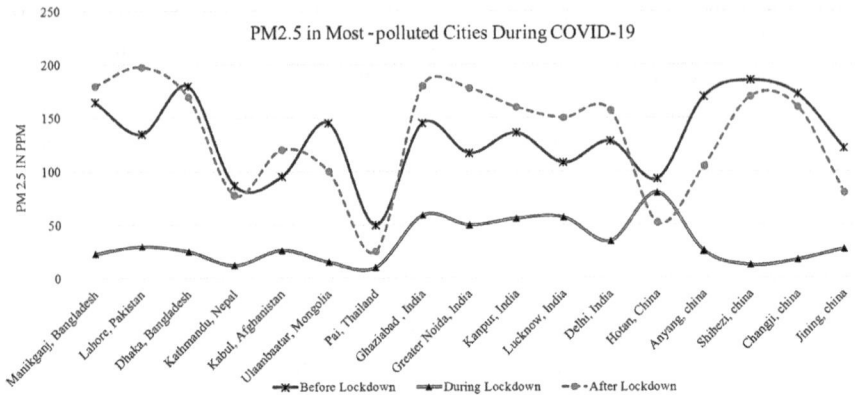

FIGURE 13.4   Global most-polluted cities $PM_{2.5}$ during pre-lockdown, lockdown and post-lockdown period.

diagnostic technique. First no new local corona virus COVID-19 transmissions after quarantine measures were imposed on March 18, 2020. After that date, the NHC will start releasing the number of symptom-free patients who tested positive for the corona virus. On 17 April 2020, Wuhan's health authorities revised their death toll by 50%. By 30 May 2021, the novel coronavirus SARS-CoV-2 had infected around 103,000 people and killed 4,846 people in China.

Overall, 74.9% of US participants, 70% of non-US participants, and 72.5% of global participants utilized social media for non-business objectives (Wang et al., 2020). In 2018–19, more than 99% of the population may report their air quality is "excellent" or "good". The AQI in January 2018–2019 was lower than in January 2017–2018.

Figure 13.4 shows the total $PM_{2.5}$ concentration in Bangladesh, Pakistan, Nepal, Afghanistan, Mongolia, and Thailand. The data clearly shows low $PM_{2.5}$ levels and high air quality during lockdown. After the lockdown, $PM_{2.5}$ levels climbed in Manikanj, Lahore, and Kabul.

13.4.2   IMPACT OF LOCKDOWN ON AIR QUALITY IN UNITED STATES OF AMERICA (USA)

In response to the pandemic, US flights were curtailed but preparations for the healthcare system, additional travel restrictions, and testing were delayed. WHO found 33,132,301 confirmed cases of COVID-19 in the United States, with 594,495 deaths starting on 3 January 2020 and ending at 5:46 pm CEST on 14 June 2021. Although the population, economy, and energy consumption have all increased over the previous five decades, the Clean Air Act has helped to decrease the levels of fine particulate matter to acceptable levels. Despite efforts over the long term in lowering pollution levels, there was a 5.5% rise in $PM_{2.5}$ levels between 2016 and 2018 (Han et al., 2021). Air pollution levels are going up, since there is an increased dependency on fossil fuels, a rise in frequency and severity of wildfires, and the complete absence of enforcement of the Clean Air Act (Ghosal and Saha, 2021).

In particular, metropolitan regions in the western US have the greatest $PM_{2.5}$ concentrations, and data suggests that low-income and persons of colour are more exposed to $PM_{2.5}$. In 2020, 38% of the cities in the database in the US failed to reach the WHO's annual mean $PM_{2.5}$ concentration limit of 10 g/m³ (Perera et al., 2021). Despite limits imposed to combat the COVID-19 pandemic, which resulted in short-term $PM_{2.5}$ reductions of 10–30%, this spike in $PM_{2.5}$ occurred. Despite the fact that automobile traffic was cut by almost 40% as a result of COVID-19 restrictions, air pollution levels were quite comparable to prior records (Zangari et al., 2020; Liu et al., 2021). This revealed that industries, refineries, power plants, and heavy-duty vehicles continue to be the primary sources of pollution. During 2020, millions of automobiles were no longer utilized on a regular basis, and the air became visibly cleaner, however the amount of pollutants suspended in that "clean" air remained the same. Readings were gathered from over 900 monitoring locations around the US, and it was discovered that ozone levels were around 15% lower than at similar periods in prior years (Pereira and De Mello, 2021; Li, 2021). This finding alone shows that lowering exhaust emissions will not be sufficient to reduce ozone levels significantly.

13.4.3 PROPOSED FRAMEWORK FOR THE AIR POLLUTION MONITORING AND MODELING

A framework system is proposed in Figure 13.5. The primary components of the framework are: A network of sophisticated sensors for adaptive environmental monitoring that automatically collects timely and high-quality environmental observations at particular geographical and temporal scales. Environmental observations and predicted environmental conditions are examples of vast volumes of spatio-temporal

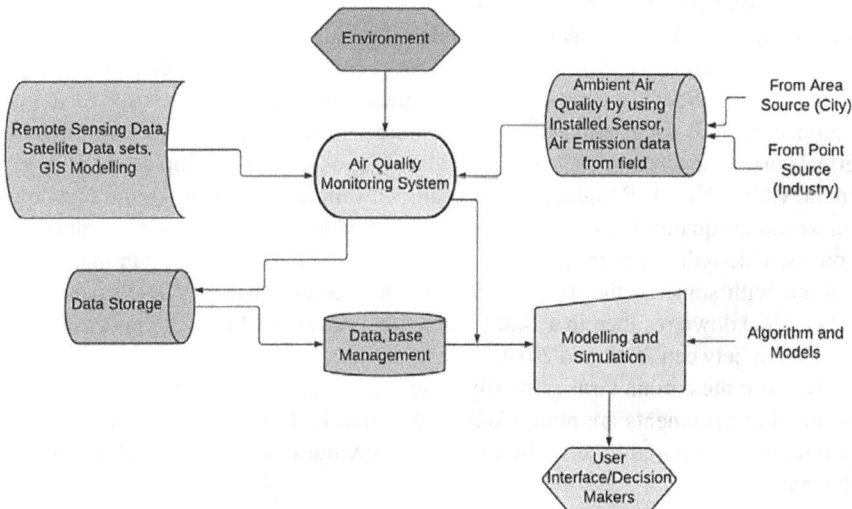

FIGURE 13.5   Proposed framework for the Air Pollution monitoring and modeling system.

data that may be efficiently stored, queried, and reported. Spatio-temporal data fusion, information extraction, knowledge discovery, and decision support are all possible. Fourier spatial data fusion, information extraction, knowledge discovery, and decision support need complex and durable algorithms and models. EnviroQuery is a set of customizable and user-friendly multimodal user interfaces for efficient data access and reporting. It also supports additional framework components including the sensor network, geographical databases, analytical and modeling engines, and user interfaces.

To monitor and manage pollution issues in cities, geoinformatics tools must be used to collect and share environmental data. This method may assist identify major sources of air pollution in a city, their particular contributions to pollutant concentrations, and how these contributions change by location. This method is expected to enable environmental managers in analyzing emissions, monitoring air pollutants, understanding dispersion and anticipating pollutants' consequences. Ultimately, this method is designed to help management avoid the worst pollutants and superfluous red tape. Participatory decision-making requires environmental modeling and information exchange. A new synthesis of important communication and information technologies, as well as geographical information sciences, must be applied to execute the framework (Liu et al., 2007). This framework could help in many ways, including identifying the most significant pollution sources, calculating each pollutant source's overall contribution to pollution concentrations at various locations, understanding and predicting pollution transport and exposure pathways, and quantifying pollution sources and exposure.

13.5 SUMMARY OF GLOBAL AIR QUALITY DURING COVID-19 LOCKDOWN

Globally, humans are responsible for 33% of COVID-19 fatalities. Human-caused air pollution deaths may have been avoided if pollution levels were reduced. While improved security, behaviour, and economic conditions led in cleaner air, new viruses like SARS-CoV-2 arose. Several studies show that people exposed to air pollution, especially long-term exposure, are more sensitive to COVID-19's detrimental effects on the respiratory and cardiovascular systems. In densely populated areas like India, China, Nepal, Bangladesh, Pakistan, Kuwait, and Afghanistan the lockdown improved air quality but disrupted the lives of hundreds of millions. Our data show considerable reductions in air pollutants during lockdown, especially in most Asian nations with some of the world's most polluted cities. The COVID-19 pandemic-related lockdown resulted in a 14.8% reduction in overall $PM_{2.5}$ readings throughout the region between 2018 and 2019.

Because the corona virus primarily attacks the respiratory system, those living in polluted environments are more likely to die from it. The majority of India's urban population is exposed to air pollution for lengthy durations, putting older people at danger.

13.6   CONCLUSION

The coronavirus pandemic has shifted the global energy pattern. Global lock-down protocols have been devised to cope with the present situation. The usage of energy by many sectors has been significantly reduced due to factory closures, flight bans, quarantine procedures, and other limitations. Alternatively, the above actions benefitted the environment by considerably improving global air quality. The virus propagated based on humidity and temperature. This research shows that several efforts taken by governments throughout the world have considerably reduced the quantity of pollutants in the air. COVID-19 processes also reduced PM_{10}, $PM_{2.5}$, CO, NO_2, SO_2, and O_3 emissions. Shutting down industries and other energy sources raises O_3 levels. Satellite images were utilized to monitor pollution levels during the lockdown. While the shutdown was good for the environment, several nations' economy suffered. COVID-19 became a pandemic in March 2020, causing severe economic losses and behavioural shifts. As a consequence, the lockdown improved worldwide air quality. Despite the costs, management, and control measures were implemented to prevent COVID-19 spread. Policymakers, on the other hand, are particularly worried about the cost and benefits of reducing emissions. Thus, future study should examine the cost-effectiveness of air pollution control techniques.

ACKNOWLEDGEMENTS

Authors are thankful to Civil Engineering Department, Institute of Technology, Nirma University for providing an opportunity to do research work. Authors are also thankful to Central Pollution Control Board (CPCB) and MoEF, Govt of India for their valuable support in data provision as well as guidance in this project. We would also like to extend our gratitude to all those who have directly and indirectly funnelled us in this research work.

REFERENCES

Agarwal, N., Swaroop, C., Raj, B. P., & Saini, L. (2021). Indoor air quality improvement in COVID-19 pandemic: Review. *Sustainable Cities and Society*, 70(April), 102942.

Ahmad, E., & Ahmad, E. (2020). Multilevel responses to risks, shocks and pandemics: lessons from the evolving Chinese governance model lessons from the evolving Chinese governance model. *Journal of Chinese Governance*, 0(0), 1–29.

Ambade, B., Kumar, T., Amit, S., Alok, K., & Gautam, S. (2021). COVID-19 lockdowns reduce the Black carbon and polycyclic aromatic hydrocarbons of the Asian atmosphere: source apportionment and health hazard evaluation. *Environment, Development and Sustainability.*

Balasubramaniam, D., Kanmanipappa, C., & Saravanan, M. (2020). Environmental effects assessing the impact of lockdown in US, Italy and France – What are the changes in air quality? *Energy Sources, Part A: Recovery, Utilization, and Environmental Effects*, 00(00), 1–11.

Barnes, J. H., Chatterton, T. J., & Longhurst, J. W. S. (2019). Emissions vs exposure: Increasing injustice from road traffic- related air pollution in the United Kingdom. *Transportation Research Part D*, 73, 56–66.

Brodawka, E., Korzeniewska, A., Szczurowski, J., & Zar, K. (2021). LCA and economic study on the local oxygen supply in Central Europe during the COVID-19 pandemic, *Science of the Total Environment*, 786.

Claeys, M. J., Argacha, J., Collart, P., Carlier, M., Caenegem, O. Van, Sinnaeve, P. R., Hanet, C. (2020). Impact of COVID-19-related public containment measures on the ST elevation myocardial infarction pandemic in Belgium: a nationwide, serial, cross-sectional study Impact of COVID-19-related public containment measures on the ST. *Acta Cardiologica*, 0(0), 1–7.

Cole, M. A., Elliott, R. J. R., & Liu, B. (2020). The Impact of the Wuhan Covid-19 Lockdown on Air Pollution and Health: A Machine Learning and Augmented Synthetic Control Approach. *Environmental and Resource Economics*, 76(4), 553–580.

Collett, R.S. and Oduyemi, K. (1997), Air quality modeling: a technical review of mathematical approaches. *Met. Apps*, 4: 235–246.

El-Sayed, M. M. H., Elshorbany, Y. F., & Koehler, K. (2021). On the impact of the COVID-19 pandemic on air quality in Florida, 285, *Environmental Pollution*, 117451.

Garg, A., Kumar, A., & Gupta, N. C. (2021). Impact of lockdown on ambient air quality in COVID-19 affected hotspot cities of India: Need to readdress air pollution mitigation policies impact of lockdown on ambient air quality in. *Environmental Claims Journal*, 33(1), 65–76.

Ghosal, R., & Saha, E. (2021). Impact of the COVID-19 induced lockdown measures on PM 2.5 concentration in USA, *Atmospheric Environment*, 254.

Han, L., Zhao, J., & Gu, Z. (2021). Assessing air quality changes in heavily polluted cities during the COVID-19 pandemic: A case study in Xi an, China. *Sustainable Cities and Society*, 70(April), 102934.

Kabiraj, S., & Vyankat, N. (2020). Impact of SARS – CoV – 2 pandemic lockdown on air quality using satellite imagery with ground station monitoring data in most polluted city Kolkata, India. *Aerosol Science and Engineering*, 4(4), 320–330.

Liu, Q., Harris, J. T., Chiu, L. S., Sun, D., Houser, P. R., Yu, M., Yang, C. (2021). Spatiotemporal impacts of COVID-19 on air pollution in California, USA. *Science of the Total Environment*, 750, 141592.

Othman, M., & Talib, M. (2021). Air pollution impacts from COVID-19 pandemic control strategies in. *Journal of Cleaner Production*, 291, 125992.

Patel A., Chitnis K. (2021) Application of fuzzy logic in river water quality modeling for analysis of industrialization and climate change impact on Sabarmati river. Water Supply 2021; ws2021275.

Patel A. (2020) Rainfall-Runoff Modeling and Simulation Using Remote Sensing and Hydrological Model for Banas River, Gujarat, India. Advances in Water Resources Engineering and Management. Lecture Notes in Civil Engineering, vol. 39. Springer, Singapore.

Pandya, U., Patel, A., & Patel, D. (2017). River cross section delineation from the Google Earth for development of 1D HECRAS model–A case of Sabarmati River, Gujarat, India. In *International Conference on Hydraulics, Water Resources & Coastal Engineering*, Ahmedabad, India.

Pereira, S., & Mello, C. B. S. De. (2021). Efficiency evaluation of Brazilian airlines operations considering the Covid-19 outbreak, *Journal of Air Transport Management*, 91(May 2020).

Perera, F., Berberian, A., Cooley, D., Shenaut, E., Olmstead, H., Ross, Z., & Matte, T. (2021). Potential health benefits of sustained air quality improvements in New York City: A simulation based on air pollution levels during the COVID-19 shutdown. *Environmental Research*, 193, 110555.

Sahoo, P. K., Chauhan, A. K., Mangla, S., Kumar, A., & Garg, V. K. (2021). COVID-19 pandemic: An outlook on its impact on air quality and its association with environmental variables in major cities of Punjab and Chandigarh, India. *Environmental Forensics*, 22(1–2), 143–154.

Singh, R. P., & Chauhan, A. (2020). Impact of lockdown on air quality in India during COVID-19 pandemic, *Air Quality, Atmosphere & Health*, 13 (921–928) (July).

Sreekanth, V., Kushwaha, M., Kulkarni, P., & Upadhya, A. R. (2021). Impact of COVID-19 lockdown on the fine particulate matter concentration levels: Results from Bengaluru megacity, India. *Advances in Space Research*, 67(7), 2140–2150.

Wang, L., Li, M., Yu, S., Chen, X., Li, Z., Zhang, Y., Xia, Y. (2020). Unexpected rise of ozone in urban and rural areas, and sulphur dioxide in rural areas during the coronavirus city lockdown in Hangzhou, China: Implications for air quality. *Environmental Chemistry Letters*, 18(5), 1713–1723.

Yadav, A.K., Jamal, A. (2018) Technological reviews of particulate matter and their source identification techniques, Environmental Quality Management, 27(4), 87–95

Yadav, A.K., Sahoo, S.K., Dubey, J.S., Kumar, A.V., Pandey, G., Tripathi, R.M. (2019) Assessment of particulate matter, metals of toxicological concentration, and health risk around a mining area, Odisha, India, Air Qual Atmos Hlth, 12(7), 775–783

Yadav, A.K., Sahoo, S.K., Kumar, A.V., Pandey, G. (2013) Spatial and temporal variation of particulate matter with height in residential and sand mining areas in Ganjam district of Odisha, India. International Research Journal of Environment Sciences, 2(12), 19–24

Zangari, S., Hill, D. T., Charette, A. T., & Mirowsky, J. E. (2020). Science of the Total Environment Air quality changes in New York City during the COVID-19 pandemic. *Science of the Total Environment*, 742, 140496.

14 Integration of Geospatial Techniques in Environment Monitoring Systems

S. Sreedevi, Rakesh Kumar Sinha, and T. I. Eldho

CONTENTS

DOI: 10.1201/9781003203445-17

14.1   INTRODUCTION

The environment which constitutes air, water, land, or vegetation is continuously subjected to changes due to human activities or natural processes. For the management of environment in a sustainable way and for policy making, its spatio-temporal changes are to be monitored closely. Water resources forms an integral component of the environment and is in a great threat with regard to diminishing quality/quantity. Management of water resources in the wake of shortages, surpluses and resource impairment (due to droughts, floods, or pollution) is challenging.

Modeling of environmental processes such as hydrologic processes at watershed level is thus imperative for efficient water resources management. Hydrologic models ranging from "lumped to distributed" or "conceptual to physics-based" are generally applied for this purpose. In the recent times, the immense capability of GIS in handling spatial geographic datasets has led to integration of GIS with hydrologic modeling. Distributed hydrologic models represent the spatial heterogeneity in terms of terrain characteristics, soil, land use/cover, precipitation, and meteorological parameters at a fine spatial resolution. These models provide better understanding of hydrologic processes within a catchment and also captures the impacts associated with land use and climate change. For this, extensive data such as geographical, topographical, and meteorological are required. As field level data collection is too cumbersome, time consuming, and expensive, the remotely sensed data obtained through aircraft and satellites have become so handy in all types of environmental monitoring or modeling.

In the past decade, remotely sensed data pertaining to hydro-meteorological datasets has increased substantially which has facilitated application of distributed hydrologic models even in regions where field level datasets are scarce. Direct and indirect measurements of most of the hydrological cycle components could be captured by satellite-based sensors (McCabe et al., 2017). The availability of high-resolution environmental datasets of extent varying from regional to global scale demands the need to efficiently manage, analyze, process and visualize such big data (Gebbert and Pebesma, 2014). Integration of Geographic Information System (GIS) with remotely sensed data serves as an important tool to aggregate geospatial data from various sources, present them in map form generating information about the environmental conditions of the region of interest. In this chapter, the geospatial techniques of GIS, RS, and its integrated use and its applications in hydrological modeling are discussed with a case study.

The content of this chapter is presented in six sections. The introduction section is followed by a brief description of steps involved in environment monitoring systems and, an overview of geospatial techniques, RS, and GIS are given in Section 14.2. Sections 14.3 and 14.4 deals with integration of RS and GIS, its application in environmental monitoring systems with specific focus on hydrologic modeling. In Section 14.5, a case study of application of GIS/RS techniques in two distributed hydrologic models are provided. Finally, the concluding remarks are presented in Section 14.6.

14.2   ENVIRONMENT MONITORING SYSTEMS

An environmental monitoring system (EMS) automates the process of data collection and (pre) processing to monitor the quality of the environment. Actual monitoring involves the following steps (Larsen, 1999):

1. A communication task collecting data from a sensor and communicating it to a receiving monitoring system.
2. Data pre-processing by calibration, checking, and formatting.
3. Data storage in some sort of database.
4. Displaying the data in a suitable form to users.

The EMS require large-scale data collection, processing, and assimilation. Hence an effective user interface required which demands the introduction of GIS into monitoring processes. GIS mapping functions not only enables display objects such as measurement stations but also aids in data processing tasks which involve spatial analysis. The spatio-temporal patterns in the properties of major interconnected environmental systems (atmosphere, biosphere, cryosphere, and oceans) are identified using the data derived from remotely sensed data. That way, for effective EMS, integration of RS and GIS is very important. One of the important aspect of EMS is hydrological system monitoring. Due to large-scale complexity of hydrological system, effective hydrological simulation models integrated with geospatial techniques such as GIS and RS is very essential.

14.3   OVERVIEW OF GEOSPATIAL TECHNIQUES

Data pertaining to events, or phenomena with location on Earth's surface could be referred to as geospatial data. Geospatial data are of two types: vector and raster. Geospatial technology is an evolving area of interest that comprises of Remote Sensing (RS), Global Positioning System (GPS), and Geographic Information System (GIS). In environment monitoring and modeling, most commonly used geospatial techniques are RS and GIS. In this section, a brief overview of RS and GIS are provided.

14.3.1   REMOTE SENSING TECHNIQUES

14.3.1.1   Basics of Remote Sensing

Remote sensing is a technique to detect and monitor the physical characteristics of an object or area by measuring while it is reflected and emitted radiation at a distance (from RS platforms such as satellite, aircraft, or various terrestrial platforms) (Lillesand and Kiefer, 2002). Remote sensing satellites are artificial and bear sensors to capture image of the surface of Earth. The whole globe or an assigned part of it can successively be observed by satellites for a specific time period (Guo et al., 2016). Remote sensing techniques allow taking images of the Earth's surface in various wavelength region of the electromagnetic spectrum which is one of the major attributes of a remotely sensed image (Lillesand and Kiefer, 2002). Table 14.1 provides the wavelength ranges of waves in electromagnetic spectrum for RS.

The visible, near infrared, or thermal infrared wavelength regions in the electromagnetic spectrum are mostly used by RS images. Microwave region energy is the measure of relative return from the Earth's surface and is called as active RS, since the remote sensing platform provides energy source. Passive remote sensing systems

TABLE 14.1
Electromagnetic Spectrum Divisions for Remote Sensing

Name	Wavelength
Visible Spectrum	0.4–0.7 µm
	Violet 0.4–446 µm
	Blue 0.446–0.5 µm
	Green 0.5–0.578 µm
	Yellow 0.578–0.592 µm
	Orange 0.592–0.62 µm
	Red 0.62–0.7 µm
Infrared (IR) Spectrum	0.7–100 µm
Microwave Region	1 mm–1 m
Radio Waves	(>1 m)

depend upon an external source of energy for the RS measurements (e.g., the Sun) (Lillesand and Kiefer, 2002).

Detection and discrimination of objects or surface features means detecting and recording of radiant energy reflected or emitted by objects or surface material. Depending upon the property of material, surface roughness, angle of incidence, intensity, and wavelength of radiant energy, different objects return different amount of energy in different bands of the electromagnetic spectrum, incident upon it. Different stages are involved in the processing of RS (Figure 14.1).

14.3.1.2 Remote Sensing Datasets
Different nations across the world have developed land observation satellites used for a wide range of applications such as investigation of land resources, environment

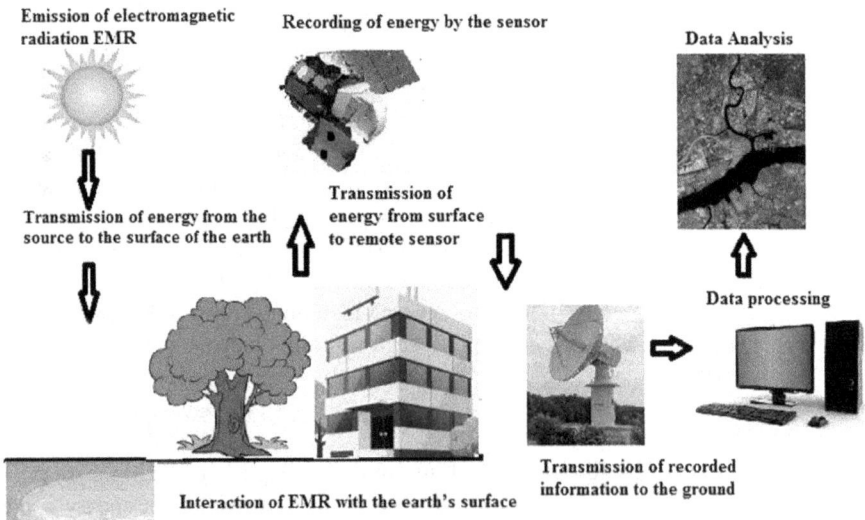

FIGURE 14.1 Different stages of remote sensing process.

TABLE 14.2
List of Major Land/Ocean/ Meteorological Satellites

Nation	Satellites
US	Landsat series, TRMM, Terra, ACRIMSAT, GRACE, Aqua, ICESat, SORCE, Suomi NPP, TIROS-I, SeaSTAR, TOPEX/Poseidon, OrbView, Jason, DMSP, NOAA, GOES
European	CryoSat-2, Sentinel 1, SPOT, CHAMP, ERS-1/2, ENVISAT, The Gravity Field and Steady State Ocean Circulation Explorer (GOCE), Meteosat series, MetOp
China	China-Brazil Earth Resource Satellites (CBERS), ZY-1 02C, HJ-1A/B, HJ-1C, GF-1 (Gaofen-1), SJ" series, "Tsinghua-1", "NS-2", and "Beijing-1", Shenzhou, Jilin-1, Beijing-2, SuperView-1, and Lishui-1, HY-1A, 1B, 2A, FY satellite series
Japan	Japan's Earth resource satellite (JERS-1), Japan's Advanced Earth Observation Satellite (ADEOS), Advanced Land Observing Satellite (ALOS), MOS-1, MTSAT
India	RESOURCESAT-1, 2, 2A CARTOSAT-1, 2, 2A, 2B, RISAT-1, and 2, OCEANSAT-2, Megha-Tropiques, SARAL and SCATSAT-1, and INSAT-3D, Kalpana & INSAT 3A, INSAT-3DR
Russia	Okean-O1, Meteor" series, GOMS,

Source: Fu et al., 2020

research related to Earth, prediction of crop condition, and observing natural disasters. A list of different satellites and the nations which developed them are listed in Table 14.2 (Fu et al., 2020). Satellite mission and the hydro-meteorological variable for water resources management are precipitation (TMPA, PERSIANN, CMORPH, GSMAP, GPM/IMERG, CHIRPS, MSWEP), Land Surface Temperature (for Evapotranspiration) (Landsat, AVHRR, ASTER, MODIS, VIIRS, Sentinel-3, ECOSTRESS), Evapotranspiration products (RS-PM, MOD16 ET, PT-JPL, GLEAM, Global ESI), Soil moisture (AMSR-E, SMOS, SMAP, Sentinel-1), Groundwater (GRACE), Vegetation (Landsat, AVHRR/GIMMS, MODIS, VIIRS, SPOT/PROBA-V, Sentinel-2). The details of spatio-temporal resolution, launch dates, and source are mentioned in Sheffield et al., 2018.

14.3.1.3   Use of Remote Sensing for Environmental Monitoring

For environmental monitoring, the RS satellites are effectively used for data collection since 1980s. Data scarcity issue for many of the environmental applications such as hydrological modeling can be solved to a certain extend using the RS data. The RS technology provides data pertaining to spatial distribution of soil and land use parameters, initial conditions, water bodies inventories etc. (Lillesand and Kiefer, 2002). The RS data can map conditions of snow, ice cover, atmospheric and climatic, sea surface, and determine the water quality parameters (Schummgge and Gurney, 1988). The advantage of remote sensing is its ability to observe several hydrological variables on the basis of the electromagnetic energy measured at different wavelengths as given in Table 14.3 (Lillesand and Kiefer, 2002).

Further evapo-transpiration can also be measured by using remote sensing data. From the RS, we can estimate various parameters required in the evapotranspiration process such as the incoming solar radiation, surface albedo, vegetation cover, surface temperature, atmospheric temperature, water vapour, and soil moisture.

TABLE 14.3
The Environmental Variables and Their Interpretable Wavelength

Sl. No	Environmental Variables	Applicable Wavelength Range
1	Surface temperature on Earth	Thermal infrared wavelengths
2	Surface soil moisture	Microwave lengths
3	Snow cover	Microwave lengths and visible wave lengths
4	Land use/ Land cover/ Vegetation	Visible, infrared, and micro wavelengths
5	Energy balance Components	Visible and thermal wavelengths

The soil moisture can be determined through the use of sensors operating in microwave range (wavelength 3 to 30 cm) based on the principle of strong sensitivity of soils dielectric properties on its moisture content.

The few areas where RS can be used for environmental monitoring/ modeling include (Singh and Woolhiser, 2002):

1. Real-time flood forecasting.
2. Monitoring and prediction of snow melt runoff.
3. Changes in land use and land cover.
4. Developing watershed management strategies to make conservation plans.
5. Agricultural developments.
6. Mapping of ground water potentialities.
7. Environmental impact assessment.
8. Inventorying coastal and marine processes.

14.3.2 GEOGRAPHIC INFORMATION SYSTEMS (GIS)

14.3.2.1 Basics of GIS

GIS is a system of capturing, storing, manipulating, analyzing, and representing the diverse sets of geo-referenced data. On the other hand, a software package that efficiently relates graphical information to attribute data stored in a database and vice-versa (Malczewski, 1999). The first GIS technique was developed in the Canada during the 1960s for the need to handle the government data., three categories of computer platforms used to run GIS include: mainframes which are oldest, personal computers (PCs), and most recently, workstations. Although PCs are relatively inexpensive and easy to use, they may be insufficient for handling large datasets with topological complexity encountered in water resources management. Some of the important packages of GIS include: Arc/Info, GRASS, QGIS, Intergraph, MapInfo, Gram++.

Spatially referenced GIS data can be represented by vector and raster forms and attribute tables in tabular forms (Malczewski, 1999). Vector data again can be grouped into three categories: point, line, and polygon types. Point data representing nonadjacent features and discrete data points such as hydrologic gauging station, place name, locations etc. and having zero dimensions. Line is used to represent linear features such as rivers and streets. Polygons are used to represent areas such as the river basin and sub-basin boundary, city/village boundary, lake, forest areas etc.

Raster data represented as grids are of two types: continuous and discrete. Population density is an example of discrete raster data whereas temperature and elevation are continuous data. There are also three types of raster datasets: thematic data, spectral data, and pictures (imagery). Thematic raster dataset is called a Digital Elevation Model (DEM). Spatial hydrology modeling such as extracting watersheds and flow lines also uses a raster-based system. The spectral signatures of each feature are classified to gather vegetation/geologic information from aerial or satellite imagery (Malczewski, 1999).

Data processing in GIS interface can be very useful and complex, since it usually involves the analysis of large amounts of spatial and non-spatial data from various diverse applications. In hydrological modeling, GIS software's facilitate construction and import of Digital elevation models (DEM) and triangulated irregular networks (TIN's) from which slope and aspect can be derived using specific functions (Malczewski, 1999).

14.3.2.2   Recent Advances in GIS Techniques

For more than two decades, integration of time in GIS has been an active field of research since the first milestone set by Langran (1992). A few approaches in this direction are: The snapshot approach (Armstrong, 1988), the amendment vector approach Langran (1992), Event-Based Spatio-Temporal Data Model (Peuquet and Duan, 1995), Modeling of Application Data with Spatio-temporal features (Parent et al., 1999), and Extended Dynamic GIS (Pultar et al., 2010).

14.3.2.3   Use of GIS for Environmental Monitoring

GIS has many applications in environmental monitoring, planning, and management. Some of the application areas of GIS in environmental monitoring and modeling are:

1. Geographical data management in a particular area.
2. Mapping and monitoring of surface water bodies.
3. Locating and mapping Ground water potential zones and rechargeable areas.
4. Watershed delineation and management.
5. Irrigation command area management.
6. Hydrological modeling.
7. Water quality monitoring and analysis etc.

In hydrological modeling, GIS integrated with computer models perform functions to design, calibrate, modify, evaluate, and compare the hydrological/ watershed models. The use of GIS in integrated watershed modeling can be enumerated as following (Singh and Woolhiser, 2002):

1. Delineation of watershed and subdivision into hydrologically homogeneous sub-areas.
2. Hydrologic properties require to be categorized based on application types for which many spatial overlay combinations could be conducted.
3. Incorporation of spatial details beyond which hydrologic models can represent.

4. A better resolution of features pertaining to land and drainage areas enhances the ability to generate more suitable grid layers for computational watershed model.

14.4   INTEGRATION OF GEOSPATIAL TECHNIQUES OF RS AND GIS

Remote sensing data and GIS tool have played a crucial role for environmental monitoring, planning and management, such as river basin planning, management, assessment of river basin condition, and visualization of human activities. For instance, GIS and RS can help to find the location and suitability of areas to build up different project for development. Various methods for the integration of RS, GIS, and GPS exists. Four models were conceptualized and summarized by Gao (2002): linear, interactive, hierarchical and complex. In view of environmental monitoring say in hydrology, satellite remote sensing data is a good source of LULC change detection, monitoring of water quantity and quality for river/lake/pond and other hydrological processes within a river basin. An example of initial level of RS and GIS tool is overlay of digitized imagery (photos) with cartographic dataset (road, line, etc.) derived from GIS, producing a combined product that allow to visualize information derived from both. Recently, space-born microwave especially Synthetic Aperture Radar (SAR) which work on all weather conditions can provide important spatially distributed flood information, mapping and measurement of river network, lakes, reservoir, inundated areas.

14.5   APPLICATION OF INTEGRATION OF GEOSPATIAL TECHNIQUES IN ENVIRONMENT MONITORING SYSTEM

For environmental impact assessment, geospatial technology is an important part as environmental resources are directly impacted by anthropogenic activities. Application and integration of spatial techniques like GIS, RS, and GPS in environmental monitoring has enhanced visualizing, including map making capabilities (Satapathy et al., 2008). With the help of geospatial techniques agriculture sector, air, water, rainfall variations, vegetation cover, soil erosion, identification of ecological sensitive zone and hotspot, landscape change, dispersion of pollutants, etc. can be managed in a sustainable manner. A spatial decision support system in GIS can add spatial/non-spatial data, provide graphical display of the monitoring status, and assess impacts using modeling and analytical techniques.

14.5.1   APPLICATION IN HYDROLOGICAL MODELING

Four different approaches have been widely used to integrate GIS with hydrological modeling (Sui and Maggio, 1999): Embedding GIS-like functionalities into hydrological modeling packages, Embedding hydrological modeling into GIS packages, loose coupling, or tight coupling. Loose coupling approach uses a package of GIS (e.g., Arc/Info) and hydrological/hydraulic modeling programs or a statistical package (e.g., SAS or SPSS). Tight coupling approach embeds certain hydrological models within a commercial GIS software package via either GIS macro or conventional programming. A well-defined interface to the data structures held by the GIS is required in this approach. A combination of loose and tight coupling is used in many empirical studies. Examples of Loosely coupled models include TOPKAPI, VIC, DREAM, DHVSM, WaSiM, HYDROTEL, and Wet Spa. Some of the tightly coupled models are WMS (HEC-HMS, WEPP, AGWA, SWAT and PIHMgis (PIHM). HEC-1 (HEC-HMS), TR-20, TR-55, Rational Method, NFF, MODRAT, and HSPF), SHE, CASC2D, LISFLOOD, ANSWERS are models which comes under the embedded coupling category (Fortin et al., 2001).

In hydrological modeling, the application of integration of geospatial techniques include:

1. Measuring the essential variables using remote sensing and conventional methods and processing in GIS.
2. Design of layers for the considered hydrologic system within GIS including data layers and attribute information.
3. Building of database to manage all RS data.
4. Data processing (by image processing, analyses such as regression and correlation analyses) for acquiring significant model variables.
5. Input the variables into the hydrological model, do the model computations and error analysis.
6. Output presentation and comparison of the results from model computations with observed data.

14.6   CASE STUDY – APPLICATION OF INTEGRATION OF GEOSPATIAL TECHNIQUES IN HYDROLOGIC MODELING

Here a case study is presented to show the application of integration of geospatial techniques such as RS and GIS for environment monitoring. A case study of hydrological modeling involving the use of tools in GIS and RS datasets specifically in two distributed hydrologic models i.e., Soil and Water Assessment Tool (SWAT) and SHETRAN are discussed in this section. The SWAT model uses semi-empirical equations to represent hydrological processes while SHETRAN is a fully distributed model based on relationships derived from physical laws. Both these models are integrated with GIS. The application of GIS techniques in SWAT and SHETRAN model setup for streamflow and sediment load simulation in Netravathi River basin, India is presented. The remotely sensed data is used for hydrological modeling.

14.6.1   Description of Study Area

The Netravathi River basin belongs to the humid tropical western Ghats region in the southern state of Karnataka, India. The length of river is 107 km draining an area of 3,657 km² and a streamflow/sediment load gauging station is located at Bantwal. The different types of spatio-temporal datasets used for this case study are given in Table 14.4. The SHETRAN model is set up at 1km grid resolution whereas SWAT model operates at Hydrologic Response Units (HRUs) with uniform soil/land use

TABLE 14.4

Input Data for Hydrologic Modeling for Netravathi River Basin

Input data	Scale/Resolution	Description of Data/Source
Digital Elevation Model	30m	ASTER DEM from Earth explorer
Soil	1:5,000,000	Harmonized World Soil Database (HWSD) (FAO, 2009)
Land use	30m	LANDSAT 7 ETM+ (Acquisition Date: 20 Dec 2000)
Rainfall (SWAT model)	Daily (0.25°)	Indian Meteorological Department
Rainfall (SHETRAN model)	Daily (0.05°)	Directorate of Economics and Statistics, Karnataka
Temperature (SHETRAN model) (Max. and Min.)	Daily	Indian Meteorological Department
Meteorological data (solar radiation, relative humidity, and wind velocity) (SWAT)	Daily	Climate Forecast System Reanalysis
Streamflow and sediment load	Daily	Central Water Commission

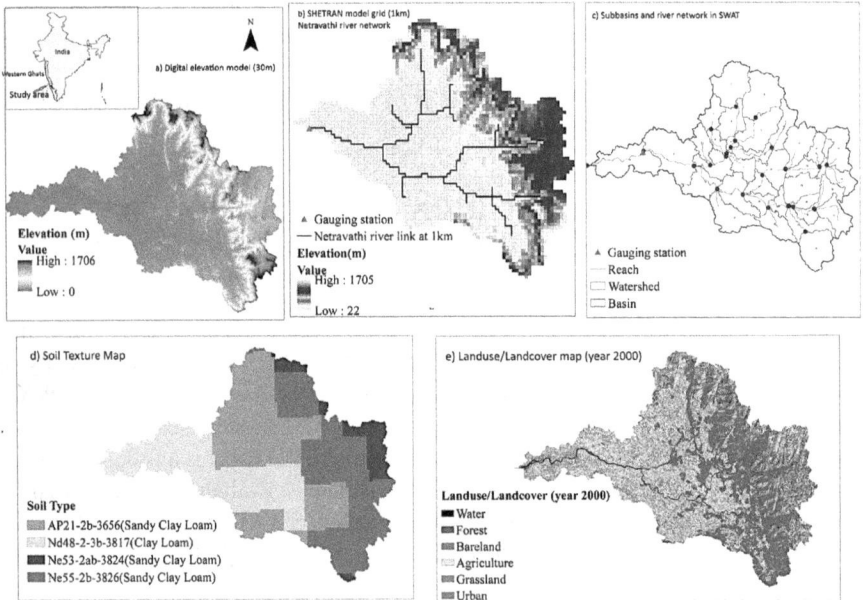

FIGURE 14.2 (a) Location map of Netravathi River basin and DEM at 30 m resolution. (b) SHETRAN model grid. (c) SWAT subbasins. (d) Soil Texture map. (e) Land use/Landcover map (year 2000).

types (Figures 14.2b and 14.2c). The rainfall/meteorological datasets used in the two hydrologic model runs are different and hence direct comparison of streamflow/ sediment load simulation is not intended in this study. The main purpose is to demonstrate processing of different datasets in GIS and use of RS data for the setup of hydrologic models in selected study region.

14.6.2 Description of Hydrologic Model SHETRAN and SWAT

14.6.2.1 SWAT Model

For runoff simulation at river basin scale, a semi-distributed, hydrological model SWAT (version 2012) which is physically based and available as open source, interfaced with ArcGIS was used. Originally the SWAT model development was intended for simulations in river basins with no gauging stations (Arnold et al., 1998). SWAT model has been applied effectively to assess the land cover and climate change effects on runoff (Sinha et al., 2020b; Sinha and Eldho, 2021). The SWAT model calibration could be subjected to large calibration uncertainty from the measured or observed data. SWAT-Calibration and Uncertainty Program with Sequential and Uncertainty Fitting (SUFI-2) algorithm is chosen because it's simple and effective to use (Sinha and Eldho, 2018). SWAT model operates at the level of hydrological response units (HRUs), based on land use, soil, and slope uniqueness in the model. The soil conservation service (SCS) curve number (CN) method (USDA, 1972) considers input data on a daily scale based on the soil hydrological groups, antecedent soil moisture, and land cover characteristics (Arnold et al., 1998) and is used for computing surface runoff. The technical details of the SWAT model regarding surface runoff and sediment yields are described by Neitsch et al. (2011). Figure 14.3a shows the typical modeling procedure using SWAT, which is fully integrated with GIS.

14.6.2.2 SHETRAN Model

SHETRAN is a fully distributed physically based model for flow, sediment and contaminant transport in river basins (Ewen et al., 2000). Finite difference method is used to solve the partial differential equations for flow and transport in SHETRAN. The model has a grid-column structure with a river link network running along edges of grid square. The details of equations which govern the hydrological process representation are provided in SHETRAN manual (SHETRAN, 2013a, 2013b). The SHETRAN model can be interfaced with GIS platforms and data from GIS/ RS can be directly used as data input files (Sreedevi and Eldho, 2020; 2021).

The SHETRAN modeling framework in the present study with its different modules and corresponding input files required to be supplied to the model are presented in Figure 14.4a. The version of SHETRAN (V4.4.5) water flow and sediment component were used in this study.

14.6.2.3 Processing of Data in GIS for SWAT and SHETRAN Model Setup

In SWAT model, the DEM is processed for obtaining useful derivatives by performing series of operations like Fill sink, Flow direction, Flow accumulation, catchment polygon processing, drainage line processing, river basin delineation etc. in ArcGIS. The delineated river basin is shown in Fig. 14.2. The river basin was discretized to polygons and taken note of the elevation of the edges to route the surface flow to the outlet of the river basin within GIS environment. The information pertaining to land use of river basin and soil parameters are also incorporated in the form of spatial maps. Within the GIS environment, the soil, land use and slope map are overlaid to develop Hydrologic Response Units (HRUs). The input parameters generated using

(a)

(b)

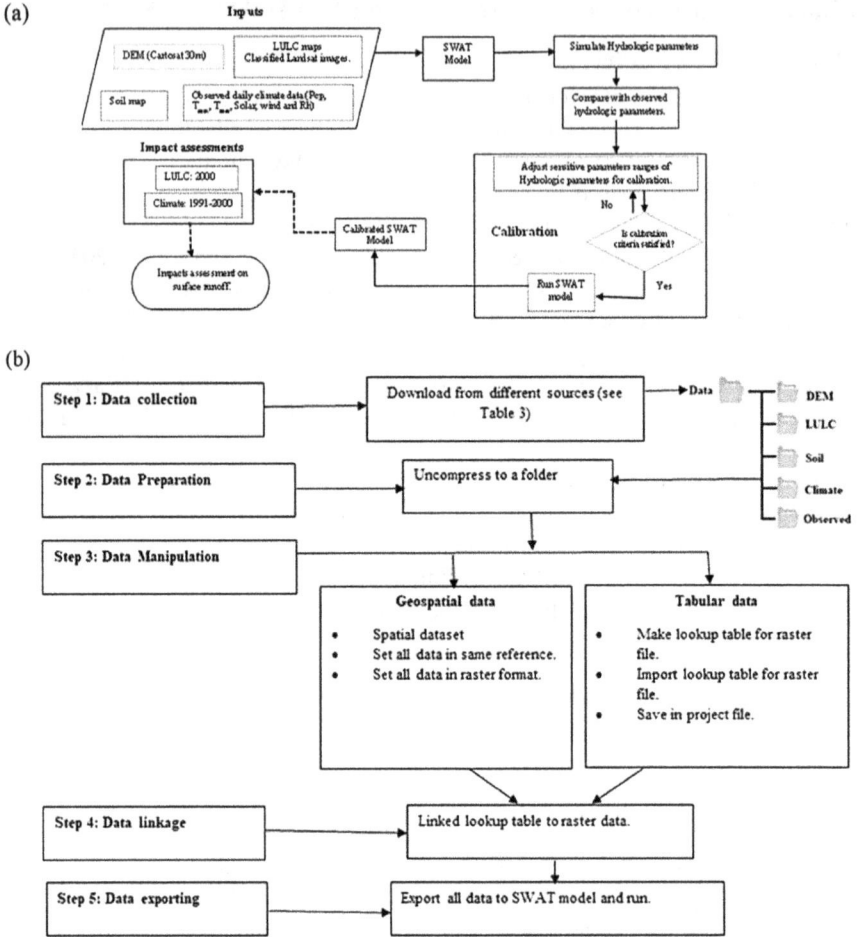

FIGURE 14.3 (a) Typical modeling flow diagram of SWAT model. (b) Procedure for input file preparation using GIS for SWAT model setup model.

GIS is connected with the runoff generation model, which gives the output in the form of runoff rate from input rainfall. The stepwise procedure in preparing the necessary input files for SWAT model setup is presented in Figure 14.3b. Model parameter values are to be provided corresponding to each of the spatially varying soil and land use types for the HRUs in the study area.

In SHETRAN model, the study area is discretized into orthogonal grids of specific size to represent spatial heterogeneity. The physiographic factors like topography, soil, and land use are prepared using ArcMap 10.1 and aggregated to the required grid size (1km in this case study) and supplied to SHETRAN model. The meteorological data, i.e., rainfall and potential evapotranspiration, are used at the daily time scale. The rainfall at ten rain gauge stations was distributed spatially to 5 km by 5 km grids (Sreedevi and Eldho, 2021) using inverse distance weight (IDW) method

(a)

DEM | Meteorological data (spatial and temporal distribution- daily/subdaily) | LULC Map | Soil Map

Discharge gauging site

Soil data (layer wise)

Catchment Mask

Soil hydraulic properties

Grid, channel, Bank discretization | Strickler overland coefficient | Channel geometry (Cross section and Bed elevation) | Vegetation parameters- Canopy storage, AET/PET LAI, root density etc. | Sediment size and number, Sediment parameters

Frame Module (frd file) | Prd , epd files | Overland/Channel Module (ocd file) | Variably Saturated Subsurface Module (vsd file) | Evapotranspiration module (etd file) | Sediment erosion and transport module (syd file)

SHETRAN Model

Adjust parameter values — Calibration/Validation

No

Acceptable? — Observed streamflow, sediment load

Yes

Simulated Streamflow, Evapotranspiration, Sediment load

(b)

Aster DEM | Satellite imageries | Soil Texture Map

Elevation Map | Clip for watershed in ArcMap | Supervised classification in ArcMap

Landuse type Map

Rainfall time series dataset (Raingauge derived or gridded)

Rainfall and PET spatial distribution → Input as ascii files of required grid size

Temperature time series dataset (Metereological station derived or gridded)

Potential evapotranspiration time series data — SHETRAN Model

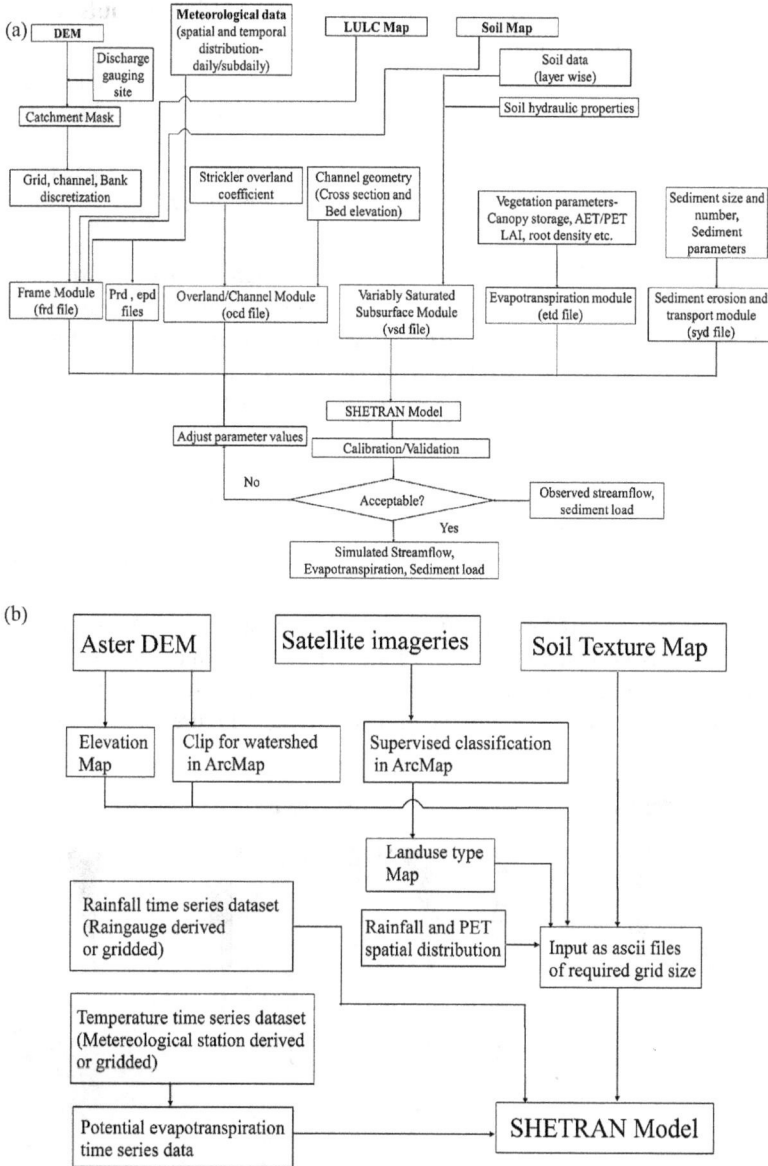

FIGURE 14.4 (a) Framework of SHETRAN model. (b) Procedure for input file preparation using GIS for SHETRAN model setup model.

(Shepard, 1968). The main tools in ArcGIS such as clipping, extract by mask, raster to ASCII data conversion, supervised classification of LANDSAT image for land use map preparation are used for data processing. The stepwise procedure in preparing the necessary input files for SHETRAN model setup is presented in Figure 14.4b. Model parameter values are to be provided corresponding to each of the spatially varying soil and land use types in the study area.

14.6.2.4 Post Processing of SWAT and SHETRAN Model Outputs in GIS

Spatial variation of SWAT model outputs is obtained at sub-basin levels in the form of a table. Further, in ArcMap using the "Join" function the attributes are joined corresponding to each sub-basin. Steps followed in processing SWAT model is as follows:

Import to database(using SWAT Editor) >> Copy to excel >> Calculate average >> Join Table in GIS >> Calculate field >> Set up symbology.

Spatial distributed SHETRAN model outputs of variables such as overland flow, phreatic surface level, sediment load, soil erosion/deposition are presented in the form of HDF—Hierarchical Data Format. ArcGIS supports the raster type dataset format of HDF5 and the tool "Add Rasters to Mosaic Dataset" can be used to display and visualize the results of spatially varying model outputs.

14.6.2.5 Hydrologic Simulation

The details of parameter sensitivity analysis and calibrated model parameter values for SWAT model is presented in Sinha and Eldho (2018) and for SHETRAN model is presented in Sreedevi and Eldho (2019; 2021), SHETRAN and SWAT model setup for Netravathi River basin are run on a daily scale to simulate streamflow and sediment load. Discharge hydrograph indicates better match of peak flows simulated using SHETRAN compared to SWAT (Figure 14.5). This may be attributed to the use of a physically based model, as in SHETRAN model. Sediment load simulated by both models matched well with the observed values (Figure 14.6).

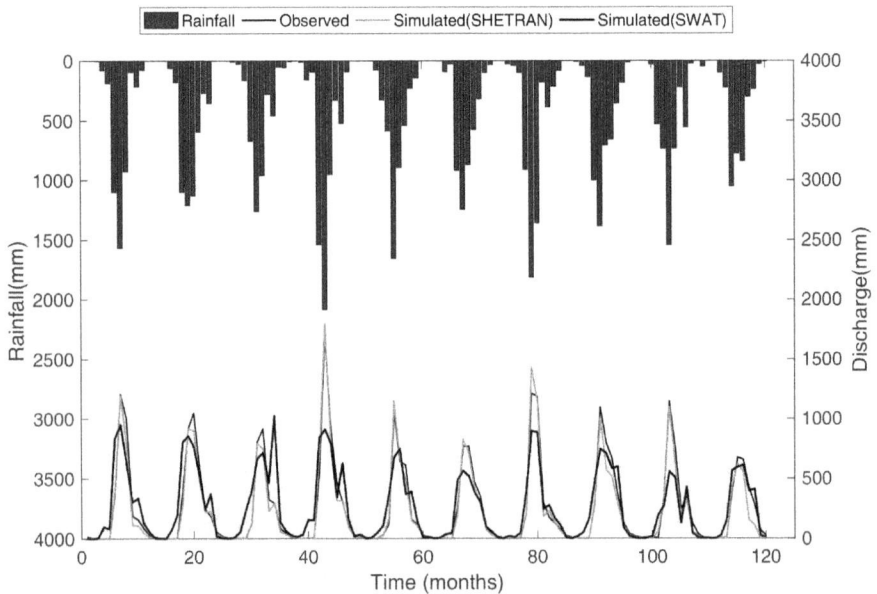

FIGURE 14.5 Monthly streamflow simulated in SHETRAN and SWAT for the years 1991 to 2000.

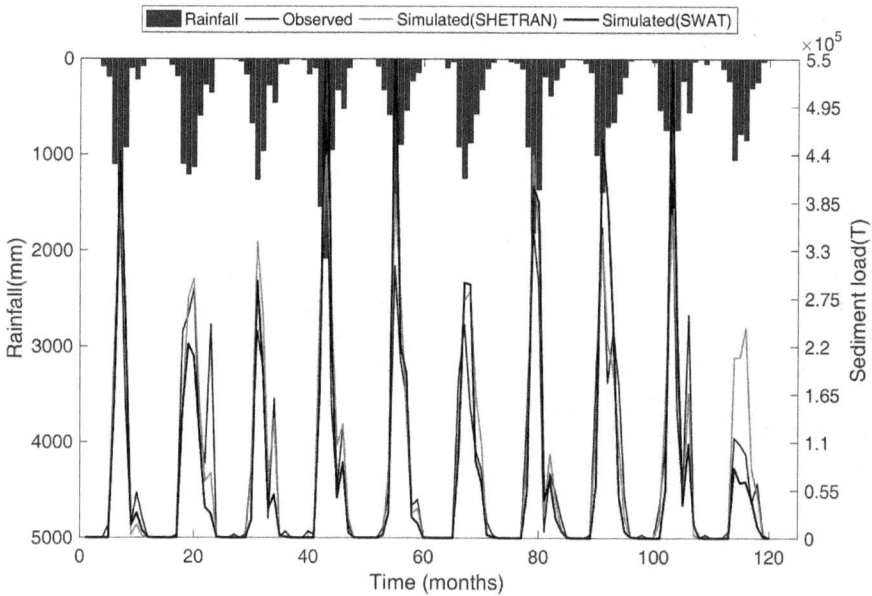

FIGURE 14.6 Monthly sediment load simulated in SHETRAN and SWAT for the years 1991 to 2000.

The objective functions that were considered to assess model performances are: Nash-Sutcliffe efficiency (NSE) (Nash and Sutcliffe, 1970) capable of determining the relative magnitude of the residual variance compared to the measured data variance, RMSE-Observations standard ratio (RSR) and percentage bias (PBIAS). According to Moriasi et al. (2007), the streamflow and sediment simulations obtained as shown in Table 14.5 can be regarded as very good (Streamflow: NSE > 0.5, PBIAS < (±) 10% or REVF <0.1 and RSR < 0.5 for monthly scale; Sediment load: PBIAS < ±15).

Although the Soil Conservation Service-Curve Number (CN) method used in the SWAT model is simple and efficient for runoff amount estimation from a rainfall, catchment characteristics representation using single CN parameter value is a main limitation. In the SWAT model, several empirical and quasi-physical equations, such

TABLE 14.5
Model Performance for Years 1991 to 2000 (Monthly Scale)

	SWAT		SHETRAN	
	Streamflow	**Sediment Load**	**Streamflow**	**Sediment Load**
NSE	0.75	0.71	0.96	0.85
RSR	0.50	0.54	0.19	0.39
PBIAS	−2.82	2.04	9.18	−8.30

as the SCS-CN method and MUSLE method (Neitsch et al., 2011), were developed based on climatic conditions in the USA, which may not be appropriate for India (Sinha et al., 2020a). The SCS-CN method does not reflect flow assessment based on events and the MUSLE method does not account processes such as landslides which may cause soil erosion. This may lead to some uncertainty in runoff and sediment yield simulation, mainly at the daily time scale. Only average annual soil loss can be estimated using USLE, which does not compute gully erosion, resulting in underestimation of the actual soil loss (Benavidez et al., 2018).

SHETRAN model uses Richards's equation for the estimation of infiltration in the unsaturated zone enabling detailed knowledge about processes related to infiltration and a more accurate estimation of excess rainfall. However, some of the limitations of this model includes its inability to account preferential flow and high computational demand. Erosion and sediment transport processes are represented individually in SHETRAN helping to identify, the separate effects of hillslope and riverbank sediment supply (Ewen et al., 2000).

14.7 CONCLUDING REMARKS

The developments in GIS and RS technologies in the recent years have made environmental monitoring and modeling much easier. The integration of GIS and RS datasets with environmental and hydrologic models has facilitated a wide range of application of distributed models for environmental monitoring and modeling. In this chapter, brief overview and explanation of geospatial techniques of GIS and RS and its applications are discussed.

The capability to handle large environmental datasets which are spatio-temporal in nature and the visualization tools has made GIS, an integral component of most of the environmental monitoring systems. The advancements in observation satellites have enabled the availability of environmental datasets from regional to global scale. Although a realistic representation of the spatial heterogeneity within a watershed is possible with distributed hydrologic models, the demand for large amounts of spatial datasets often constraints its widespread application. However, the availability of RS datasets has facilitated the application of distributed models even in regions where direct procurement of datasets might be difficult, as demonstrated in the case study. The tools in GIS aids in the processing of different RS datasets in the specific format demanded by hydrologic models as inputs and represents hydrologic model outputs. There has been a steady increase in the number and variety of functions incorporated in GIS that are suited to environmental and water resource applications. Although, integrating GIS, RS and hydrologic models have proved to be useful in solving various water related issues further research in this direction is needed to construct easy-to-use simulation models and decision support systems that help to identify and solve real-world water resource problems. However, the integrated geospatial tools will be useful for modellers to handle different geospatial datasets to set up distributed hydrologic models or environment monitoring systems.

REFERENCES

Arnold, J.G., et al., 1998. Large area hydrologic modeling and assessment part I: model development. JAWRA Journal of the American Water Resources Association, 34 (1), 73–89.

Armstrong, M.P., 1988. Temporality in spatial databases. Proceedings: GIS/LIS, 88 (2), 880–889.

Benavidez, R., Jackson, B., Maxwell, D., & Norton, K. 2018. A review of the (Revised) Universal Soil Loss Equation ((R)USLE): With a view to increasing its global applicability and improving soil loss estimates. Hydrology and Earth System Sciences, 22(11), 6059–6086. https://doi.org/10.5194/hess-22-6059-2018

Ewen, J., Parkin, G., Connell, P.E., 2000. SHETRAN: Distributed river basin flow modeling system. Journal of Hydrologic Engineering 5, 250–258.

Food and Agriculture Organization (FAO) of the United Nations. 2009. Harmonized world soil database, version 1.1. pp. 1–43. https://www.fao.org/3/aq361e/aq361e.pdf [Accessed on April 10, 2022].

Fortin, J. P., Turcotte, R., Massicotte, S., Moussa, R., Fitzback, J. E., and Villeneuve, J. P. 2001. Distributed watershed model compatible with remote sensing and GIS data. I: Description of model. Journal of Hydrologic Engineering, 6(2), 91–99.

Fu W., Ma J., Chen P., Chen F. 2020. Remote Sensing Satellites for Digital Earth. In: Guo H., Goodchild M.F., Annoni A. (eds) Manual of Digital Earth. Springer, Singapore. https://doi.org/10.1007/978-981-32-9915-3_3s

Gao J. 2002. Integration of GPS with remote sensing and GIS: reality and prospect. Photogrammetric Engineering and Remote Sensing 68:447–453

Guo H, Dou C, Zhang X et al. 2016 Earth observation from the manned low Earth orbit platforms. ISPRS Journal of Photogrammetry Remote Sensing 115:103–118

Langran G. 1992, March. Time in geographic information systems. Taylor & Francis.

Gebbert, S., Pebesma, E., 2014. A temporal GIS for field based environmental modeling, Environmental Modeling & Software, 53 (C): 1–12.

Larsen, L., 1999. GIS in environmental monitoring and assessment. Geographical information systems, 2, pp. 999–1007.

Lillesand, T.M., Kiefer, R.W. 2002. Remote sensing and image interpretation. John Wiley & Sons (Asia) Ltd., Singapore.

Malczewski, J. 1999. GIS and multicriteria decision analysis. John Wiley & Sons, Inc., New York.

McCabe, M. F., Rodell, M., Alsdorf, D. E., Miralles, D. G., Uijlenhoet, R., Wagner, W., Lucieer, A., Houborg, R., Verhoest, N., Franz, T., Shi, J., Gao, H., & Wood, E. F. 2017. The future of Earth observation in hydrology. Hydrology and Earth System Sciences Discussions, 21, 3879–3914.

Moriasi, D.N., Arnold, J.G., Van Liew, M.W., Bingner, R.L., Harmel, R.D. and Veith, T.L., 2007. Model evaluation guidelines for systematic quantification of accuracy in watershed simulations. Transactions of the ASABE 50(3), 885–900. doi:10.13031/2013.23153

Nash, J.E. and Sutcliffe, J.V., 1970. River flow forecasting through conceptual models part I—A discussion of principles. Journal of Hydrology 10(3), 282–s290.

Neitsch, S.L., et al., 2011. Soil and water assessment tool theoretical documentation version 2009. Texas Water Resources Institute, Texas A&M University, College Station, TX.

Parent, C., Spaccapietra, S., and Zimányi, E., 1999. Spatio-temporal conceptual models: data structures + space + time. In: *Proceedings of the 7th ACM International Symposium on Advances in Geographic Information Systems, GIS '99*, Kansas City, Missouri, USA, New York, NY, USA: ACM, 26–33

Peuquet, D. and Duan, N., 1995. An event-based spatiotemporal data model (ESTDM) for temporal analysis of geographical data. International Journal of Geographical Information Science, 9 (1), 7–24. doi:10.1080/02693799508902022

Pultar, E., et al. 2010. Edgis: A dynamic GIS based on space time points. International Journal of Geographical Information Science, 24 (3), 329–346. doi:10.1080/13658810802644567

Satapathy, D. R., Katpatal, Y. B., & Wate, S. R. 2008. Application of geospatial technologies for environmental impact assessment: an Indian Scenario. International Journal of Remote Sensing, 29(2), 355–386.

Schummgge, T., and Gurney, R.J. 1988. Applications of remote sensing in hydrology. Developments in water science, Computational methods in water resources, Vol-1, Modeling surface and subsurface flows, *Proceedings of VII international conference*, MIT, USA, June 1988, M.A. Celia, L.A. Ferrand, C.A. Brebia, W.G. Gray, G.F. Pinder ed., Computational Mechanics Publications, Amsterdam, Netherlands.

Sheffield, J., Wood, E.F., Pan, M., Beck, H., Coccia, G., Serrat-Capdevila, A. and Verbist, K., 2018. Satellite remote sensing for water resources management: Potential for supporting sustainable development in data-poor regions. Water Resources Research, 54(12), pp. 9724–9758.

Singh, V.P., and Woolhiser, D.A. 2002. "Mathematical modeling of watershed hydrology." Journal of Hydrologic Engineering, ASCE, 7(4), 270–292.

Sinha, R. K., & Eldho, T. I. 2018. Effects of historical and projected land use/cover change on runoff and sediment yield in the Netravati river basin, Western Ghats, India. Environmental Earth Sciences, 77(3), 1–19.

Sinha, R. K., Eldho, T. I., & Subimal, G. 2020a. Assessing the impacts of land cover and climate on runoff and sediment yield of a river basin. Hydrological Sciences Journal, 65(12), 2097–2115.

Sinha, R. K., Eldho, T. I., & Subimal, G. 2020b. Assessing the impacts of land use/land cover and climate change on surface runoff of a humid tropical river basin in Western Ghats, India. International Journal of River Basin Management, 1–12.

Sinha, R. K., & Eldho, T. I. 2021. Assessment of soil erosion susceptibility based on morphometric and landcover analysis: A case study of Netravati River Basin, India. Journal of the Indian Society of Remote Sensing, 1–17.

Shepard, D. 1968. A two-dimensional interpolation function for irregularly-spaced data, in: 23rd Natl Conf of American Computing Machinery, Princeton, NJ. 517–524.

SHETRAN. 2013a. Data requirements, data processing and parameter values. Available from: http://research.ncl.ac.uk/shetran/SHETRAN%20V4%20Data%20Requirements.pdf

SHETRAN. 2013b. Shetran water flow component equations and algorithms. Available from: http://research.ncl.ac.uk/shetran/water%20flow%20equations.pdf

Sreedevi, S., Eldho, T.I., 2019. A two-stage sensitivity analysis for parameter identification and calibration of a physically based distributed model in a river basin. Hydrological Sciences Journal 64(6), 1–19. doi:10.1080/02626667.2019.1602730.

Sreedevi, S, Eldho, T. I., 2020. "Effects of grid size on effective parameters and model performance of SHETRAN for estimation of streamflow and sediment yield," International Journal of River Basin Management, doi.org/10.1080/15715124.2020.1767637

Sui, D.Z. and Maggio, R.C., 1999. Integrating GIS with hydrological modeling: practices, problems, and prospects. Computers, Environment and Urban Systems, 23(1), pp. 33–51

USDA-SCS (United States Department of Agriculture – Soil Conservation Service), 1972. National engineering handbook, Section 4 hydrology. Chapter 4–10. Washington, USA: USDA-SCS.

15 Agent-Based Modeling for Integrated Urban Water Management

Satya Prakash Maurya and Ramesh Singh

CONTENTS

15.1 INTRODUCTION

"Water is the Driving Force of All Nature" quoted by Leonardo da Vinci outlines the essence and importance of this natural resource for plant, animal, and human. Contrary to the availability of abundance water on earth, the freshwater available for human use is only 0.5% of the total freshwater. According to a report published by United Nations (UN-WWDR, 2018) by 2050, more than half of the global population (57%) will live in areas that suffer water scarcity at least one month each year. Furthermore, it is being seen as a threat of water stress region (UN-WWDR, 2020) to the areas where water resources are presently, in abundance, into a because of climatic changes which may affect the quality and availability of freshwater. Therefore, the water sector is facing a big challenge, especially urban sustainable water development planning (SWDP). Various technological and institutional frame works were developed to improve efficiency of water consumption – or preferably,

DOI: 10.1201/9781003203445-18

enhance the capacity of water resources which aims for fulfilling the present water requirement needs of all stakeholders without hampering the need of future generation water demand. Chapter 18 of Agenda 21 sets the basis for integration of water management which includes different aspects of water availability and consumption practices (GWP-TAC, 2000).

Some critical aspects of sustainable water resource management such as supply–demand, non-conventional water reuse, ecology of water resources and its complex inter-relationship draw the attention of the researchers to move forward for spatial-temporal behaviour and new attributes with significant change in approach and develop models to address the complexity and diversification in urban water management.

15.2 BACKGROUND

Various theoretical and policy frame works were developed by the different researchers for understanding the qualitative and quantitative behaviour at various scales. The pressure-state-response (PSR) framework is a widely accepted for the compilation of sustainability performance indicators. The framework links the causes of environmental changes (pressure) to their effects (state) and finally to the projects, actions, and policies (response) designed and undertaken to tackle these changes (Mega and Pedersen, 1998). Organization for Economic Development and Co-operation (OECD, 1998) included two sets of indicators i.e., water quality and water resources. Water quality concerns runoff of untreated wastewater and other polluted water which pollutes the existing surface water. Water resources related to quantitative measure of intensity of water exploitation. Moreover, to retain desired sustainability there is need for a quantitative measure of water resources on the spatial and temporal scale within the boundary. There are several approaches applied to evaluate the sustainability of water resources among which most of them implemented to evaluate three aspects: social, technical, and economic. However, the criteria applied, and the results of each category could not be measures as a single index effectively. Urban water management (UWM) frameworks are flexible process that respond to changes and predict the impact of intervention, usually demand-supply, waste, and degradation of water resources. The major components of such frameworks are identified as (a) water demand; (b) water supply; (c) water users; (d) rainfall runoff; (e) evapo-transpiration; and filtration (f) pollution threat.

Maurya et al., 2020 attempted to establish Water Development Planning Index (WDPI), single measure index based on PSR framework consisting of seven indicators: 1) Water Security; 2) Investment Scope; 3) Water Quality; 4) Water Quantity; 5) Infrastructure; 6) Reuse, Recycle and Recharge; and 7) Governance under PSR (Figure 15.1). Twenty-two measures have been defined under seven indicators. Earlier, this framework has been formulated using identical weight for each category, but to make it a dynamic and accurate predictive model, we need to intervene a technique which can be more tightly coupled to clarify the impacts of PSR on the objective function WDPI.

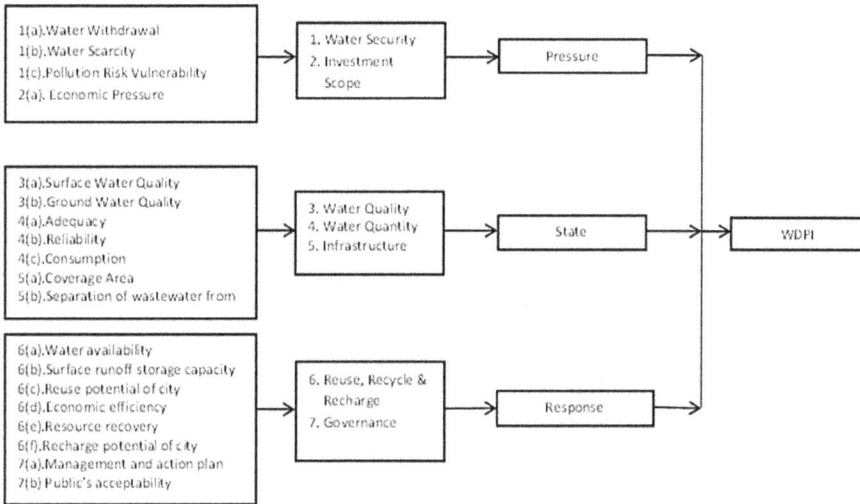

FIGURE 15.1   Framework of Water for Development Planning Index (WDPI) (Maurya et al., 2020).

15.3   MODELING AND SIMULATION IN WATER RESOURCE MANAGEMENT

Establishing inter-relationship of various measures within the frameworks and simultaneously plotting its impacts on urban water sustainability is a rigorous complex computational task in which computer applications may play a vital role. The numerical models in water sector started in late nineteenth century to address the optimization in design of urban water and sewerage system, land reclamation drainage system and reservoir spillway (Todini, 1988). However, with spatial-temporal variations, it is difficult to fit all the components into a single framework as the various parameters have complex inter-relation among themselves.

The conceptual model (Figure 15.2) of sustainable water resource system is the initial unit to analyze the problem. In this PSR-based framework, tools and approaches are essential components to define a sustainable water resource system. PSR framework provide a platform where various approaches and tools can be utilized to optimize the problems of water resources to achieve the sustainability.

ABM in water resources is applied to control agricultural water pollution Hare (2000), to manage agricultural land use and water resource Berger (2001), to develop water supply system Tillman et al. (2001), and for water resource allocation and watershed management Bars et al. (2002). To analyze the threat management strategies in water distribution systems Zechman (2007) proposed a multi-agent framework modeling that combined agent-based, mechanistic, and dynamic method to stimulate contamination events. To the existing water resources management institutions in a Socio-hydrological system, Agent-Based modeling was used by Kock (2008) for Albacete, Spain. Chu et.al (2009) worked out a Residential Water use Model (RWUM) that can be used as a tool for UWM policies, to estimate potential

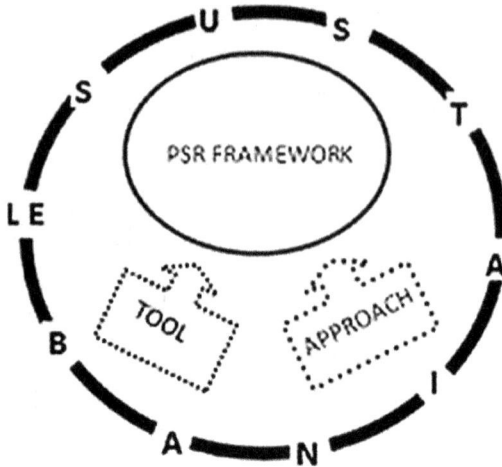

FIGURE 15.2 Conceptual framework of sustainable water resource model (Maurya and Singh, 2021).

water saving opportunity for future infrastructure development plan. Barthel et al. (2010) applied the concepts of multi-actor-based model in water sector to specify critical regions for which adaptation strategies are required for water supply due to the effects of climate change. Nikolic et al. (2013) studied integrated system dynamics simulation with ABM to analyze the spatial and temporal dynamics of water resource management system.

In the process of optimal allocation of the alternative water resources under integrated urban water management (IUWM) framework, it was observed that the models neglect uncertainties of the water demands frequently (Díaz et al., 2016; Vakilifard et al., 2019), whereas previous literature indicates the effects of change in water use patterns which cannot be ignored. Contemporarily, the development of computational models such as agent-based models (ABM) to simulate the behaviour of the society, and social choices and optimization methods has empowered decision makers to evaluate the outcome of different social rules and their influence on the applicability of adopted IUWM strategies (Castonguay et al., 2018). Water sustainability cannot be improved only focusing to implement either optimal water allocation or demand reduction rather an efficient IUWM framework must be exercised. Factors such as population growth, rainfall time series, and consumption patterns may significantly affect the IUWM, while their prediction is subject to immense uncertainties. Several studies proved that consideration of these uncertainties and their modeling methodology can significantly improve the applicability of the IUWM (Furlong et al., 2017). Moreover, water stakeholders may play a significant role in water supply–demand balance due to its complex adaptive nature. An agent-based model for domestic water management in the metropolitan area of Valladolid, Spain was developed by Galán et al. (2009). Various studies indicate that ABM may be a game changer in water development for existing and future water related activities in an urban setup.

15.4   AN AGENT-BASED INTEGRATED URBAN WATER MANAGEMENT FRAMEWORK

15.4.1   Basic Framework and Issues Need to Be Addressed

An artificial intelligence-based approach for problem solving involves three steps; *learning i.e., cognitive skills, reasoning,* and *self-correction.* Learning deals with acquiring data and establishing rules to turn data into actionable information. Such information supports computing devices with certain constraints to solve the specified problem. The reasoning step provides the justification for choosing the right rule to reach a desired outcome. A self-correction step is useful for amplifying the rules and ensures the accuracy of results. In complex environmental problems, fuzzy logic is also useful to apply knowledge from one domain to another and find a solution autonomously.

In an agent-environment approach, PSR component may be considered as an agent for an integrated urban water resource management because of its potential to determine sustainability measure. A schematic framework of ABM is shown (Figure 15.3a). Here, the major task is to design an agent program that implements the agent function i.e., the mapping from precepts to actions.

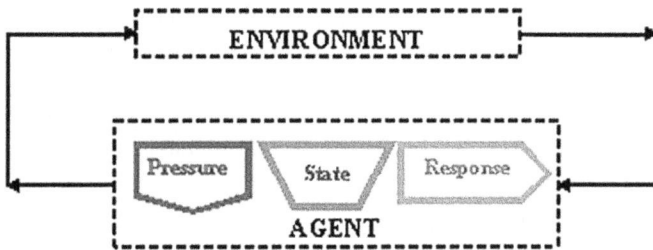

(a)

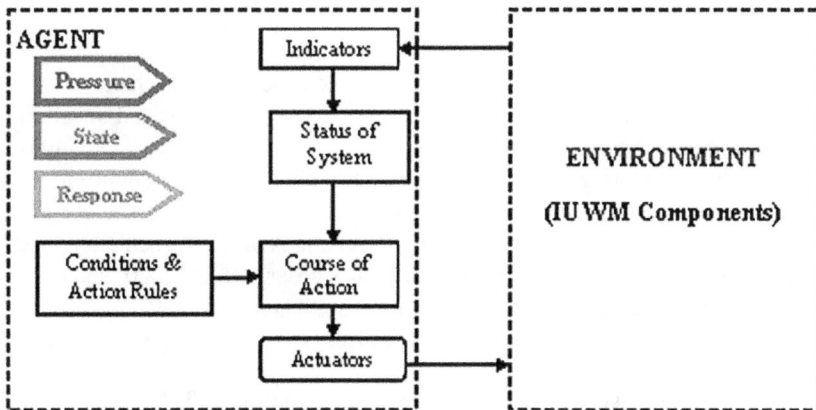

(b)

FIGURE 15.3   (a) Basic ABM framework for IUWM; (b) extended agent-environment framework for sustainable urban water management.

An agent framework consisting of sensors and actuators may combine with a software application to implement a decision support framework. This basic concept of artificial intelligence can be much useful in the complex problems of water resources management, decision-making, policy planning, and sustainable solutions. However, explaining the solution with rationales and under strict regulatory compliance requirements still needs some representative devices.

ABM has ability of asynchronous interactions among agents and between agents and their environments. The agent-based model for urban water sustainability with focusing the socio-economic aspects can be an effective tool to analyze the use of sustainable water resources by several stakeholders within the urban boundary. A modern dynamic computational system envisaged of modeling paradigm of interacting agents of an integrated urban water management (IUWM) framework can be modelled as a collection of basic entities 'agent' for decision-making which is autonomous in nature (Figure 15.3b).

This model simulates the effects of availability of water resources for recognized system actors, as well as it also examines the effects that actions of system actors have on the environment. The spatially explicit agent-based model can be designed to address two main questions:

1. How does the individual entities develop which simulate an urban water system and what are the factors that play a role in the sustainability measuring the changing physical existence and usability of the water resources?
2. How does this model can affect the urban water cycle and dynamic adaptive socio-economic aspects to achieve sustainable development goals (SDGs)?

15.4.2 Agent Model-Based Framework

To evaluate the effective use of water resource, the water managers considered inter-relation among the competing demands of resource from different stakeholders. Water resource management framework considers quantity, quality of water resource, use pattern, wastewater generation and its reuse, but this framework was limited to water supply and wastewater infrastructures i.e., the physical characteristics of urban water cycle. However, the inter-dependencies, complexities and uncertainties of the water resource pollution from discharge of waste from domestic and industrial sources and climate change which is a key issue for freshwater availability for future generation. This fact attracted attention of researchers to develop a more generic integrated water resource management model. Some ABM approach attempted for water resources planning and management (Darbandsaria et al., 2020) and complex mechanism adoption for water conservation (Rasoulkhani et al., 2017). Some of the ABMs approaches in water sector are briefly discussed in the following paragraphs.

15.4.2.1 ABM for Allocation of Total Water Availability (TAW)

Akhbari and Griggs (2013) proposed an agent-based model for estimation of total available water (TAW) for agriculture within a defined area. Total inflows (storm water, upstream inflow, and tributary inflows) over the specified area may be estimated by Equation 15.1. Further, Equation 15.2 is used to estimate the TAW for the diversion to *defined agricultural land (agent?)* within the study area. The agent's behaviour will be identified based on the quantity of water demand requested by agent *i*. If it is more than the available water for this agent, the behaviour is considered as non-cooperative; otherwise, the agent is cooperating (Equation 15.3).

$$TAW = \sum_{in=1}^{N} Q_{in} - Q_{min}; \quad \forall y, m \tag{15.1}$$

$$AW = \left[TAW / \sum_{i=1}^{I} LA_i \right] \times LA_i \quad \forall y, m \tag{15.2}$$

$$\begin{cases} \text{if } AW_i \langle D_{max,i} =\rangle i \to NC \\ \text{if } AW_i \geq D_{max,i} => i \to C \end{cases} \tag{15.3}$$

where (Akhbari and Griggs (2013))

TAW is the total available water (cms);
Q_{in} is the inflow to the river from the upstream and all tributaries (cms);
Q_{min} is the minimum river water flow rate required for environmental purposes (cms);
AW_i is the amount of available water for diversion *i* (cms);
LA_i is the area of the land belong to diversion *i* (hectare);
$D_{max,i}$ is the maximum water demand for water user *i* (cms);
m is the number of months (from 1 to 12); and,
y is the number of years in the time series.

15.4.2.2 ABM for Water Resource Management

Berglund (2015) developed an active-reactive based agent model for water resource management. Although this model was developed for a complex adaptive system, but it was specifically parameterized for the *San Joaquin watershed, California*. It was also claimed that management scenarios may be defined for encouraging different stakeholders to cooperate in the trade-off. Here, Active agents are defined with proposed behavioural rules that can update a decision-making system to satisfy the individual goals. Objective functions are responsible to select and evaluate decisions. Active agent models a set of agents that to optimize individual goals in making decisions of water uses agents are allowed to use optimization algorithm or heuristic rules to select an action rather than a predefined logical rule. The reactive agent explores the previously described water demand and uses to build the reactive agent's rules.

15.4.2.3 Conceptual ABM Frame for Urban Water Development (UWD)

The integrated urban water agent-based model identifies the entities that influence and are influenced by the urban water resources in the city boundary. In the pressure sensor, fresh water withdrawal monitoring, piped supply, and social awareness towards water uses guidelines have been considered whereas freshwater extraction reduction, vigilance of ration of water demand and water balance of the area considered as action rules. Similarly other sensors have been derived from the original WDPI framework. ABM creates a scope for evaluation of various management scenarios which deals with complex formulas. Conceptualization of ABM with a PSR framework for IUWM may change the paradigms of sustainable urban water simulation model that may help to designate effective social and institutional enhancements in improving the evaluation quality and sensitivity analysis, which results in reduced conflict levels. Such models can be a powerful tool that helps to set up rules-based scenario evaluations, especially focusing on water demands, fresh water availability, and environmental concerns.

An agent modeling based conceptual framework for urban water sustainability evaluation has been developed by Maurya and Singh, 2021 (Figure 15.4).

In this model, to develop an algorithm for interaction within the agents, a single measure, WDPI, is considered as primary agent and this agent will further interact with PSR indicators. The individual weight of indicators based on their importance is worked out within the environment of the PSR system using ABM. If the value of WDPI satisfies the sustainability criteria several scenarios based on the given problem will be generated and presented to decision makers else WDPI will be calculated

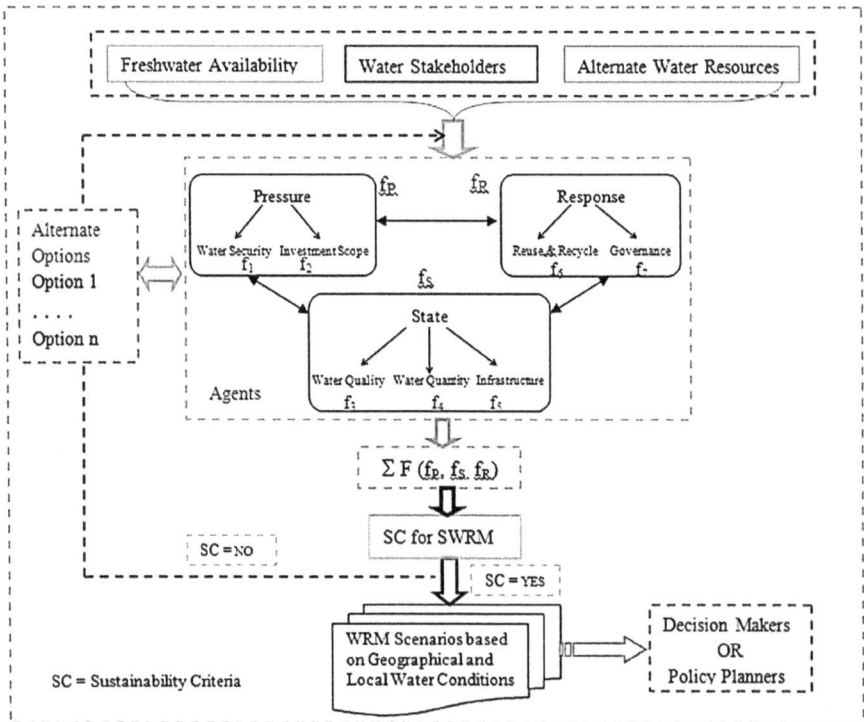

FIGURE 15.4 Conceptual framework of agent-based sustainable water management (Maurya and Singh, 2021).

again followed by the sustainability correction alternate options. In this model the objective function (F) and Sustainability Criteria (SC) plays the key role for evaluation and further water use scenario generation. Pressure function (f_P) derived from Water Security (f_1) and Investment Scope (f_2); State function (f_S) from Water Quality (f_3), Water Quantity (f_4) and Infrastructure (f_5); Response function (f_R) from Reuse and Recycle (f_6) and Governance (f_7).

However, a major challenge before the researcher is how to define the attribute and action rules for each sensor so that the basic property of the framework is undisturbed. In the present model, WDPI is considered as objective function so it can be identified as an agent that senses the facts in the environment through sensors. In this case sensors are Pressure, State, and Response functions which derived from the indicators.

The attributes and action rule for an ABM–IUDM model is depicted in a framework (Figure 15.5) State sensor has attributes as less preferred that do not prefer expensive strategies such as new construction of advanced wastewater treatment plants, and water transferring project over less expensive strategies such as modernization of existing water systems. However, action rules consist of project investment and infrastructure improvement (e.g., distribution network coverage, storm water collection coverage, wastewater treatment structures), some constraints imposed like adequacy of water quality and quantity and if it is adequate then improve the response system.

Response sensor has attributes focuses on augmentation of alternate water strategies (i.e., storm water and treated wastewater) for water supply with constraint that storm water is preferred over advanced treatment of wastewater (expensive strategies), also acceptability of alternate water uses for smart water practices. For the purpose, reuse of treated wastewater and utilization of storm water strategies and improvement in governance have been considered for action rules.

15.5   KEY ISSUES OF AGENT-BASED MODEL FOR IMPLEMENTATION

15.5.1   Mathematical Formulation

Basically, old-fashioned differential equations are used for such models. For example, the model based on PSR assumes the following mathematical equations:

$$dP/dt = \lambda PS - \gamma P$$

$$dS/dt = -\lambda PS$$

$$dR/dt = \gamma P$$

Here, P(t) is the pressure function, S(t) is the state of infrastructure, and R(t) is the response function of the system. The λ is the pressure rate and γ is the response rate. The intrinsic dimension of this model is 2 (λ and γ) which enables analysts to explore the simulation space by varying these two parameters, and studying how the response (say, duration of study timescale) depends on these values.

```
┌──────────────────────────────┐
│  WDPI   (Agent)              │
└──────────────────────────────┘

   (Sensor)        ┌──────────────────────────────────────────────────────┐
                   │ Attributes:                                          │
   ┌──────────┐    │ • Monitoring of fresh water withdrawal               │
   │ Pressure │    │ • Piped supply is more preferred than private        │
   └──────────┘    │   groundwater extraction                             │
                   │ • Raising social awareness to change water use       │
                   │   patterns is the most preferred strategy            │
                   │ • pollution risk vulnerability                       │
                   └──────────────────────────────────────────────────────┘

                   ┌──────────────────────────────────────────────────────┐
                   │ Action Rules:                                        │
                   │ • Reduce Fresh Water Extraction                      │
                   │   (If restricted ground water extraction is          │
                   │   implemented, the surface /alternate water          │
                   │   become more desirable for supply).                 │
                   │ • If the ratio of water demand and supply is more    │
                   │   than 90%, and urban water balance is +ve no        │
                   │   strategy will be adopted for existing system.      │
                   └──────────────────────────────────────────────────────┘

                   ┌──────────────────────────────────────────────────────┐
                   │ Attributes:                                          │
   ┌──────────┐    │ • Expensive strategies like construction of advanced │
   │  State   │    │   wastewater treatment plants, water transferring    │
   └──────────┘    │   project are less preferred whereas modernization   │
                   │   of water distribution systems are preferred.       │
                   │ • Water quantity and quality monitoring              │
                   └──────────────────────────────────────────────────────┘

                   ┌──────────────────────────────────────────────────────┐
                   │ Action Rules:                                        │
                   │ • Project Investment & Infrastructure Improvement    │
                   │   (Distribution network coverage, Storm water        │
                   │   collection coverage, Wastewater treatment          │
                   │   structures)                                        │
                   │ • If adequate water quality & quantity is not        │
                   │   supplied improve the infrastructure and plan new   │
                   │   projects.                                          │
                   │ • If water quality & quantity is adequate and        │
                   │   coverage network is efficient then improve         │
                   │   response of the system.                            │
                   └──────────────────────────────────────────────────────┘

                   ┌──────────────────────────────────────────────────────┐
                   │ Attributes:                                          │
   ┌──────────┐    │ • Augment alternate water i.e. storm water and       │
   │ Response │    │   treated wastewater for water supply. Where storm   │
   └──────────┘    │   water is preferred over expensive strategies like  │
                   │   advanced treatment of wastewater.                  │
                   │ • Public participation for smart water practices and │
                   │   acceptability of alternate water uses.             │
                   └──────────────────────────────────────────────────────┘

                   ┌──────────────────────────────────────────────────────┐
                   │ Action Rules:                                        │
                   │ • Reuse Treated Wastewater                           │
                   │ • Utilize Storm Water                                │
                   │ • Improvement in Governance & Public Participation   │
                   └──────────────────────────────────────────────────────┘
```

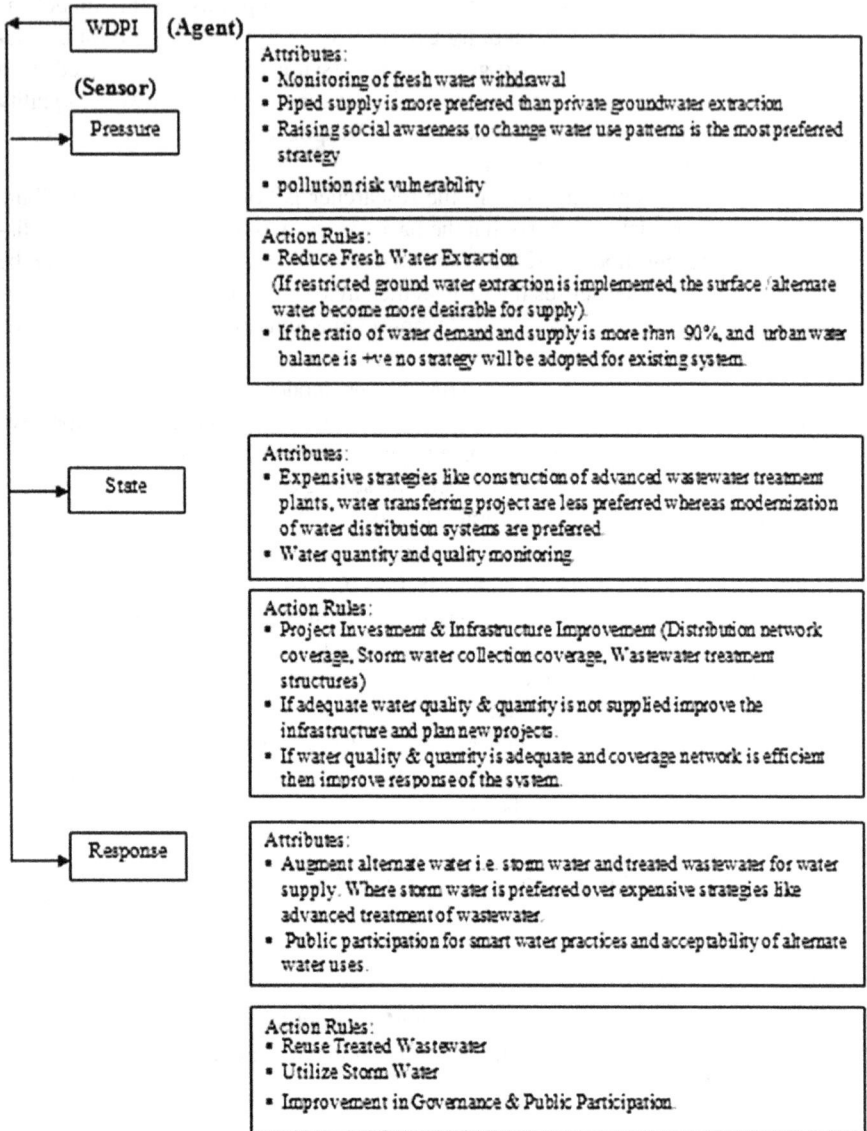

FIGURE 15.5   Framework of agent-based model for sustainable urban water planning.

For the same problem, an ABM simulation is not completely equivalent to the differential equation model – the sustainability change rate may vary drastically due to climate change before measurement within the assigned period. Nonetheless, qualitatively, the two models are extremely similar. ABM users do not realize that their model is essentially two-dimensional. The dimensionality of a model is a key property that drives inference and governs complexity. In general, an ABM analysis does not know the intrinsic dimension of the data. It is related to the rule set, but that relationship is usually unclear. Therefore, a close look participation of each factor must be considered based on correlation and their uncertainty.

15.5.2 STATISTICAL SIMULATION

For statistical simulation, there are several models, but statisticians did not yet accept a general theory and methodology for estimating their parameters from data or for making quantified statements of uncertainty about ABM predictions. In inferential strategies, Stochastic ABMs are models just like other statistical models but, with some exceptions, we cannot write down their likelihood functions. Therefore, we often have little theoretical guidance in estimating the parameters needed to fit the model to data, or in making quantified statements of uncertainty about model forecasts. But when the likelihood function is intractable, statisticians generally must use one of two tools. The first is statistical emulation? and the second is Approximate Bayesian Computation (ABC).

15.5.3 GEOSPATIAL SIMULATION

Since the urban geographical area changing invariably especially in developing countries, the agent-based modeling (ABM) which allows one to simulate the individual actions of diverse agents, measuring the resulting system behaviour and outcomes over time. With increasing power of computation and large and refined datasets of high-resolution information now exist for initializing ABM for urban water resource management simulations. Geospatial technology can encode these datasets into forms that provide the foundations for such simulations along with providing spatial methods for relating these objects based on their proximity, intersection, adjacency, or visibility to each other. However, usability of the secondary data to support the accuracy of the existing model needs to be assured.

15.6 CONCLUDING REMARKS

SDG-6 covers clean water and sanitation which also ensures the urban water sustainability under the integrated framework must face a complex system adaption. The number of management frameworks has been worked out in last several decades, but a few are applicable and implementable. However, these frameworks are also targeting specific goals of sustainability with limited stakeholders. In such situation, the ABM approach may play a paradigm changing role to achieve the urban water sustainability. In this chapter, we attempted to illustrate the current scenario of ABM in water resource management. In this chapter, the author has attempted to come forward with an ABM rule-based modeling with combinatory framework of PSR based single measured Index WDPI that can be an effective tool for water sustainability scenario evaluation and confident decision-making for policy-makers. We also discussed issues and challenges in the formulation of ABM framework for water resource management to ensure its adaptability for practical implications.

REFERENCES

Akhbari, M., & Grigg, N. S. (2013) A framework for an agent-based model to manage water resources conflicts. Water Resource Management 27:4039–4052. doi:10.1007/s11269-013-0394-0

Bars, M. L., Attonaty, J. M., & Pinson, S. (2002) An agent-based simulation for water sharing between different users. In: *International Proceedings of the First International Joint Conference on Autonomous Agents and Multi-Agent Systems*, Bologna, Italy.

Barthel, R., Janisch, S., Nickel, D., Trifkovic, A., & Hörhan, T. (2010) Using the multifactor-approach in GLOWA-Danube to simulate decisions for the water supply sector under conditions of global climate change. Water Resource Management 24:239–275. doi:10.1007/s11269-009-9445-y

Berglund, E. Z. (2015). Using agent-based modeling for water resources planning and management. Journal of Water Resources Planning and Management. doi:10.1061/(ASCE)WR.1943-5452.0000544

Berger, T. (2001). Agent-based spatial models applied to agriculture: a simulation tool for technology diffusion, resource use changes and policy analysis, Agricultural Economics, 25, 245–260.

Castonguay, A. C., Urich, C., Iftekhar, M. S., & Deletic, A. (2018). Modeling urban water management transitions: A case of rainwater harvesting. Environmental Modeling & Software, 105, 270–285.

Chu, J., Wang, C., Chen, J., & Wang, H. (2009). Agent-based residential water use behavior simulation and policy implications: a case-study in Beijing City. Water Resource Management 23:3267–3295. doi:10.1007/s11269-009-9433-2

Darbandsaria, P., Kerachianb, R., Malakpour-Estalakic, S., & Khorasani, H. (2020). An agent-based conflict resolution model for urban water resources management. Sustainable Cities and Society 57, 102112.

Díaz, P., Stanek, P., Frantzeskaki, N., & Yeh, D. H. (2016). Shifting paradigms, changing waters: transitioning to integrated urban water management in the coastal city of Dunedin, USA. Sustainable Cities and Society, 26, 555–567.

Furlong, C., Brotchie, R., Considine, R., Finlayson, G., & Guthrie, L. (2017) Apr 1. Key concepts for integrated urban water management infrastructure planning: lessons from Melbourne. Utilities Policy, 45:84–96.

Galán, J. M., López-Paredes, A., & Olmo, R. (2009). An agent-based model for domestic water management in Valladolid metropolitan area. Water Resources Research, 45(5): 1–17.

GWP-TAC (Global Water Partnership Technical Advisory Committee). 2000. Integrated water resources management. TAC Background Paper No. 4, Stockholm, Sweden.

Hare M. P. (2000) Agent-base integrated assessment of policies for reducing groundwater pollution by nitrates from agricultural fertilizer, part I: pilot study model description and initial results. Working Report, Swiss Federal Institute of Environmental Science and Technology.

Kock B. E. (2008) Agent-based models of socio-hydrological systems for exploring the institutional dynamics of water resources conflict. MS Thesis, Massachusetts Institute of Technology.

Maurya S. P. and Singh R., 2021. 6 – Sustainable water resources. Sustainable Resource Management, 2021:147–162.

Maurya, S. P., Singh, P. K., Ohri, A., Singh, R., 2020. Identification of indicators for sustainable urban water development planning. Ecological Indicators, 108.

Mega, V. & Pedersen, J. (1998) Urban Sustainability Indicators Luxembourg: Office for Official Publications of the European Communities.

Nikolic V. V., Simonovic S. P., Milicevic D. B. (2013) Analytical support for integrated water resources management: a new method for addressing spatial and temporal variability. Water Resource Management 27:401–417. doi:10.1007/s11269-012-0193-z

Organisation for Economic Development and Co-operation (OECD), 1998. Towards Sustainable Development – Environmental Indicators. OECD, Paris, France.

Rasoulkhani, K., Logasa, B., Reyes, M. R., Mostafavi, A., 2017. Agent-based modeling framework for simulation of complex adaptive mechanisms underlying household water conservation technology adoption. *Proceedings of the 2017 Winter Simulation Conference* W. K. V. Chan, A. D'Ambrogio, G. Zacharewicz, N. Mustafee, G. Wainer, and E. Page, eds., Las Vegas, NV, USA. doi:10.1109/WSC.2017.8247859

The United Nations World Water Development Report. 2018. Nature-Based Solutions for Water. United Nations Educational, Scientific and Cultural Organization.

The United Nations World Water Development Report. 2020. Water and Climate Change. United Nations Educational, Scientific and Cultural Organization.

Tillman, D. E., Larsen, T. A., Pahl-Wostl, C. & Gujer, W. (2001) Interaction analysis of stakeholders in water supply systems. Water Science Technology 43 (5), 319–326.

Todini, E., (1988) Rainfall-runoff modeling – Past, present, and future, Journal of Hydrology 100, 1–3, pp. 341–352.

Vakilifard, N., Bahri, P. A., Anda, M., & Ho, G. (2019). An interactive planning model for sustainable urban water and energy supply. Applied Energy, 235, 332–345.

Zechman E (2007) Agent-based modeling to simulate contamination events and to analyze threat management strategies in water distribution systems. In: *Proceeding of the World Environmental and Water Resources Congress*, Tampa, FL.

16 Data-Driven Modeling Approach in Model Rainfall-Runoff for a Mountainous Catchment

Arunava Poddar, Akhilesh Kumar, Veena Kashyap, and Sashank Thapa

CONTENTS

DOI: 10.1201/9781003203445-19

16.1 INTRODUCTION

The understanding of the flow of runoff over ground surfaces is essential for the migration of contaminants, flood forecasting, sediment transport, irrigation scheduling, moisture infiltration into the soil, and water resources management. Different physically based models necessitate a thorough explanation of the overland flow plane at the numerical model's grid-scale. A comprehensive explanation is not often available, mostly due to time and budgetary limitations. Therefore, for several practical reasons, numerical models of overland flow have limited applicability due to insufficient data (Poddar et al., 2020). Further, the numerical solution of these extremely non-linear flow equations is susceptible to problems of uncertainty and lack of conjunction. Thus, solutions with simpler approaches may be necessary for the projection of discharge, particularly for design purposes (Kumar et al., 2020).

Hydrologists are frequently confronted with challenges of forecasting and estimation of runoff using rainfall data (Kumar et al., 2020). The rainfall-runoff procedure is an extremely non-linear and dynamic physical structure that is tremendously challenging to simulate using a model. Mountain catchments are specific by the fast response of runoff to precipitation events. This phenomenon is explained by the low retention capacity of high mountain areas (Walter et al., 2020). However, in general, the hydrologic alterations involving increased impervious area, increased flooding, soil compaction, and improved drainage effectiveness generally lead to enhanced direct runoff, decreased groundwater recharge (Sharma et al., 2019; Sun et al., 2020). In the past, investigators have utilized either conceptual or systems theoretical techniques in separation. Literature provides numerous rainfall-runoff models. Jaiswal et al. (2020) have compared and evaluated the existing models. However, to the authors' knowledge, very few comparative studies of model efficiency are available in the literature. Of course, each model is assessed to some degree by its creator. These informal assessments, nevertheless, are not always objective because they often do not sustain the independence of the information used for measurement and validation (Beard et al., 2020).

The regression analysis is the most frequent and uncomplicated approach in the evaluation of rainfall-runoff relationships (Poddar et al., 2021). The value of regression constants is calculated with the help of regression analysis. The lack of sophistication of this method lays in its basic data needs, modification is performed depending on the catchment's history and the time passed since the previous storm; thus, specifically rainfall and runoff records. The usage of this approach is common during the initial stages of runoff estimation for design because in this method internal certainties are not considered given the complexity of internal watershed structures. However, inputs change over time, imperviousness levels due to urbanization and soil humidity conditions imprecise information, and measurement yield to inaccurate parameter estimation (Saadi et al., 2020). Over the past few years, data-driven modeling (DDM) has been the topic of research in hydrological modeling. DDM analyzes the data about a system, in specific finding associations between the input and output parameters of the system without explicit knowledge of its physical behaviour. Artificial neural networks (ANN) and fuzzy logic (FL) are DDM practices that have presented promising results (Kumar et al., 2020). Physical models have their restrictions because of complex processes and difficult representation. Data-driven approaches provide an alternative to these traditional physical-based models.

The ANN and FL methodologies were used, especially from the last decade, for solving a wide range of problems in water resource engineering (Tayfur and Singh, 2006; Thakur et al., 2021). Dawson and Wilby (1998) debated the use of ANN to forecasting flow in flood-prone catchments in England utilizing hourly hydrometric data.

Several researchers have investigated the utilization of the FL approach in rainfall-runoff studies. Yu and Yang (2000) gave a fuzzy multi-objective function (FMOF) to advance the performance of orthodox objective functions of root-mean-square error (RMSE) and mean percentage error (MPE), that are used in calibrating rainfall-runoff conceptual models. See and Openshaw (2000) offered artificial intelligence methodology-based prediction models to offer a solution to a river level and flood forecasting problem. The fuzzy logic approach has been applied to flood forecasting (Mukerji et al., 2009); precipitation (Tayfur and Singh 2006; Poddar et al., 2020); sediment transport (Tayfur et al., 2003), reservoir operation (Tilmant et al., 2002), and stormwater infiltration (Hong et al., 2002). Whereas ANN has found application in sediment transport, streamflow forecasting, rainfall forecasting seepage, and dispersion. In the present chapter, four regressions and two DDM approaches are applied to model the rainfall-runoff of a mountainous catchment in Himachal Pradesh, India. The chapter presents an approach that associates data and techniques to advance integrated models of the rainfall-runoff process. Specifically, results from five different models are presented and discussed.

16.2 MATERIALS AND METHODS

16.2.1 STUDY AREA

The watershed of 56 hectares is located at Palampur, in the Kangra district of Himachal Pradesh, India. It lies under 32.6 N latitude and 77.2 E Longitude with an elevation of 1300 m. The areas fall under the mid-hills sub-humid agro-climatic zone of the state (Kumari et al., 2021). The annual rainfall of the region is about 3000 mm, with more than 80% of the rainfall traced in the monsoonal season (mid-June to mid-September) and a major portion as waste runoff. The general slope of the watershed is 1–10%. The rainfall and runoff data for a period of June to September for continuously 17 years (1999–2015) was obtained.

16.2.2 PHYSICAL PROPERTIES OF THE SOIL

Various physical properties of the soil were determined for the study area. The results of the sieve analysis of the soil sample show that fine sand content was minimum (15.1%) at the surface which with soil depth increased and attained a maximum of 47.5% in a 60–75 cm layer. Coarse sand was similar in all the layers and minimum in 15–30 cm (1.4%). Clay content is more prevalent in the upper 30 cm, whereas silt content was optimal in the upper 7.5 cm, and decreased up to 30 cm depth, beyond which its concentration again increased. The textural classification of the watershed soil is silty clay loam. In general, bulk density increased at depths up to 60 cm. It decreased at 60–75 cm, and again increased below 75 cm. Maximum value bulk density (1.45 g/cm³) was obtained for the 45–60 cm layer and minimum (1.06 g/cm³) for the top layer. The aggregates < 0.25 mm diameter were maximum in 75–90 cm and minimum in 30 to 45 cm depth. A profile infiltration vs time relation of the study site showed decreasing infiltration rate with time, and arrived at a steady-state value of 0.25 m/hr after about 4 hours of study. The cumulative intake after 8 hours was about 6 cm. The water content at all depths decreased with an increase in suction while 45–60 cm depth retained the highest and 60 to 75 cm and 0 to 15 cm depths retained the lowest amount at a particular value of suction. Other depths were intermediate. At 100 m bars of suction, the water retained in samples was 0.53, 0.50, 0.47, 0.47, 0.45, 0.45, and 0.39 cm³ for corresponding depth increments of 45–60, 90–105, 30–45, 75–90, 15–30, 0–15, and 60–75 cm.

16.2.3 METHODS USED

In this chapter, one statistical regression (Polynomial) and three data-driven methods (non-linear regression, and Fuzzy logic and ANN approaches) were applied to the data, and the results from these models were then compared in the sections that follow.

16.2.3.1 Polynomial Regression

Polynomial regression uses the following type of equation:

$$Y = A_0 + A_1 X + A_2 X^2 + A_3 X^3 + \ldots + A_n X^n \tag{16.1}$$

Where $A_0, A_1, A_2, A_3, \ldots, A_n$ are constants.

16.2.3.2 Linear Regression Model

The equation for straight line regression between rainfall (P) and runoff (R):

$$R = A + BP' \tag{16.2}$$

The values of coefficients A and B are given by

$$A = \frac{\Sigma R - B \Sigma P}{n} \tag{16.3}$$

$$B = \frac{n \Sigma PR - \Sigma P \Sigma R}{n \Sigma P2 - (\Sigma P)2} \tag{16.4}$$

The coefficient of correlation, r:

$$r = \frac{n\sum PR - \sum P \sum R}{sqrt\left[n\sum P^{\wedge}2 - (\sum P)^{\wedge}2\right]\left[\left(n(\sum R^{\wedge}2) - (\sum R)^{\wedge}2\right)\right]} \qquad (16.5)$$

16.2.3.3 Quadratic Regression Model

The equation for quadratic regression between (R) and (P) is

$$R = AP^2 + BP + C \qquad (16.6)$$

The values of the coefficients A, B, and C can be computed using Equations (16.3) and (16.4), and similarly, the correlation coefficient is calculated using Equation (16.5).

16.2.3.4 Non-Linear Regression

For the determination of rainfall-runoff correlations, two non-linear regression methods, exponential and logarithmic, have been used.

16.2.3.5 Exponential Regression

For an exponential regression, data should not contain zero or negative values. We have used Equation (16.7) to calculate the least squares fit through points.

$$R = A.e^{B.P} \qquad (16.7)$$

where e is expressed as the base of the natural logarithm. The value of A and B are obtained with the help of Equations (16.3) and (16.4), and the value regression coefficient by Equation (16.5).

16.2.3.6 Logarithmic Regression

For a logarithmic regression, data can contain negative and positive values. We have used the following equation to calculate the least squares fit through points:

$$R = A.lnP + B \qquad (16.8)$$

where ln is the natural logarithm function.

16.2.3.7 Fuzzy Logic Approach

Many investigators have used the fuzzy approach to various engineering problems. Rather than numerical uncertainty estimates, the fuzzy logic approach is based on linguistic uncertain expressions. In this chapter, P and R variables were categorised into five partial subgroups named low (L), medium-low (ML), medium (M), medium-high (MH), and high (H). A small number of fuzzy subgroups choice leads to unreliable predictions whereas a large no. implies redundant calculations. In the preliminary stage, the number of subgroups is selected as five subgroups in each variable imply that there are $5 \times 5 = 25$ different partial relationship pairs that may be considered between the R and P variables. However, many of these relations are not possible. For instance, if the P is High, it is not possible to state that the R is Low or Medium positions of the Fuzzy words employed one shows the relative to the model (Figure 16.1).

Each of the middle fuzzy words is exposed as a triangle with maximum membership degree at its apex. It is significant to reflect those neighbouring fuzzy subsets interfere with each other, providing the fuzziness in the model simulation. Since the rainfall-runoff relationships, in general, have a direct proportion feature, it is probable to write the following five rule-bases for the description of fuzzy rainfall-runoff

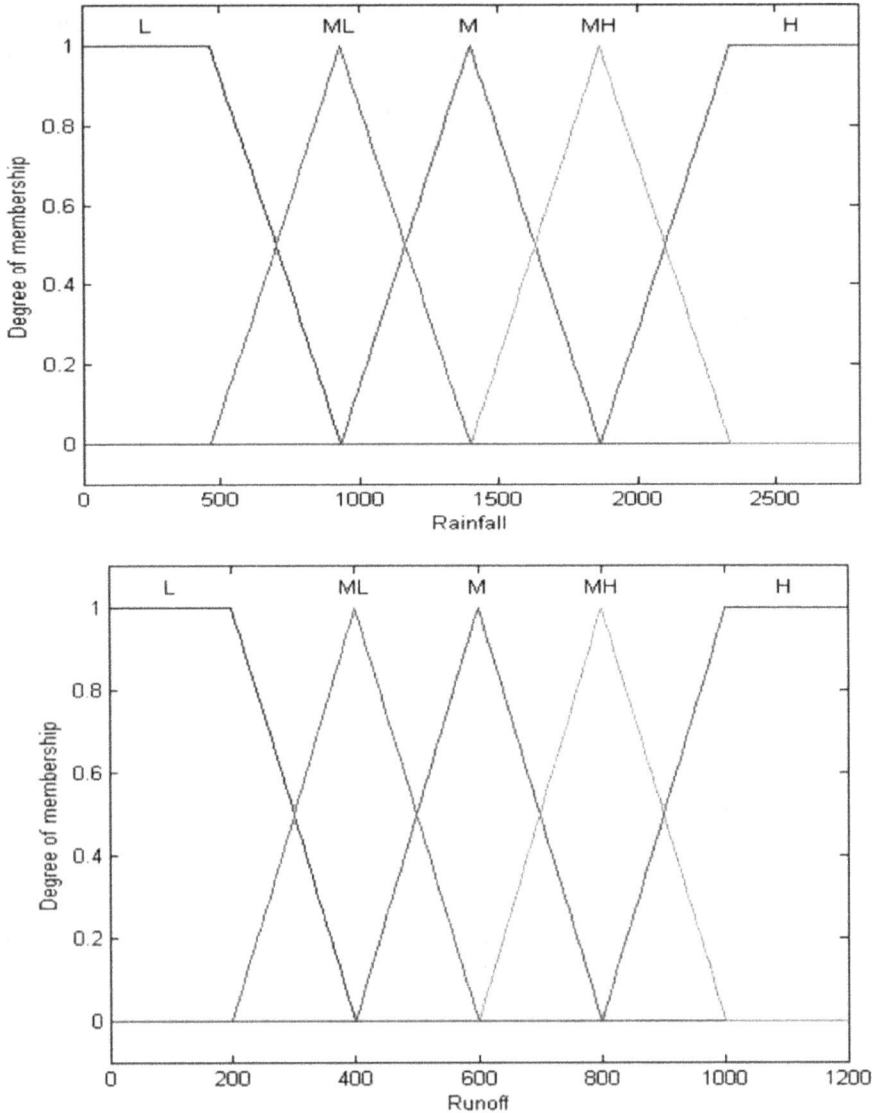

FIGURE 16.1   Fuzzy subgroups of rainfall and runoff.

modeling. In Figure 16.2(a), rainfall-runoff relationships rule with fuzzy interference sets have been shown.

1. Rule 1: IF rainfall is L THEN runoff is L, or
2. Rule 2: IF rainfall is ML THEN runoff is ML, or
3. Rule 3: IF rainfall is M THEN runoff is M, or
4. Rule 4: IF rainfall is MH THEN runoff is MH, or
5. Rule 5: IF rainfall is H THEN runoff is H.

For the fuzzy logic-based approach for rainfall-runoff correlation determination, a computer program was written on MATLAB software which provides the value of runoff when rainfall is treated as the input value. In Figure 16.2(b), the layout of the developed fuzzy logic model has been shown.

An ANN can be defined as "a data processing system consisting of a large number of simple, highly interconnected processing elements (artificial neurons) in an architecture inspired by the structure of the cerebral cortex of the brain". The ANN tries mimicking the operation of the human brain, which contains billions of neurons and their interconnections. The objective of ANN is to process the information in a way that is previously trained, to generate satisfactory results. Neural networks learn from experience, generalize from previous examples to new ones, and abstract essential characteristics from inputs containing irrelevant data.

In this chapter, a computer program is written to develop an artificial neural network-based model for the rainfall-runoff correlation determination. The computer program has been written in MATLAB. We have utilized the Feedforward Backpropagation Network for the present chapter. In this network, the training

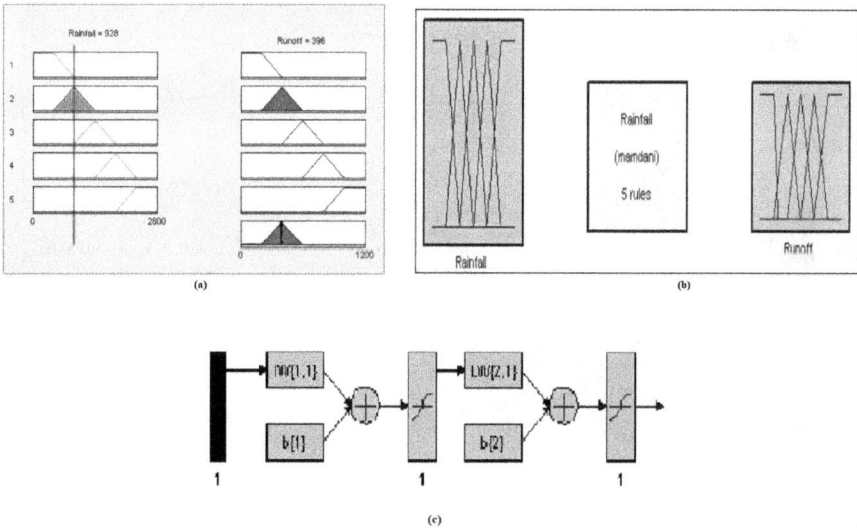

FIGURE 16.2   (a) Rainfall-runoff relationships rules with fuzzy interference set, (b) Fuzzy logic model developed Artificial Neural Network Approach, and (c) Feedforward Back Propagation network for the rainfall-runoff correlation.

function TRAINLM is used, adaption learning function LEARNGDM, performance function MSE.

And for layer 2, the hidden layer, the number of neurons selected is eight, and the transfer function used is TANSIG. In Figure 16.2(c) the Feedforward Back propagation network is shown which was created using the computer program.

16.3 RESULTS AND DISCUSSION

Based on the method discussed above, calculations of the rainfall-runoff correlations have been done. Conclusions from the comparison of the results obtained for correlation coefficient and sum square errors (SSE) of different methods have been summarized below.

16.3.1 POLYNOMIAL REGRESSION

In polynomial regression, linear, and quadratic regression methods have been used, and the results for the linear and quadratic regression methods are as follows:

16.3.2 LINEAR REGRESSION (LR) METHOD

For LR, the assumed rainfall-runoff data were utilized for a straight line curve fitting and the curve obtained is shown in Figure 16.3.

The values of the norms of residuals (S) which were the alteration between the given values of the Runoff and the values obtained from the LR model have been calculated. The curve attained for S for LR is shown in Figure 16.4(a).

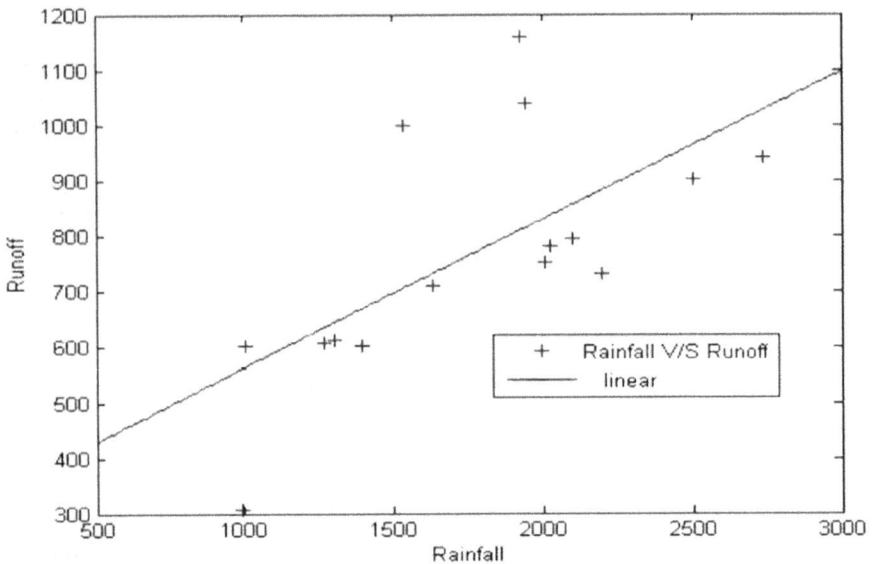

FIGURE 16.3   Straight line curve fitting.

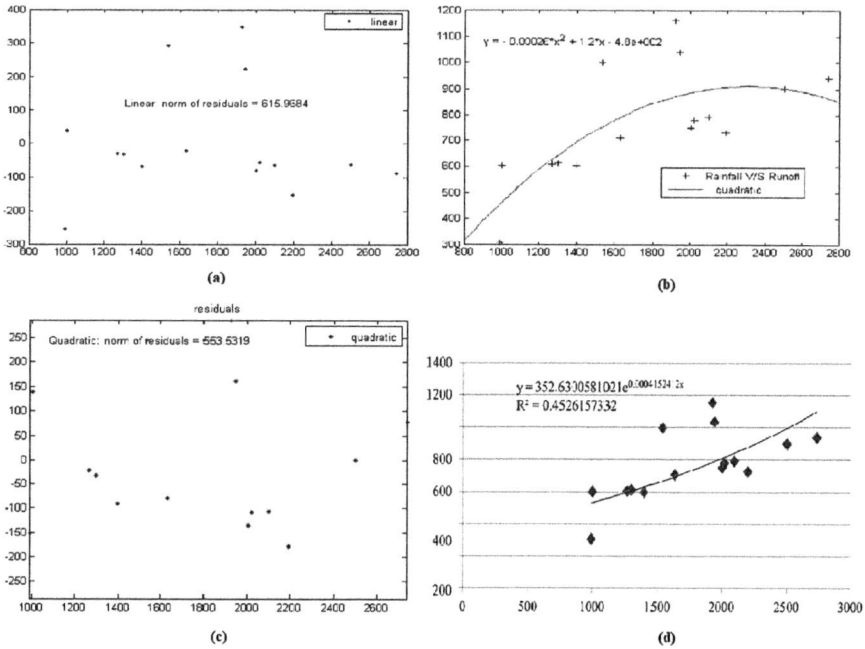

FIGURE 16.4 (a) Norms of residuals for linear regression, (b) Quadratic curve fitting, (c) Norms of residuals for quadratic regression, and (d) Exponential curve fitting.

The values for the constants A, B, R, and S are 295.6, 0.266, 0.645755, and 615.968 respectively. In Table 16.1 the values of the runoff and SSE using LR have been shown.

The total rainfall for the years 1995, 1996, and 1997 are 1827.3 mm, 1387.4 mm, and 1517.7 mm respectively, whereas the calculated runoff values for the given rainfall for the years 1995, 1996, and 1997 are 786 mm, 666 mm, and 701 mm.

16.3.3 QUADRATIC REGRESSION (QR) METHOD

For QR, the given rainfall-runoff data was used for a quadratic curve fitting and the curve was obtained as shown in Figure 16.4(b).

The curve obtained for norms of residuals for linear regression has been shown in Figure 16.4(c).

The values for the constants A, B, and C are −0.00025915, 1.2, and −478.21 respectively. While the values of R and S are 0.727324 and 553.53 respectively. In Table 16.2 the values of the runoff and SSE using linear regression have been shown.

The total rainfall for the years 1995, 1996, and 1997 are 1827.3 mm, 1387.4 mm, and 1517.7mm respectively, whereas the calculated runoff values for the given rainfall for the years 1995, 1996, and 1997 are 849 mm, 688 mm, and 746 mm.

TABLE 16.1

Calculated Values of Runoff and SSE Using LR

Year	Rainfall (mm)	Given Runoff (mm)	Calculated Runoff (mm)	SSE
1979	1535	998	705	86048
1980	1923	1158	809	121849
1981	1943	1037	814	49894
1982	1270	605	634	828
1983	1631	709	731	455
1984	1397	600	668	4576
1985	2023	780	835	2984
1986	2196	729	881	23292
1987	992	305	560	64933
1988	2740	939	1026	7577
1989	2500	900	963	3907
1990	2100	792	856	4039
1991	2005	750	830	6467
1992	1300	610	642	1009
1993	1001	602	562	155
			Total	**378013**

TABLE 16.2

Calculated Values of Runoff and SSE Using QR

Year	Rainfall (mm)	Given Runoff (mm)	Calculated Runoff (mm)	SSE
1979	1535	998	753	60184
1980	1923	1158	871	82272
1981	1943	1037	875	26393
1982	1270	605	627	486
1983	1631	709	789	6422
1984	1397	600	692	8425
1985	2023	780	888	11649
1986	2196	729	907	31749
1987	992	305	457	23023
1988	2740	939	864	5711
1989	2500	900	902	2325
1990	2100	792	899	11331
1991	2005	750	886	18427
1992	1300	610	644	1105
1993	1001	602	463	19212
			Total	**308714**

16.3.4 Exponential Regression (ER) Method

For the ER, the assumed rainfall-runoff data was used for an exponential curve fitting, and subsequently, the resultant curve was thus obtained as shown in Figure 16.4(d).

The values for the constants A, B, and R are 352.630058102, 0.0004152412, and 0.672309, respectively. In Table 16.3 the values of the runoff and SSE using linear regression have been shown.

The total rainfall for the years 1995, 1996, and 1997 are 1827.3 mm, 1387.4 mm, and 1517.7mm respectively. Whereas the calculated runoff values for the given rainfall for the years 1995, 1996, and 1997 are 753.0937 mm, 627.3628 mm, and 662.2419 mm.

16.3.5 Logarithmic Regression (LoR) Method

For the ER, the assumed rainfall-runoff data was used for an exponential curve fitting and thus the curve is obtained as shown in Figure 16.5.

The values for the constants A, B, and R are 479.5, 2797, and 0.691375. In Table 16.4 the values of the runoff and SSE using linear regression have been shown.

The total rainfall for the years 1995, 1996, and 1997 is 1827.3 mm, 1387.4 mm, and 1517.7 mm, respectively, whereas the calculated runoff values for the given rainfall for the years 1995, 1996, and 1997 are 804.3302 mm, 672.2721 mm, and 715.3142 mm.

TABLE 16.3
Calculated Values of Runoff and SSE Using ER

Year	Rainfall (mm)	Given Runoff (mm)	Calculated Runoff (mm)	SSE
1979	1535	998	706	85653
1980	1923	1158	796	130518
1981	1943	1037	801	55660
1982	1270	605	650	1960
1983	1631	709	727	317
1984	1397	600	676	5692
1985	2023	780	821	1676
1986	2196	729	867	19115
1987	992	305	596	84447
1988	2740	939	1027	7680
1989	2500	900	953	2825
1990	2100	792	841	2424
1991	2005	750	817	4481
1992	1300	610	656	2054
1993	1001	602	597	1912
			Total	**406414**

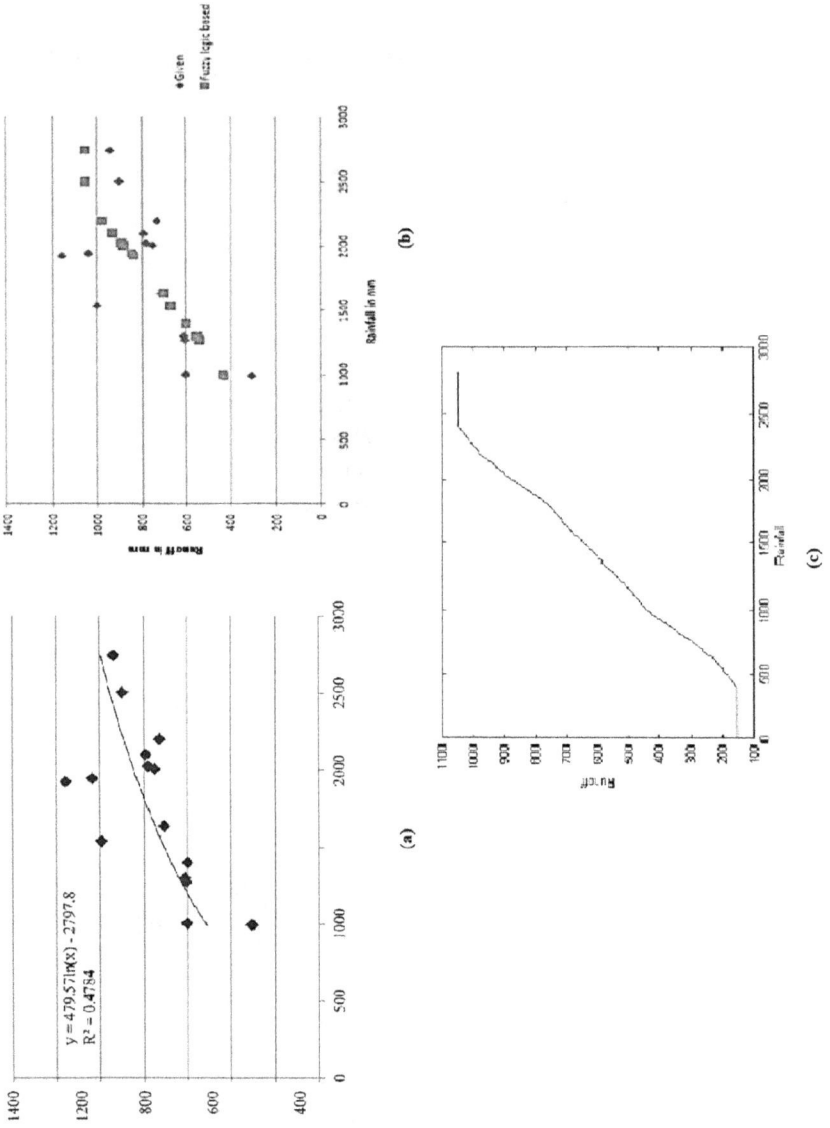

FIGURE 16.5 (a) Logarithmic curve fitting, (b) Comparison of results between given values and values from fuzzy logic, and (c) Surface between rainfall and runoff.

TABLE 16.4
Calculated Values of Runoff and SSE Using LoR

Year	Rainfall (mm)	Given Runoff (mm)	Calculated Runoff (mm)	SSE
1979	1535	998	720	77202
1980	1923	1158	829	108322
1981	1943	1037	833	41530
1982	1270	605	629	580
1983	1631	709	749	1622
1984	1397	600	675	5597
1985	2023	780	853	5212
1986	2196	729	892	26685
1987	992	305	511	42400
1988	2740	939	998	3451
1989	2500	900	954	2940
1990	2100	792	871	6160
1991	2005	750	848	9704
1992	1300	610	641	920
1993	1001	602	515	7449
			Total	**339774**

TABLE 16.5(a)
Calculated Values of Runoff and SSE Using Fuzzy Logic Approach

Year	Rainfall (mm)	Given Runoff (mm)	Calculated Runoff (mm)	SSE
1979	1535	998	665	111415
1980	1923	1158	835	104412
1981	1943	1037	846	36695
1982	1270	605	537	4732
1983	1631	709	699	114
1984	1397	600	598	882
1985	2023	780	889	11666
1986	2196	729	975	60432
1987	992	305	433	16230
1988	2740	939	1050	12133
1989	2500	900	1050	22326
1990	2100	792	927	18074
1991	2005	750	880	16816
1992	1300	610	549	3842
1993	1001	602	437	27324
			Total	**447093**

16.3.6 Fuzzy Logic Method

In this method, a computer program was used using MATLAB. In this method, five rules were evaluated based on fuzzy sets interference. The program uses the graphic user interface (GUI) of MATLAB and gives output in the fuzzy logic Toolbox. Results obtained from the fuzzy logic approach are shown in Table 16.5(a).

TABLE 16.5(b)
Comparisons of the Values of Coefficient of Correlation and SSE

Method	Coefficient of Correlation	Value of SSE
Linear regression	0.64	379417
Quadratic regression	0.72	306399
Exponential regression	0.67	404528
Logarithmic regression	0.69	339781
Fuzzy Logic Approach	0.71	320567

The total rainfall for the years 1995, 1996, and 1997 is 1827.3 mm, 1387.4 mm, and 1517.7 mm, respectively, whereas the calculated runoff values for the given rainfall for the years 1995, 1996, and 1997 are 777.13 mm, 592.66 mm, and 658.24 mm. In Figure 16.5(b), the comparison of the given runoff data and the runoff values obtained from the fuzzy logic method have been shown.

In Figure 16.5(c), surface generated between rainfall and runoff from the developed program has been illtreated. The value of the coefficient of correlation for the fuzzy logic method for the assumed rainfall-runoff data was 0.71126.

16.4   COMPARISON OF RESULTS

Comparison of the results was conducted based on the values obtained for SSE and coefficient of correlation. The value of the coefficient of correlation lies between 0.6 to 1 gives the best results. The higher the value of the correlation coefficient higher will be the accuracy of the results. In our chapter, the highest value of the coefficient of correlation is obtained for the QR and subsequently for the fuzzy logic. Also, the value of SSE was calculated and the value of SSE tending to 0 gives the higher value of goodness of the curve. So, for the case of SSE also the lowest value is obtained for the QR and thereafter for the fuzzy logic method. In Table 16.5(b), the values of coefficient of correlation and SSE have been compared for the different methods used.

16.5   CONCLUSIONS

This chapter concludes that for the given rainfall-runoff data of a watershed near the Palampur region of Kangra in Himachal Pradesh, the best correlation method of the rainfall-runoff is dependent upon the obtained value of the coefficient of correlation. In our chapter, the best results for both coefficients of correlation and SSE were obtained for the QR (Polynomial regression of second order). The value of the coefficient of correlation for the fuzzy logic approach is also nearly equal to the QR method.

For the study area (watershed near Palampur (H.P.)), it is concluded that the best method for the rainfall-runoff correlation is the QR model. For this model, the obtained value of the correlation coefficient is 0.727324, and the value of SSE is

306399.012. However, the fuzzy logic approach-based method also proves good for the rainfall-runoff correlation.

It is recommended that in case of rainfall-runoff records existence, it is preferable to apply a QR approach for the runoff estimations from the given rainfall measurements. To adjust the uncertainties in the runoff estimation for a hilly catchment area, one can use the fuzzy logic approach.

REFERENCES

Beard, S., Rowe, T., and Fox, J. 2020. An analysis and evaluation of methods currently used to quantify the likelihood of existential hazards. *Futures*, 115, 102469.

Dawson, C. W., and Wilby, R. 1998. An artificial neural network approach to rainfall-runoff modeling. *Hydrological Sciences Journal*, *43*(1), 47–66.

Hong, Y. S., Rosen, M. R., and Reeves, R. R. 2002. Dynamic fuzzy modeling of storm water infiltration in urban fractured aquifers. *Journal of Hydrologic Engineering*, *7*(5), 380–391.

Jaiswal, R. K., Ali, S., and Bharti, B. 2020. Comparative evaluation of conceptual and physical rainfall-runoff models. *Applied Water Science*, 10(1), 1–14.

Kumar, N., Poddar, A., Shankar, V., Ojha, C. S. P., and Adeloye, A. J. 2020. Crop water stress index for scheduling irrigation of Indian mustard (*Brassica juncea*) based on water use efficiency considerations. *Journal of Agronomy and Crop Science*, *206*(1), 148–159.

Kumari, S., Poddar, A., Kumar, N., and Shankar, V. 2021. Delineation of groundwater recharge potential zones using the modeling based on remote sensing, GIS and MIF techniques: a study of Hamirpur District, Himachal Pradesh, India. *Modeling Earth Systems and Environment*, 1–12.

Mukerji, A., Chatterjee, C., and Raghuwanshi, N. S. 2009. Flood forecasting using ANN, neuro-fuzzy, and neuro-GA models. *Journal of Hydrologic Engineering*, *14*(6), 647–652.

Poddar, A., Kumar, N., and Shankar, V. 2021. Evaluation of two irrigation scheduling methodologies for potato (Solanum tuberosum L.) in north-western mid-hills of India. *ISH Journal of Hydraulic Engineering*, 27(1), 90–99.

Poddar, A., Kumar, N., Kumar, R., Shankar, V., and Jat, M. K. 2020. Evaluation of non-linear root water uptake model under different agro-climates. *Current Science*, 119(3), 485.

Saadi, M., Oudin, L., and Ribstein, P. 2020. Beyond imperviousness: the role of antecedent wetness in runoff generation in urbanized catchments. *Water Resources Research*, 56(11), e2020WR028060.

See, L., and Openshaw, S. 2000. A hybrid multi-model approach to river level forecasting. *Hydrological Sciences Journal*, *45*(4), 523–536.

Sharma R. K., Kaur A., and Kumar A., 2019. Slope stability analysis by bishop analysis using MATLAB program based on particle swarm optimization technique. In: Singh H., Garg P., Kaur I. (eds) *Proceedings of the first International Conference on Sustainable Waste Management through Design. ICSWMD 2018. Lecture Notes in Civil Engineering*, 21. Springer, Cham.

Sun, A., Yu, Z., Zhou, J., Acharya, K., Ju, Q., Xing, R., ... and Wen, L. 2020. Quantified hydrological responses to permafrost degradation in the headwaters of the Yellow River (HWYR) in High Asia. *Science of the Total Environment*, 712, 135632.

Tayfur, G., and Singh, V. P. 2006. ANN and fuzzy logic models for simulating event-based rainfall-runoff. *Journal of Hydraulic Engineering*, *132*(12), 1321–1330.

Tayfur, G., Ozdemir, S., and Singh, V. P. 2003. Fuzzy logic algorithm for runoff-induced sediment transport from bare soil surfaces. *Advances in Water Resources*, *26*(12), 1249–1256.

Thakur, M. S., Pandhiani, S. M., Kashyap, V., Upadhya, A., & Sihag, P. 2021. Predicting bond strength of FRP bars in concrete using soft computing techniques. *Arabian Journal for Science and Engineering*, 46(5), 4951–4969.

Tilmant, A., Vanclooster, M., Duckstein, L., & Persoons, E. 2002. Comparison of fuzzy and nonfuzzy optimal reservoir operating policies. *Journal of Water Resources Planning and Management, 128*(6), 390–398.

Walter, F., Amann, F., Kos, A., Kenner, R., Phillips, M., de Preux, A., ... and Bonanomi, Y. 2020. Direct observations of a three million cubic meter rock-slope collapse with almost immediate initiation of ensuing debris flows. *Geomorphology,* 351, 106933.

Yu, P. S., and Yang, T. C. 2000. Fuzzy multi-objective function for rainfall-runoff model calibration. *Journal of Hydrology,* 238(1–2), 1–14.

17 Geospatial Technology-Based Artificial Groundwater Recharge Site Selection for Sustainable Water Resource Management
A Case Study of Rajkot District, Gujarat

Jaysukh Chhaganbhai Songara and Jayantilal Naginbhai Patel

CONTENTS

DOI: 10.1201/9781003203445-20

17.1 INTRODUCTION

Groundwater is one of the most valuable natural resources, accounting for about 34% of the globe's freshwater resources (Shekhar and Pandey, 2015). It is in high demand in recent years worldwide, as its application is not confined to agricultural activity, and it has numerous applications in household purpose, various industrial demand, and many more. In countries like India, on average 50% of the urban and 90% of rural populations rely on groundwater to fulfil their daily needs. According to report of GEC 2015, approximate 70% of total groundwater is used for agricultural purposes in India. Groundwater potential zones and recharge sites are depending on various factor including soil, geomorphology, rainfall, existing groundwater levels, lineament density, and drainage density.

Nowadays, GIS- and RS-based technology play significant roles in assessment of different worldwide problems, as this technology has emerged over the last few years, and has capacity to handle massive amounts of satellite data. Moreover, this technology has more advantages than the traditional methods of assessment as it saves various resources like personnel, cost, and time. To assessment of groundwater potential zones, various researchers have used numerous GIS integrated techniques. Index models have been used in a number of studies to analyse groundwater prospective mapping (Sahoo and Khaoash, 2020; Pathak and Hiratsuka, 2011; Jaiswal et al., 2003).

Rainwater harvesting sites and potential groundwater zone (PGWZ) are found and suggested to increase the groundwater and surface water for agricultural activity (Contreras et al., 2013; Singha et al., 2021). The assistance of Landsat satellite picture and DEM extricated from SRTM information. This information was utilized to deliver eight thematic maps of the volume of yearly floods, slope, area of the basin, length catchment, maximum flow distance in the catchment, drainage density of the area, lineament density of the catchment, and stream order (Mugo and Odera, 2019).

The artificial recharge is used to replenish the groundwater reservoir and ensure the long-term viability of groundwater development. Artificial recharge schemes are being implemented in a variety of hydro geological situations. The ability to identify the characteristics of different rock types, rainfall pattern, evaporation losses, climatological features, natural topography, soil types, drainage pattern, sources of water availability, detailed knowledge of geological and hydrological features, help in the selection of sites and the design of artificial recharge structures. The area with a high yield obviously has more storage potential, and thus there is potential for more groundwater recharge.

National awareness to the use of artificial recharge to support ground water supplies is one result of the growing competition for water. Simply put, artificial recharge is the process of guiding excess surface water into the ground to replenish an aquifer either by spreading on the surface, using recharge wells, or altering natural conditions to increase infiltration. Artificial recharge (also known as scheduled recharge) is a method of storing water underground during periods of excess rainfall in order to meet demand during periods of scarcity. Water managed to recover from recharge projects can be used for non-potable purposes, such as landscape irrigation, or for potable purposes, which is less common.

Artificial recharge can be used to reduce the cost of ground water pumping by trying to control seawater intrusion in coastal aquifers, controlling land subsidence due to declining ground water levels, maintaining base flow in some streams, and increasing water levels to control seawater intrusion in coastal aquifers. As per the reports of CGWB (2000; 2007), they have applied artificial recharge in Ahmedabad, Mehsana, Kachchh, and Coastal Saurashtra region using different techniques such as injection well, farm pond, check dam, and many more. Moreover, Weirs/check dams are considered feasible in challenging rock areas with moderate slope, whereas percolation tanks are deemed appropriate to hard rock plateau and the plain regions. In addition, weirs/check dams are considered feasible in semi-consolidated formations. On the other hand, percolation tanks are considered appropriate in alluvium-affected areas (CGWB, 2020).

Selection of rainwater harvesting sites depend on hydrological and hydro-morphometric criteria (Shazwani Muhammad et al., 2017). For the water supervisors, primarily on greater scales, the assessment of RWH prospective and the decision of reasonable goals for RWH structures are incredibly trying and for evaluating RWH prospective and recognizing sites/zones for different RWH structures using geospatial and multi-criteria decision examination (MCDA) strategies using ArcGIS programming remote identifying data and common field data were used to design needed topical layers (Adham et al., 2018).

The Rajkot district is the major district of Saurashtra region of Gujarat state, India. The district comprises many land use patterns. However, the agriculture area is major in the district. The majority agricultural fields of the district are depending on rainwater. Due to uncertainty and irregularity of the rainfall, Rajkot district is facing water crisis over the last few years (CGWB, 2013). Additionally, issues such as lack of awareness, improper planning of groundwater extrusion, salinity in groundwater, improper pumping of water, and plummeting groundwater levels, have increased the water instability in the district (CGWB, 2013). Rajkot district has contributed more than 20% of the country's overall GDP rate. Furthermore, different industries like power industries, construction, countrywide, and worldwide trade have increased the importance of study area. There has been much research conducted on groundwater potential zones with numerous techniques. In this research authors have contributed towards groundwater potential zone mappings and the proposed recharge sites, as few studies have been found regarding groundwater potentiality and recharge methodology in Rajkot district. Moreover, CGWB have provided information about groundwater recharge. According to the hydrologic and hydro-geologic conditions, authors have suggested techniques for the groundwater recharge in the Rajkot district.

17.2 METHODOLOGY

1) **Data collection**: The collection relevant to this work will be collected from the GWRDC, SWDC, and Bahumali Bhavan Rajkot.
2) **Data analysis:** Analysis of the data will be done using Microsoft Excel and GIS.

3) **Map preparation**: Develop thematic layers such as Drainage, Lineament, Rainfall map, LULC map, geology, and soil map, using remote sensing.
4) **Delineation of the map**: Development of lineament map using Geomatica software.
5) **Rainfall mapping**: Study and mapping of rainfall data using ArcGIS.
6) Find a Rainwater harvesting site in the catchment area with the help of ArcGIS.
7) Various thematic layers were prepared for data analysis and integration in ArcGIS.
8) Determination of groundwater potential zones using ArcGIS technology for the study area.
9) The flow chart showing the methodology used in the present study is shown in Figure 17.1.

Figure 17.1 shows a flowchart of the methodology. Visual interpretation of satellite images was used for site selection, based on the lineament using Geomatica software, soil texture, LU/LC, and geology from LISS Image in ArcGIS. These features permitted site selection to be supported by the natural features of the region. The lineaments were extracted by interpreting aerial photographs from 2013 using ArcGIS and geological maps. As input data, satellite imagery and a 30 m digital elevation model (DEM) from the Shuttle Radar Topographic Mission (SRTM) were applied. A total of 32 suitable RWH locations were chosen for hydrological factor computations. As a result, both ground control points and feature heights enhanced the SRTM DEM.

ArcGIS software-defined the basins and afforded numerous features for each watershed. Eight thematic maps – the volume of the yearly flood, catchment region, catchment length, catchment slope, maximum flow distance, drainage frequency density, lineament frequency density, and stream order – were extracted to determine prospective sites for rainwater collection.

17.2.1 STUDY AREA

Rajkot can be said as important district of central Saurashtra peninsula, Gujarat state, India. The district geographically lies between the coordinates 20°30′ and 23°12′ North and 70°00′ and 71°45′ East. The boundary of the district touches Junagadh and Amreli, Jamnagar, small desert of Kachchh district and Surendranagar at south, west, north, and east, respectively. The approximate area of the district is 7,586 km² during the winter season.

The study is of Rajkot district shown in Figure 17.2. According to the Indian Census 2011, the district population was 38 lakhs. Bhadar, Aji, Machhu, and Demai are the main rivers flowing through the district. The average rainfall of the district is 710 mm. And temperature varies between 11 °C to 40 °C. Moreover, the district falls under subtropical climate region, comprises three seasons summer (April–June), Monsoon (July–September) and winter (October–March). The major dams in the districts are Aji I, Aji II and Aji III on river Aji and Nyari dam I and II on the River Nyari. Most of the region of Rajkot district is covered with agriculture, with an area

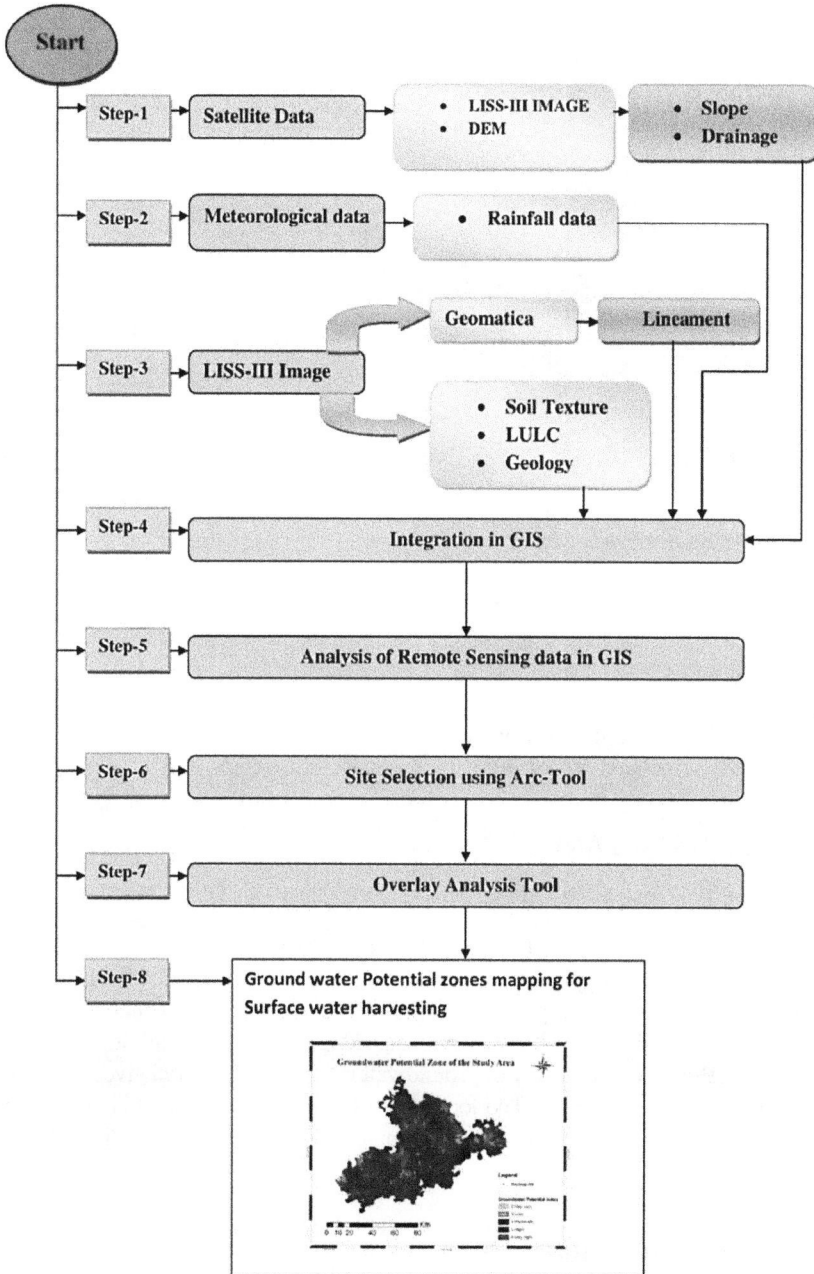

FIGURE 17.1   Methodology used in the present study.

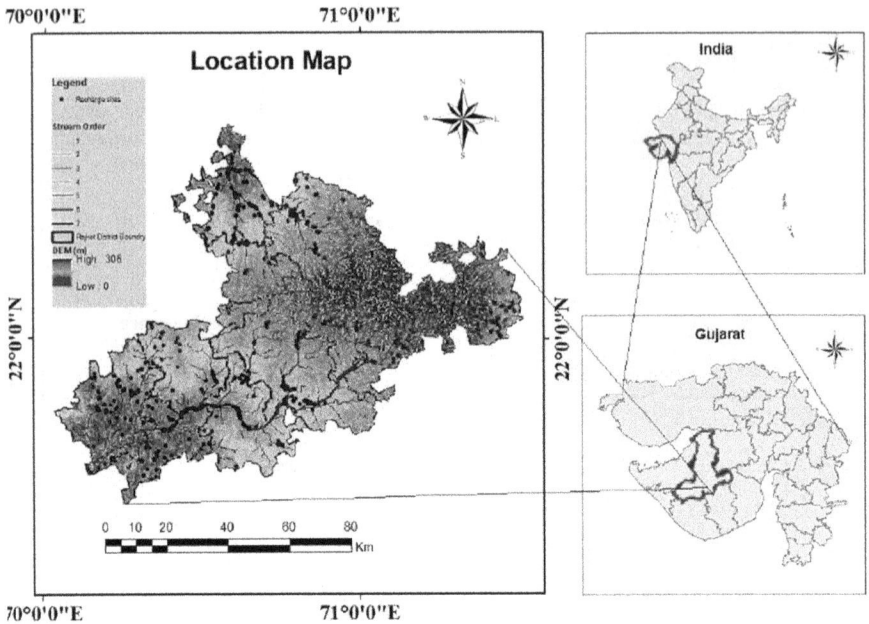

FIGURE 17.2 Location map of the study area.

of 5,685.19 km2. The major crops in the district are groundnut and cotton in monsoon season. Moreover, wheat and cumin are also cultivated.

17.3   MATERIALS AND METHOD

17.3.1   GEOLOGY

The development of groundwater relies upon the geographical setting. A visual interpretation of the satellite picture has been utilized for an outline of land highlights. Figure 17.3(a) shows geographical investigation territory has been characterized into five classes: Flood Plains, Pediplain, Plateau, Educational Hills, and Structural Hills. Pediplain yields a good quantity of groundwater; the Pediment area gives moderately good amount of groundwater. The location of the highly dissected Plateau, denudation hill, and the lateritic Plateau were poor for groundwater prospecting (Terêncio et al., 2017).

17.3.2   RAINFALL PATTERN

The examination region gets precipitation from the southwest storm beginning amid the long stretch of June and reaching out to September. October precipitation in a few areas inside the bowl Rainfall appropriation alongside the slant inclination specifically influences spill over water penetration rate. The minimum rain is 390 mm in the examination zone, and the most extreme precipitation is around 1950 mm, shown in Figure 17.3(b).

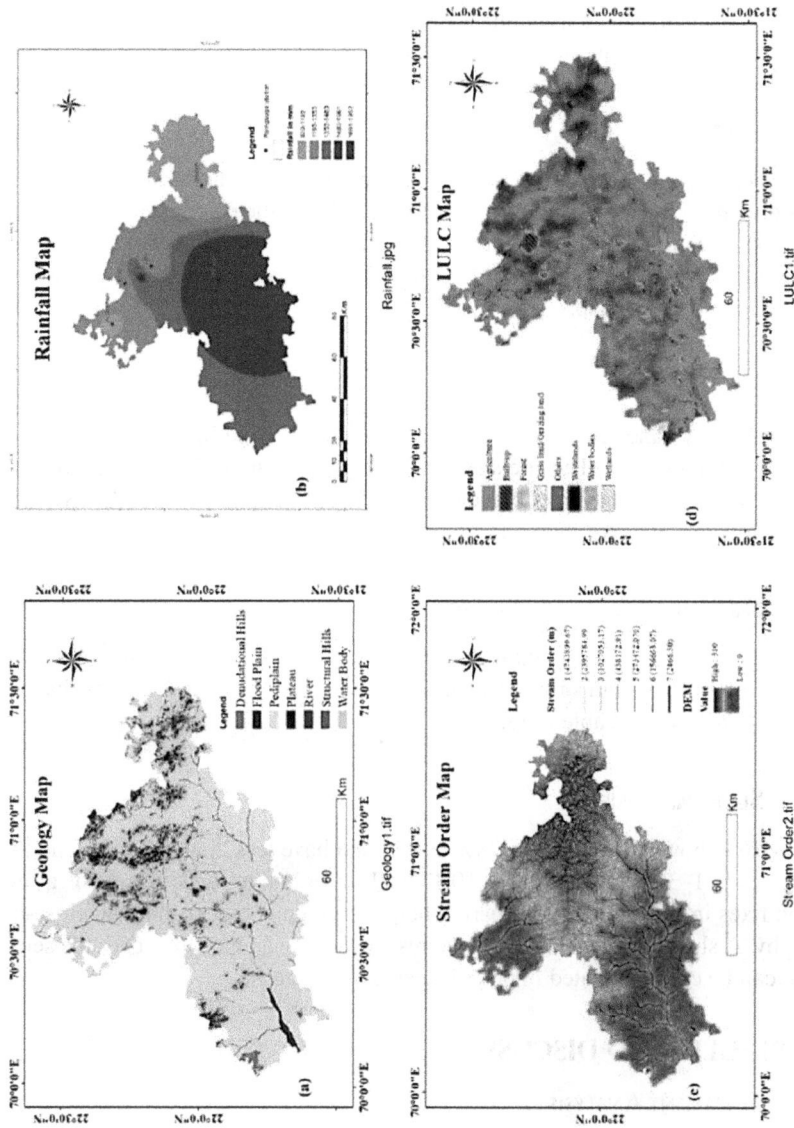

FIGURE 17.3 (a) Geology map, (b) Rainfall map, (c) Stream order map, and (d) LULC map.

17.3.3 MORPHOMETRIC ANALYSIS

Stream arranges to rely upon the relationship of tributaries. The stream analysis organizes fundamental essentialness for mapping of rainwater collection prospective, while a lower stream sets results in higher defencelessness and interruption (Glendenning et al., 2012). The overall length of the dumbest founding stream arranges it was 2 km, shown in Figure 17.3(c). The aggregate length of the plain low class (first and second request) was 7,139 km.

17.3.4 LAND USE

Land use depends on natural and anthropogenic activities. These maps have been developed using ArcGIS software. Figure 17.3(d) shows the catchment area including agriculture, buildings, forest, grassland/grazing land, wastelands, water bodies, wetlands, towns/cities (urban), villages (rural) have been identified and delineated the land use/ land cover map.

17.3.5 SOIL TEXTURE

Five types (clayey sand, clayey skeletal, fine clay, loamy, and loamy skeletal) have been identified in the area. The site's statistical analysis with the help of ArcGIS has been done, which gave the values clayey sand as 53.29% and fine clayey 40.18% and other type soil is 6.53% are shown in Figure 17.4(a).

17.3.6 DRAINAGE FREQUENCY DENSITY

The drainage density is 2.93 km/km^2, stream frequency is 3.21 streams/ km^2 shown in Figure 17.4(b). The high stream frequency and drainage density indicates that the rapid runoff and impermeable lithology.

17.3.7 SLOPE ANALYSIS

Figure 17.4(c) shows that about 86.78% of the land have less than 1% slope and the remaining area 1–3% approximately 10.17%. The region verified except for a few restricted fixes in the Northeastern part, where the grade is higher than the close-by topography. A slope is the measure of progress in surface an incentive over the separation. It can be communicated in percentages or as degrees.

17.4 RESULTS AND DISCUSSIONS

17.4.1 LINEAMENT ANALYSIS

Lineaments are land joint/crack surfaces that display straight components like straight or nearly straight, as seen in Figure 17.4(d).

They expect a fundamental occupation in seeing fitting objectives for common vivification of groundwater in light of how they judge artificial structures. Water can immerse and affect an outing up to a few metres.

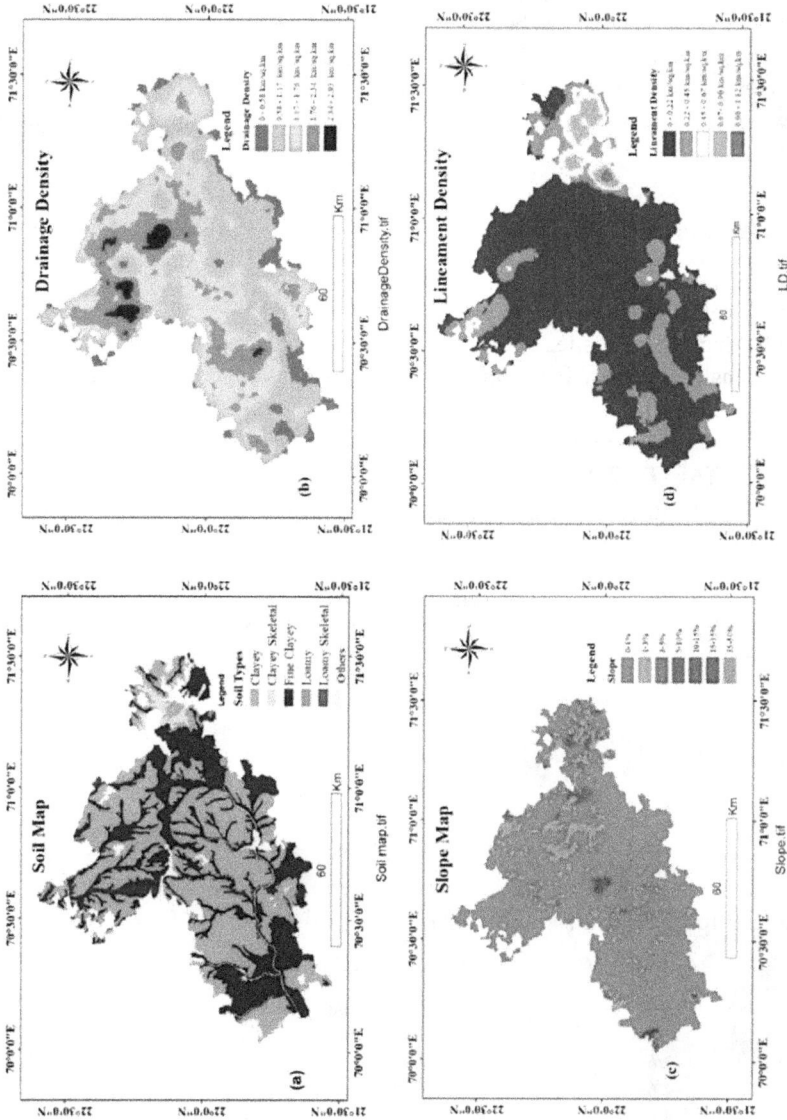

FIGURE 17.4   (a) Soil map, (b) Drainage density, (c) Slope map, and (d) Lineament density.

17.4.2 Normalized Weights

Table 17.1 summarized the weights assigned to various thematic layers. Table 17.2 appeared in the normalized weight for particular topics utilizing the AHP and eigenvector system.

Normalized class weights are associated with rename each subject's classes, and the renamed topical layers have been facilitated into the ArcGIS. The assistance of overlay examination device standardized subject loads outlines the potential groundwater zones in the investigation region. Table 17.3 reported that the normalization weights of various classes of themes had been resolved in this way, which is correct.

17.4.3 Artificial Recharge Sites

Water reaping potential guide of the examination region was created by crossing point the topical layers of drainage of the investigation zone, and lineament and processing water collecting potential site. The site is distinguished to give a proportion of the area's rejuvenate capability for taking up water protection measures by developing

TABLE 17.1
Weights Assigned to the Seven Thematic Layers

Serial	Theme	Weight
1	Rainfall	6
2	Geology	5
3	LULC	2
4	DEM	4
5	LD	4
6	DD	3
7	Soil	4

TABLE 17.2
Pairwise Comparison Matrix and Normalized Weights of the Seven Themes

Theme	GM(6)	GLG(5)	LULC(2)	DEM(4)	LD(4)	DD(3)	SL(4)
RF(6)	6/6	6/5	6/2	6/4	6/4	6/3	6/4
GLG(5)	5/6	5/5	5/2	5/4	5/4	5/4	5/4
LULC(2)	2/6	2/5	2/2	2/4	2/4	2/4	2/4
DEM(4)	4/6	4/5	4/2	4/4	4/4	4/3	4/4
LD(4)	4/6	4/5	4/2	4/4	4/4	4/3	4/4
DD(3)	3/6	3/5	3/2	3/4	3/4	3/3	3/4
SL(4)	4/6	4/5	4/2	4/4	4/4	4/3	4/4

LULC, Landuse/Landcover; DD, Drainage Density; GLG, Geology; DEM, Digital Elevation Model; LD, Lineament Density; SL, soil; RF, Rainfall; NW, Normalized Weights of a Feature Class.

TABLE 17.3

Assigned and Normalized Weight of the Different Classes of Each Theme

Theme	Feature Classes	Potential	Assigned Weight	Normalized Weight
Geology	Structural Hill	Nil	1	0.05
	Denudational hill	Nil	1	0.05
	Plateau	Moderate	3	0.16
	Flood Plain	Good	4	0.21
	Pediplain	Good to very good	5	0.26
	Alluvium	Very Good	5	0.26
	Total		19	
Soil type	Clayey	Very Poor	1	0.07
	Clayey Skeletal	Poor	2	0.13
	Fine	Moderate	3	0.20
	Loamy Skeletal	Good	4	0.27
	Loamy	Very Good	5	0.33
	Total		15	
Land Cover	Built-up area	Very Poor	1	0.03
	Wasteland	Very Poor	2	0.06
	Wetland	Poor	4	0.13
	Grassland	Moderate	5	0.16
	Agricultural	Good	6	0.19
	Forest	Good	6	0.19
	Water bodies	Very Good	7	0.23
	Total		31	
Drainage Density	2.41–3.00	Very Poor	1	0.07
	1.81–2.40	Poor	2	0.13
	1.21–1.80	Moderate	3	0.20
	0.61–1.20	Good	4	0.27
	0–.60	Very Good	5	0.33
	Total		15	
Lineament Density	0–.60	Very poor	1	0.06
	0.61–1.20	Poor	2	0.13
	1.21–1.80	Moderate	3	0.19
	1.81–2.40	Good	4	0.25
	2.41–3.00	Very Good	6	0.38
	Total		16	
Slope	35–50	Very poor	1	0.07
	15–35	Poor	2	0.13
	8–15	Moderate	3	0.20
	3–8	Good	4	0.27
	0–3	Very Good	5	0.33
	Total		15	
Rainfall	920–1195	Very Poor	1	0.07
	1195–1353	Poor	2	0.13
	1353–1483	Moderate	3	0.20
	1483–1661	Good	4	0.27
	1661–1952	Very good	5	0.33
	Total		15	

artificial recharge structures, such as Check Dams/ Cement Plug / Nala Bund, Farm Pond/Khet talavadi, Permeation/Percolation tanks, and Gabion structures, recharge wells/pits, Recharge Shaft, Injection Wells etc. are shown in Figure 17.5(a). Check dams are built across small streams with a gentle slope and can be built in both hard rock and alluvial formations. Gabion structure is a type of check dam that is commonly used to conserve stream flows across small streams with little to no submergence beyond the stream's course. The boulders that are economically feasible are kept in a steel wire. By grounding it to the streamside, this is hung across the stream's mesh to act as a small dam. A percolation tank is an artificially induced surface water body that submerges highly permeable land areas in its reservoir, allowing surface runoff to percolate and recharge ground water storage.

17.4.4 STATUS OF SAMPLE RECHARGE SITES

The sample sites in the research area are classified into two categories based on the data collected from latitude and longitude in the Google Earth Pro and Google Maps survey regarding these sites' existing status. The two types are:

1. Areas suitable for artificial recharge.
2. Sites with existing recharge structures.

The distribution can be classified as suitable recharge sites and existing recharge structures. It may be observed from Figure 17.6 that the majority of the sample sites indicate the suitability for artificial recharge (Abdulla Umar Naseef and Thomas, 2016). The Table 17.4 further reveals the accuracy of the proposed artificial recharges sites in Hingolgadh, Gangajal, Amargadh, and Atkot. The site is recognized to give a proportion of the areas energize capability and development of (artificial recharge) structures, such as Pond/Khet talavadi: 64, check dams: 06, gabion structures: 06, recharge trenches: 01 and percolation tanks etc. These are distributed throughout the study area. The most concentration convergence of such locales with existing

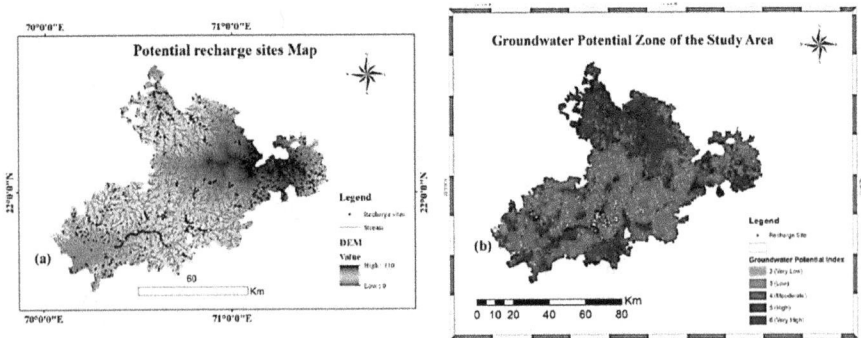

FIGURE 17.5   (a) Potential Recharge Sites, and (b) Identified groundwater potential zones and suitable recharge sites.

FIGURE 17.6 Proposed location sites suitable for artificial recharge of groundwater (GoogleEarthPro).

recharge structures is found in the northeast part of the investigation zone. Figure 17.6 reveals the current status of some sample sites.

These are the proposed artificial recharge sites visited during latitude and longitude in Google Earth Pro in the study area and are considered appropriate for the artificial recharge of groundwater. Figure 17.7 shows the artificial recharges structures constructed by the government.

These sites are categorized as sites with existing recharge structures. This is lessening the good destinations for artificial recharge of groundwater which is nearly needed for maintainable groundwater advancement in the examination zone. Table 17.5 demonstrated further uncovers that there are 12 test locales (Reservoirs: 08, Check dam: 03 and Lake: 01) where the government has developed some artificial recharge structure.

17.4.5 GROUNDWATER POTENTIAL ZONING

The zoning index for groundwater resources was divided into five categories: very low, low, moderate, high, and very high, as shown in Figure 17.5(b). The depiction was noteworthy in clearly delineating the predicted groundwater potential zones. They outline the zones where the scene is most appropriate for groundwater stockpiling furthermore exhibit the accessibility of water. Topographically, these zones are arranged in southern, north-east, and a few settlers on the zone's western parts. The site verified by the moderate and extraordinary (high) groundwater potential zoning index is about (83%). The focal part and a few fragments of the north and southwest fall under the moderate groundwater potential class and identified potential groundwater zones shown in Figure 17.5(b). The lineament and natural drainage pattern overlay with the help of Intersect tools in ArcGIS output gets as recharge

TABLE 17.4

Status of Proposed Artificial Recharge Sites in Rajkot district, Gujarat

Serial	Geographical Coordinates		Location	Suggested Location	Serial	Geographical Coordinates		Location	Suggested Location
	Latitude	Longitude				Latitude	Longitude		
1	21.64989	70.31343	Velariya	Pond/Khet talavadi	40	22.01472	71.4127	Gadhala	Pond/Khet talavadi
2	21.65188	70.22432	Meli	Pond/Khet talavadi	41	22.02161	71.18943	Jasdan	Pond/Khet talavadi
3	21.65908	70.23567	Meli	Pond/Khet talavadi	42	22.01616	71.21123	Jasdan	Pond/Khet talavadi
4	21.70665	70.33099	Bholgamda	Check dam	43	22.02642	71.40424	Gadhala	Pond/Khet talavadi
5	21.73712	70.14313	Rajkot	Check dam	44	22.02439	71.14534	Atkot	Pond/Khet talavadi
6	21.73073	70.39219	Rajkot	Pond/Khet talavadi	45	22.0369	71.35501	Vadod	Gabion Structure
7	21.72217	70.79279	Bantva-Devli	Pond/Khet talavadi	46	22.05581	71.38473	Somalpar	Pond/Khet talavadi
8	21.73059	70.83342	Barvala Baval	Pond/Khet talavadi	47	22.08348	71.47008	Sartanpar	Pond/Khet talavadi
9	21.73198	70.66574	Jetpur	Pond/Khet talavadi	48	22.08689	71.39804	Modhuka	Pond/Khet talavadi
10	21.74611	70.49874	Mota Gundala	Check dam	49	22.09877	71.48437	Belda	Pond/Khet talavadi
11	21.7635	70.21426	Nagvadar	Pond/Khet talavadi	50	22.12935	71.43691	Veraval Bhadli	Pond/Khet talavadi
12	21.73321	70.96195	Sanali	Pond/Khet talavadi	51	22.13434	71.41399	Janada	Pond/Khet talavadi
13	21.76771	70.09348	Pransla	Pond/Khet talavadi	52	22.13148	71.45901	Sanali	Pond/Khet talavadi
14	21.80031	70.57207	Panchpipla	Pond/Khet talavadi	53	22.13602	71.46275	Sanali	Pond/Khet talavadi
15	21.78242	70.45742	Bhukhi	Pond/Khet talavadi	54	22.15507	71.39317	Janada	Pond/Khet talavadi
16	21.77109	70.22187	Kolki	Pond/Khet talavadi	55	22.15537	71.3352	Hingolgadh	Recharge trenches
17	21.79759	70.17327	Rajpara	Check dam	56	22.16975	71.37252	Hathasani	Pond/Khet talavadi
18	21.80364	70.41274	Adval	Pond/Khet talavadi	57	22.19447	71.35402	Revaniya	Pond/Khet talavadi
19	21.80061	70.87253	Dhudashiya	Pond/Khet talavadi	58	22.21204	71.30228	Dadli	Pond/Khet talavadi
20	21.81793	70.9027	Mota Sakhpar	Pond/Khet talavadi	59	22.20805	71.22849	Dhokalva	Pond/Khet talavadi
21	21.82527	70.83325	Shivrajgadh	Pond/Khet talavadi	60	22.21469	70.92647	Aniyala	Pond/Khet talavadi
22	21.82741	70.90697	Shrinathgadh	Gabion Structure	61	22.23662	71.39049	Samadhiyala	Pond/Khet talavadi
23	21.8465	70.14863	Hariyasan	Gabion Structure	62	22.24347	71.35679	Kharachiya Jas	Pond/Khet talavadi
24	21.87902	70.4283	Padariya	Gabion Structure	63	22.24104	71.32214	Ori	Pond/Khet talavadi
25	21.87686	70.42678	Padariya	Pond/Khet talavadi	64	22.25024	71.27559	Chhasiya	Pond/Khet talavadi
26	21.84677	70.42969	Rodhel	Pond/Khet talavadi	65	22.25475	71.29484	Chiroda	Pond/Khet talavadi
27	21.88428	70.25888	Bhayavadar	Pond/Khet talavadi	66	22.25784	71.39593	Gangajal	Pond/Khet talavadi
28	21.89124	70.2788	Arni	Gabion Structure	67	22.28985	70.89342	Amargadh	Pond/Khet talavadi
29	21.89333	70.31988	Timbadi Jam	Pond/Khet talavadi	68	22.33595	70.91955	Targhadia	Pond/Khet talavadi

No	Latitude	Longitude	Village	Structure
30	21.87237	70.96408	Dodiyara	Pond/Khet talavadi
31	21.92952	71.05097	Pratappur	Pond/Khet talavadi
32	21.92924	71.13345	Garani	Pond/Khet talavadi
33	21.96787	71.09346	Jasapar	Check dam
34	21.98112	71.2746	Gokhalana	Pond/Khet talavadi
35	21.99309	71.14772	Atkot	Pond/Khet talavadi
36	21.98291	71.19721	Khanpar	Pond/Khet talavadi
37	21.98504	71.35404	Kaskoliya	Pond/Khet talavadi
38	22.00682	71.3922	Gadhala	Pond/Khet talavadi
39	22.01099	71.13614	Atkot	Pond/Khet talavadi
69	22.34973	70.85131	Nakaravadi	Check dam
70	22.37349	70.73077	Khandheri	Pond/Khet talavadi
71	22.37721	70.93525	Jhiyana	Pond/Khet talavadi
72	22.38157	70.74261	Anandpar	Pond/Khet talavadi
73	22.40261	70.87194	Nagalpar	Pond/Khet talavadi
74	22.46473	70.51916	Sanosara	Pond/Khet talavadi
75	22.45187	70.68681	Kothariya	Gabion Structure
76	22.47416	70.64297	Ukarda	Pond/Khet talavadi
77	22.54209	70.56917	Modpar	Pond/Khet talavadi

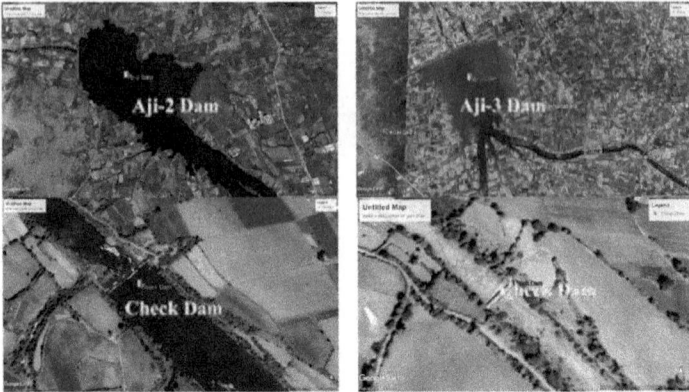

FIGURE 17.7   Sites having existing structures for artificial recharge sites (GoogleEarthPro).

TABLE 17.5
Status of Existing Artificial Recharge Sites in Rajkot District, Gujarat

| Serial | Geographical Coordinates | | Location | Existing Location |
	Latitude	Longitude		
1	21.70437	70.73994	Charaniya	Reservoir
2	21.73681	70.38907	Bhadar River	Bhadar Dam
3	21.76351	70.41611	Bhukhi	Bhadar-II
4	21.81321	70.78659	Bhadar Reservoir	Bhadar Dam
5	21.84088	70.52074	Dudhivadar	Fofal dam
6	21.88850	70.15669	Satvadi	Paneli lake
7	21.90025	71.01315	Sanathali	Check dam
8	22.00388	71.16666	Atkot	Check dam
9	22.06971	71.49011	Sanala	Reservoir
10	22.09909	71.43009	Patiyali	Check Dam
11	22.35998	70.78929	Aji River	Aji Dam II
12	22.40002	70.71713	Anandpar	Anandpar Dam

points in the high groundwater potential zones. Moreover, the overlaying of these points on groundwater potential zone, to identify some location for check dam, Gabion structure and Farm pond which is shown in Figure 17.6 and more location indicates in Table 17.4.

17.5   CONCLUSIONS

To develop prospective recharge areas and ideal recharge site structures for artificial recharge to groundwater, assorted topical layer maps were prepared using remote detecting frameworks or data assembled by conventional techniques for the study area. RS and geographic information system techniques provide very accurate results quickly for delineating the artificial recharge sites and groundwater potential zones to boost the groundwater yield and recharge. The topical layers included topography, lineaments, drainage pattern, land use and soil texture maps from remotely detected information, and incline/slope from topographic maps. The extent of each standard was controlled in geographic information system tropical maps. In light of the leakage (drainage) guide and lineament layout, a wire of all GIS criteria produces maps to perceive the sensible RWH areas. The site is recognized to give a proportion of the areas energize capability and development of (artificial recharge) structures, such as Pond/Khet talavadi (64), check dams (06), gabion structures (06), recharge trenches (01), and percolation tanks. Moreover, we suggest ideal sites for artificial recharge based on the latest Geoinformatics technology and relevant data from the Govt. of Gujarat.

Here, this model is regional model; the same technique can be used for other regions also. However, the influencing parameters are depending on the motive of the study, topography of area, and other related conditions. Therefore, to choose the affecting parameters can be the challenging task for other regions. Limitations of the study includes that overall application of this study on the field will depend on the scheme/guidelines of the CGWB-India and Govt. of Gujarat.

WEBSITES

1. http://cgwb.gov.in/Manuals-Guidelines.html
2. https://iwmpmis.nic.in/
3. https://nrega.nic.in/netnrega/guidelines.aspx
4. https://gswma.gujarat.gov.in/
5. https://swhydrology.gujarat.gov.in/

REFERENCES

Abdulla Umar Naseef, T., & Thomas, R. (2016). Identification of suitable sites for water harvesting structures in Kecheri River Basin. *Procedia Technology*, *24*, 7–14. https://doi.org/10.1016/j.protcy.2016.05.003

Adham, A., Sayl, K. N., Abed, R., Abdeladhim, M. A., Wesseling, J. G., Riksen, M., Fleskens, L., Karim, U., & Ritsema, C. J. (2018). A GIS-based approach for identifying potential sites for harvesting rainwater in the Western Desert of Iraq. *International Soil and Water Conservation Research*, *6*(4), 297–304. https://doi.org/10.1016/j.iswcr.2018.07.003

Contreras, S. M., Sandoval, T. S., & Tejada, S. Q. (2013). Rainwater harvesting, its prospects and challenges in the uplands of Talugtog, Nueva Ecija, Philippines. *International Soil and Water Conservation Research*, *1*(3), 56–67. https://doi.org/10.1016/S2095-6339(15)30031-9

Glendenning, C. J., van Ogtrop, F. F., Mishra, A. K., & Vervoort, R. W. (2012). Balancing watershed and local scale impacts of rainwater harvesting in India—A review. *Agricultural Water Management*, *107*, 1–13. https://doi.org/10.1016/j.agwat.2012.01.011

Jaiswal, R. K., Mukherjee, S., Krishnamurthy, J., & Saxena, R. (2003). Role of remote sensing and GIS techniques for generation of groundwater prospect zones towards rural development – An approach. *International Journal of Remote Sensing*, *24*, 993–1008. https://doi.org/10.1080/01431160210144543

Mugo, G. M., & Odera, P. A. (2019). Site selection for rainwater harvesting structures in Kiambu County-Kenya. *The Egyptian Journal of Remote Sensing and Space Science*, *22*(2), 155–164. https://doi.org/10.1016/j.ejrs.2018.05.003

Pathak, D. R., & Hiratsuka, A. (2011). An integrated GIS-based fuzzy pattern recognition model to compute groundwater vulnerability index for decision-making. *Journal of Hydro-Environment Research*, *5*(1), 63–77. https://doi.org/10.1016/j.jher.2009.10.015

Sahoo, S., & Khaoash, S. (2020). Impact assessment of coal mining on groundwater chemistry and its quality from Brajrajnagar coal mining area using indexing models. *Journal of Geochemical Exploration*, *215*, 106559. https://doi.org/10.1016/j.gexplo.2020.106559

Shekhar, S., & Pandey, A. C. (2015). Delineation of groundwater potential zone in hard rock terrain of India using remote sensing, geographical information system (GIS) and analytic hierarchy process (AHP) techniques. *Geocarto International*, *30*(4), 402–421. https://doi.org/10.1080/10106049.2014.894584

Singha, S., Das, P., & Singha, S. S. (2021). A fuzzy geospatial approach for delineation of groundwater potential zones in Raipur district, India. *Groundwater for Sustainable Development*, *12*, 100529. https://doi.org/10.1016/j.gsd.2020.100529

Terêncio, D. P. S., Sanches Fernandes, L. F., Cortes, R. M. V., & Pacheco, F. A. L. (2017). Improved framework model to allocate optimal rainwater harvesting sites in small watersheds for agro-forestry uses. *Journal of Hydrology*, *550*, 318–330. https://doi.org/10.1016/j.jhydrol.2017.05.003

18 Rainfall-Runoff Estimation for Rapti River Catchment Using Geospatial Technology

Suchita Pandey, Nilanchal Patel, and Ajay Kumar Agrawal

CONTENTS

DOI: 10.1201/9781003203445-21

18.1 INTRODUCTION

The quantitative assessment of surface runoff is essential for land and water resource developments and water yield potential of the watershed, planning water conservation measures, recharging groundwater zones and reducing the flooding hazard downstream (Sharma and Thakur, 2007). The accurate and direct estimation of runoff from rainfall through the Soil Conservation Services Curve Number (SCS-CN) method, known as Natural Resource Conservation (NRCS), developed by the United States Department of Agriculture, in 1954 has been done by many researchers. Due to the scarcity of past and accurate information on rainfall and runoff for ungauged watersheds in India (Sarangi et al., 2008), there is an urgent need to generate information on basin runoff.

The concept of weighted curve number is used for runoff water available for artificial recharge in the Ayyar basin, Tamil Nadu (Anbazhagan et al., 2005). The land use/land cover classes can be integrated with the hydrologic soil groups of the Palleru sub-basin in GIS, and the weighted CN can be estimated. These estimated weighted CN for the entire area can compute runoff (Viswanadh and Girdhar, 2007). Shi et al. (2009) studied the Three Gorges Area of China and showed that the Ia / S adjusted method better correlates between estimated and observed runoff than the traditional method.

In the present study, we have implemented an improved runoff evaluation method that considers the antecedent soil moisture for the subcatchment of Rapti River. The subcatchment is located in the Rapti basin, which is dominantly mountainous, and characterized by a steep slope.

18.1.1 SIGNIFICANCE OF THE RESEARCH

The various parameters that affect runoff in a watershed can be categorized into two broad groups, static and dynamic. The static parameters remain stable and unchanged over a very long period, such as soil type, slope, and major drainage patterns, whereas land use and land covers, and the various climatic parameters such as rainfall, temperature, and humidity constitute the major dynamic parameters. The soil's antecedent moisture condition (AMC) primarily determined from the five-day cumulative rainfall also significantly influences the runoff estimate, therefore, it is also important to determine the relationship between rainfall and runoff in different AMC conditions. The present study has been conducted to investigate the relationship between rainfall and runoff in different AMC conditions in the individual hydrological response units and for different LULC categories present in the study area.

18.1.2 OBJECTIVES

The primary objectives of the present study are as follows:

1. To estimate runoff for each hydrological response unit (HRU) of the study area using the SCS-CN method.

2. To estimate runoff generated within individual LULC categories covered under the influence of different rainfall stations employing area-weighted SCS-CN method.
3. To estimate runoff in different AMC conditions within each HRU and within each LULC category.
4. To determine the agreement between the runoff estimated from the different HRUs and the corresponding LULC categories falling within individual rainfall stations.

18.2 DATA AND METHODS

18.2.1 STUDY AREA

The study area comprises the subcatchment of the Rapti River, with 95% of its area covered in Nepal and 5% in India. It comprises the Middle Hill region, the Siwalik Range, the Terai region in Nepal portion with Indo–Gangetic plain of Shrawasti District. Bhinga is the main entry point of the Rapti River in India. The total areal extent of the subcatchment is 5.77 lakh hectares (57,76.09 sq. km.). The latitudinal and longitudinal extent of the subcatchment is from 27° 39′ 11.63″ N to 28° 33′ 42.34″ N and 81° 53′ 27.23″ E to 82° 32′ 9.22″ E respectively (Figure 18.1). The whole subcatchment is divided into 21 HRUs. Six rainfall stations are lying within and nearby the subcatchment. The location of the rainfall stations is shown in Figure 18.2(a).

The drainage map of the study area is shown in Figure 18.2(b). The soils of different regions of the area have different characteristics, but most of the soils are coarse-textured, excessively drained to well-drained, very deep to moderately deep with high transmission capacity. The annual rainfall varies between 1,500 and 2,500 mm, with about 80% of the precipitation confined to the monsoon period (June–September) (www.thamel.com).

18.2.2 DATA AND SOFTWARE USED

The following datasets were used in the present study.

1. PAN data of IRS P6 – of subscene A of the scene (P101 – R52) acquired on 10 February 2003, subscene B (P101 – R52) acquired on 7 March 2003 and subscene D (P100 – R51) acquired on 23 February 2004 and AWiFS data with a 56-metre spatial resolution of P101 – R53 acquired on 21 October 2004 were used to delineate LULC and drainage of the study area.
2. The SRTM data (90 m) downloaded from the website www.srtm.csi.cgiar. org was used to generate slope, contour, and upstream drainage maps.
3. The soil map for the upper portion of the subcatchment was downloaded from the website www.icimod.org, and the thematic layer of soil for the Indian portion was procured from Soil Division, UPRSAC, Lucknow, India.

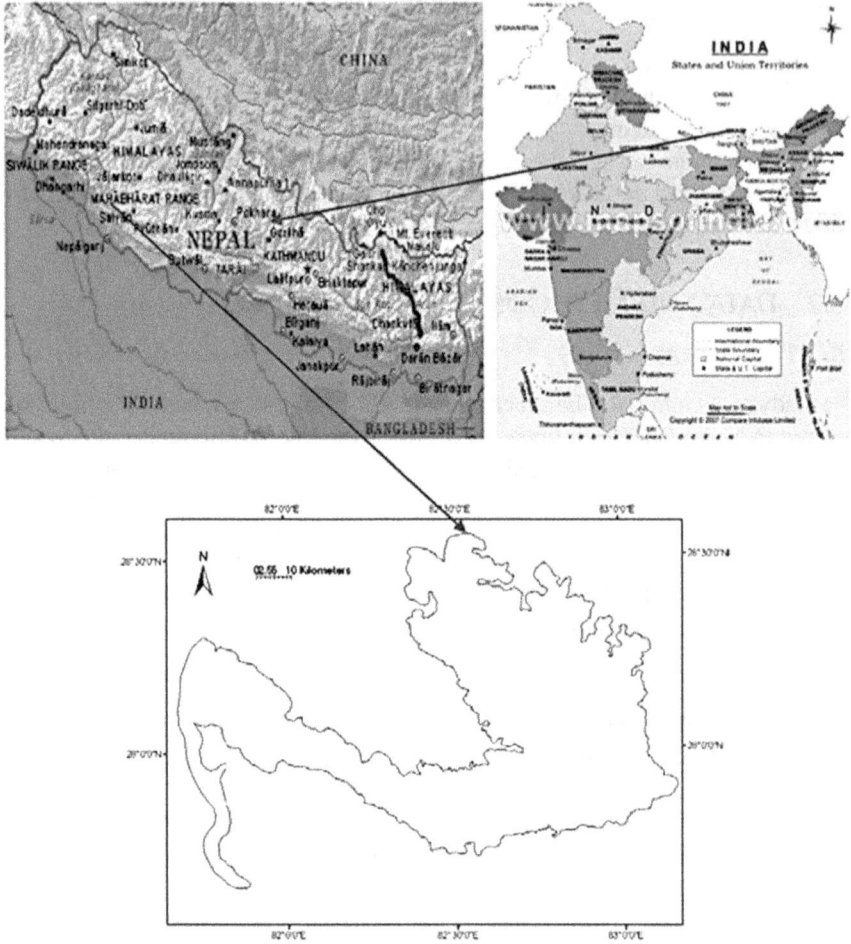

FIGURE 18.1   Location map of study area.

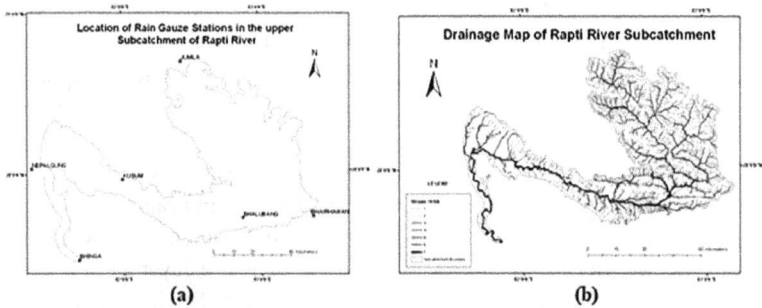

FIGURE 18.2   Rainfall stations and Rapti River Subcatchment in the study area.

The information regarding soil properties for the generation of Hydrological Soil Group was extracted from the book "Soil Series of India" (Seghal et al., 1994).

4. Rainfall data for all six rainfall stations were procured from the Irrigation Department, Central Water Commission, Lucknow.

18.2.3 METHODOLOGY

The Survey of India (SOI) topographic maps, soil map, and rainfall data were used to generate Hydrological Soil Group (HSG) map and hydrological database for runoff estimation. By integrating all the databases such as LULC, HSG, drainage density, average slope percentage and rainfall records, runoff estimation was done through the SCS-CN method. The flow chart of methodology is shown in Figure 18.3. The entire subcatchment is divided into 21 major HRUs (Figure 18.4).

18.2.3.1 Database

18.2.3.1.1 Land Use / Land Cover

The study area comprises six major LULC categories. Dense forest comprises 4,933 km^2 area, which amounts to 85.7%, followed by agricultural land (7.1%), open forest (2.3%), built-up land (0.55%). Flood plain and river comprise nearly 151 and 98 km^2 area respectively (Figure 18.4). The area of each LULC category falling under different HRUs is presented in Table 18.1.

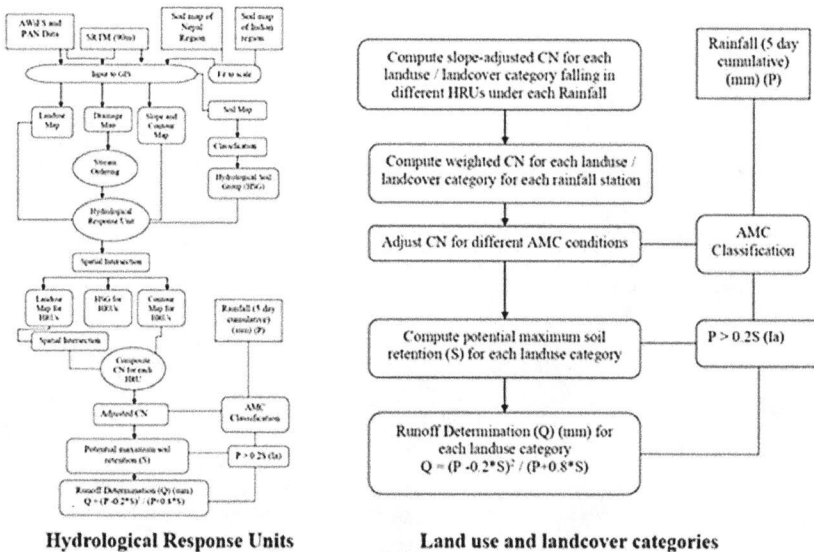

Hydrological Response Units   **Land use and landcover categories**

FIGURE 18.3 Flow chart for determination of runoff (Q).

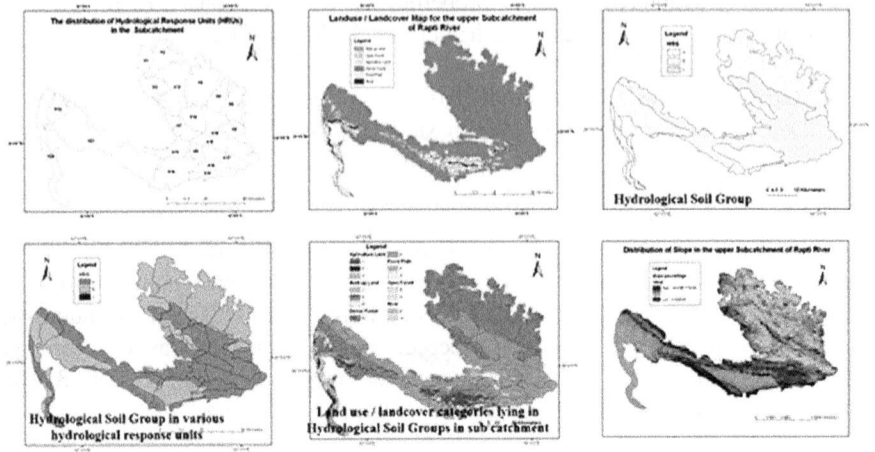

FIGURE 18.4 Hydrological, land use, and land cover status in the study area.

TABLE 18.1
Area of Different Land Use/Land Cover within Each HRU

Hydrological Response Unit	Built Up	Open Forest	Agriculture Land	Dense Forest	Flood Plain	River
K 1				157.97		
K 2				219.24		
K 3				192.07		
K 4				469.96		
K 5				166.92		
K 6				235.31		
K 7				194.99		
K 8				143.38		
K 9				66.71	31.48	0.000462
K 10		17.41	4.79	106.75	4.14	3.02
K 11	0.466	1.21	5.76	175.46	8.48	0.022
K 12	0.1796	15.59	1.91	154.71		0.446
K 13				443.95		
K 14				195.51		
K 15		3.79		120.78	0.52	0.434
K 16				377.61		0.196
K 17		6.62	6.89	204.96	12.32	5.59
K 18	0.245746	9.91	5.41	89.06	5.85	2.73
K 19	8.628	34.71	117.26	198.66	31.24	11.99
K 20	17.83	33.73	166.76	129.99	33.73	31.26
K 21	5.29	9.33	100.13	889.258261	23.36	42.78

18.2.3.1.2 Hydrological Soil Group

The HSG map of the study area has been prepared by adopting the criteria provided by Anbazhagan et al. (2005) (Figure 18.4). LULC categories lying in different hydrological soil groups covered within the individual HRUs are shown in Figure 18.4.

18.2.3.1.3 Slope

The slope of a terrain reveals the runoff and infiltration condition in a basin (Daofeng et al., 2004). A basin with less slope will induce more infiltration if all other parameters are constant. The slope in the study area varies from 2° to 89° from the lower Indo–Gangetic plain to the upper hill region. The slope percentage map for the entire subcatchment is shown in Figure 18.4. The average slope percentage for different HRUs varies from 0.019 to 0.0435.

18.2.3.1.4 Rainfall

The rainfall data is procured for six rain gauze stations located within the extent of the subcatchment, viz. Bhinga, Kusum, and Bhalubang while the remaining three lie close to the study area: Nepalgunj, Bhairhawan, and Jumla (Figure 18.2). For deriving the area of influence of a particular rain gauze station in the subcatchment, Theissen polygons were drawn.

18.2.3.2 Runoff Estimation from SCS-CN Method

If a basin comprises different LULC and soil, the concept of composite CN is used to obtain an accurate estimate of runoff (Anon, 1973; SCS). If the slope of a basin varies, the CN values should also be adjusted for slope (Daofeng et al., 2004). The following equations are used for the computation of runoff (USDA-SCS, 1985).

$$Q = \frac{(P-0.2S)^2}{P+0.8S} \left(\text{Geetha et al.2008}\right) \qquad (18.1)$$

where Q is the runoff volume expressed in depth (mm), P is total rainfall, S is potential maximum soil retention. S, the potential maximum soil retention, is computed using the following equation.

$$S = \frac{25400}{CN} - 254 \left(\text{Geetha et al.2008}\right) \qquad (18.2)$$

The CN values are tabulated in the National Engineering Handbook for various land cover categories and different soil textures.

18.2.3.2.1 Determination of Weighted CN for Each Hydrological Response Unit

The CN values tabulated in the National Engineering Handbook pertain to specific LULC and soil texture. These CN values are referred to as CN2 values in the equations used for runoff estimation. If an HRU is characterized by a single LULC and homogeneous HSG represented by uniform soil texture, the runoff for this HSU can be estimated using the standard CN (i.e., CN2 values) pertaining to the particular LULC and HSG present therein. However, many HRUs may be associated with more than one LULC occurring within one HSG. In such situations, the runoff estimation for the HRU is performed using the weighted CN computed in four steps, as discussed below. In the first step, the composite CN of the respective LULC present within the different HSGs in the HSU is determined from CN2 values using Equation 18.3; the extent of each LULC covered by the different HSGs is obtained through their spatial intersection performed in Arc GIS. In the second step, the composite CN determined for specific LULC categories within the HRU are converted to their CN3 values using Equation 18.4. The CN2 values provided in the National Engineering Handbook for different LULC categories are assumed to be appropriate for 5% slopes. Therefore, it is required to adjust the CN3 values determined for the individual LULC categories for the average slope percentage existing in the HRU. The slope-adjusted CN values are referred to as $CN2_s$, which are determined using Equation 18.5. In the final step, the weighted CN value for a particular HRU is computed by considering the $CN2_s$ values of different LULC categories and their proportionate areal extents using Equation 18.6.

The composite CN for a particular LULC category is computed using the following equation:

$$\text{Composite CN} = \frac{\Sigma(CN_i * A_i)}{\Sigma A_i} \tag{18.3}$$

where CN_i and A_i represent the curve number and area of the particular LULC category corresponding to the ith HSG.

Computation of CN3 for each LULC category was done by adopting the following formula (Daofeng et al., 2004).

$$CN3 = CN2 * \exp\left[0.00673(100 - CN2)\right] \tag{18.4}$$

The CN3 value for each LULC category is adjusted for its average slope percentage using the following equation (Daofeng et al., 2004).

$$CN2_s = \left[(CN3 - CN2)\right]/3.\left[1 - 2.\exp(-13.86 * SLP)\right] + CN2 \tag{18.5}$$

where CN3 is derived from Equation 18.4, and SLP denotes the average slope percentage.

The weighted curve number for each HRU is determined using the following equation.

$$\text{Weighted CN} = \frac{\Sigma\left[(CN2_s)_m * a_m\right]}{A} \quad (\text{Anbazhagan et al.2005}) \tag{18.6}$$

where $(CN2_S)_m$ represents the curve number for a particular LULC m, a_m is the area of that LULC, and A is the total area of the HRU.

18.2.3.2.2 Estimation of Antecedent Moisture Condition (AMC)

The AMC of the soil influences the runoff on the surface. The AMC of soil is determined by the cumulative rainfall of the five previous days' rainfall data and is classified into three types: dry, average, and wet or saturated condition, based on USDA-SCS criteria (Table 18.2). With the help of AMC determined for different dates using USDA-SCS criteria, the weighted CN values for different HRUs were adjusted for only those dates for which the rainfall is greater than Ia.

The Ia, initial abstraction shows the water losses like plant interception, infiltration and surface storage which occur before runoff and is subtracted from the total runoff (USDA-SCS 1985). Ia for different AMCs is determined using the following equations.

$$Ia = 0.3 * S \text{ for AMC I} \tag{18.7}$$

$$Ia = 0.2 * S \text{ for AMC II} \tag{18.8}$$

$$Ia = 0.1 * S \text{ for AMC III} \tag{18.9}$$

If the rainfall is greater than Ia, runoff is possible; otherwise, runoff is zero. Rainfall events greater than Ia were considered for the computation of runoff using the SCS-CN method (Anbazhagan et al., 2005).

The weighted CN values for different HRUs are adjusted for different AMCs using the following equations.

For AMC I (dry condition):

$$CN = 0.39 * CN * \exp(0.009 * CN) \quad (\text{Anbazhagan et al.} 2005) \tag{18.10}$$

For AMC II (average condition) and III (wet condition):

$$CN = 1.95 * CN * \exp(-0.00663 * CN) \quad (\text{Anbazhagan et al.} 2005) \tag{18.11}$$

After determining the weighted CN for each HRU, the S, potential maximum retention was estimated using the Equation 18.2. Finally, runoff (Q) was

TABLE 18.2
Classification of Antecedent Moisture Condition (AMC)

Antecedent Moisture Condition	Rainfall Range (mm)
I (dry)	<36
II (normal / average)	36–53
III (wet)	>53

computed for each HRU using Equation (18.1 which uses rainfall (P) and S as the input parameters.

The correlation coefficient (r) and the regression equations have been determined between rainfall and runoff for each of the 21 HRUs of the subcatchment. The coefficient of determination (r^2) has also been determined between the estimated values of runoff obtained from the above method and its corresponding computed values obtained from the regression equation for each HRU. The correlation coefficient (r) has also been determined between rainfall and runoff for each of the 21 HRUs of the subcatchment for AMC II and AMC III, respectively, to investigate the variation in the rainfall-runoff relationship in average and wet conditions. Since P values were less than Ia under AMC I for each of the 21 HRUs, runoff estimation was not carried out for AMC I.

18.2.3.3   Runoff Estimation for each LULC Category Using SCS-CN Method

In the present study, runoff estimation has been carried out for different LULC categories within the subcatchment through a modified SCS-CN method. To perform this task, first, the HRUs falling under the area of influence of individual rain gauze stations were delineated through the Theissen polygon technique and were grouped as independent land units. Thus a total of six such independent land units were delineated corresponding to as many rain gauze stations. Subsequently, runoff estimation was carried out for the individual LULC categories falling under each land unit separately since the rainfall recorded for a particular rain gauze station is uniform in the entire area of a particular land unit. Finally, the total runoff estimated for all the LULC categories falling in a particular land unit for different dates was compared with that estimated for the HRUs falling within the same land unit through regression analysis. In addition, regression analysis was performed for the individual LULC categories for AMC II and AMC III conditions, respectively, to delineate comparison between the two different AMC conditions. These tasks were performed for all the six land units separately. The stepwise procedure to estimate runoff for different LULC categories is provided below.

18.2.3.3.1   Determination of Weighted Curve Number for Each Land Use and Land Cover Category

The same database comprising the rainfall records, HSG, LULC categories and average slope percentage used to estimate runoff for the individual HRUs were used to compute runoff for each LULC category covered by the respective land units. Since the category-wise runoff estimation was carried out for each independent land unit separately, it is desired to determine the weighted CN2s value for the individual categories within each land unit. The weighted CN2s for a particular LULC category was computed by considering the areal proportion of that category present in the different HRUs falling within the land unit under consideration using the following equation.

$$\text{Weighted CN2s} = \frac{\Sigma\left[(CN2_S)_i * a_i\right]}{A} \tag{18.12}$$

where $(CN2_S)_i$ represents the CN value for a particular LULC category for the ith HRU determined from Equation 18.5, a_i is the area of this LULC category in the ith HRU, n is the number of HRUs, and A is the total area of the LULC categories present in all the HRUs lying under the area of influence of a particular rainfall station.

18.2.3.3.2 Computation of AMC-adjusted CN Values for Different Land Use and Land Cover Categories

With the help of AMC determined for different dates using USDA-SCS criteria (Table 18.2), the weighted CN2s values for different LULC categories were adjusted for only those dates for which the rainfall is greater than Ia. Ia values for AMC I, AMC II and AMC III are determined using Equations 18.7, 18.8, and 18.9 respectively. The weighted CN2s values computed for different LULC categories using Equation 18.12 are adjusted for different AMCs using the following equations (Sharma and Thakur, 2007).

For AMC I

$$CN\,I = \frac{4.2 * CN\,II}{(10 - 0.058 * CNII)} \tag{18.13}$$

For AMC II and AMC III

$$CN\,III = \frac{23 * CN\,II}{(10 + 0.13 * CN\,II)} \tag{18.14}$$

18.2.3.3.3 Computation of Runoff for Different Land Use and Land Cover Categories

The AMC-adjusted CN values computed for the individual LULC categories were used in Equation 18.2 for determining the S, potential maximum soil retention for the respective LULC categories, which was further used for the estimation of runoff (Q) using Equation (18.1 for different dates of rainfall records.

18.2.3.3.4 Types of Analyses Performed

The following analyses were performed for runoff estimation of the individual LULC categories in the different land units delineated as the area of influence under respective rainfall stations.

1. (a) Determination of coefficient of correlation (r) between P (rainfall) and Q (runoff)

   (b) Determination of coefficient of determination (r^2) between estimated Q using SCS-CN method and regressed Q determined from the regression equation obtained at 1(a).
2. Determination of correlation coefficient (r) between P and Q for AMC II and AMC III. This task was not performed for AMC I since P was found to be less than Ia.

3. (a) Determination of coefficient of determination (r^2) between the aggregate runoff ($\sum Q$) of all the LULC categories and runoff ($\sum Q$) estimated from the land unit comprising the various LULC categories.

(b) Determination of coefficient of determination (r^2) between aggregate runoff ($\sum Q$) of all the LULC categories and ($\sum Q$) of the land unit comprising the various LULC categories for AMC II and AMC III, respectively.

18.3 RESULTS AND DISCUSSION

In the present study, runoff estimation has been carried out for the individual HRUs and individual LULC categories in order to delineate comparison between the runoff estimated from both the sources. The results of the investigation performed are presented in three different parts, as discussed below.

1. Rainfall-runoff relationship for the individual HRUs.
2. Rainfall-runoff relationship for the individual LULC units.
3. Comparative analysis between the cumulative runoff estimated from the HRUs and that from the corresponding LULC categories in the individual land units covered under the influence of each rain gauze station.

18.3.1 RAINFALL-RUNOFF RELATIONSHIP FOR THE INDIVIDUAL HRUs

The graphical plots between 5-day cumulative rainfall and runoff for 21 different HRUs covering the subcatchment are shown in Figure 18.5. The r values of the HRUs exceed 0.99, which signifies a perfect linear relationship between the rainfall and runoff in all the HRUs. This situation results in the prevalence of significantly high values of coefficient of determination (r^2) between the runoff values estimated from the above and runoff computed using the regression equation for the individual HRUs with r^2 values greater than 0.99.

AMC Wise: Further, there also occur very high very correlations between the rainfall and runoff estimated through the SCS-CN method in the average and saturated AMC conditions that is AMC II and AMC III, respectively (Figure 18.6). This relationship was not determined for AMC I condition since P was found to be less than Ia.

The occurrence of r and r^2 values of 0.99 in each HRU can be attributed to the constant value of S (maximum potential soil retention) for a particular HRU that makes the runoff directly proportional to rainfall.

18.3.2 RAINFALL-RUNOFF RELATIONSHIP FOR THE INDIVIDUAL LAND USE AND LAND COVER UNITS

As per the procedure discussed in section 18.2, runoff has been estimated for different LULC categories lying within the land unit under the influence of each rain gauze station. The value of r is found to be uniform (0.99) for all the LULC categories lying under the influence of different rain gauge stations.

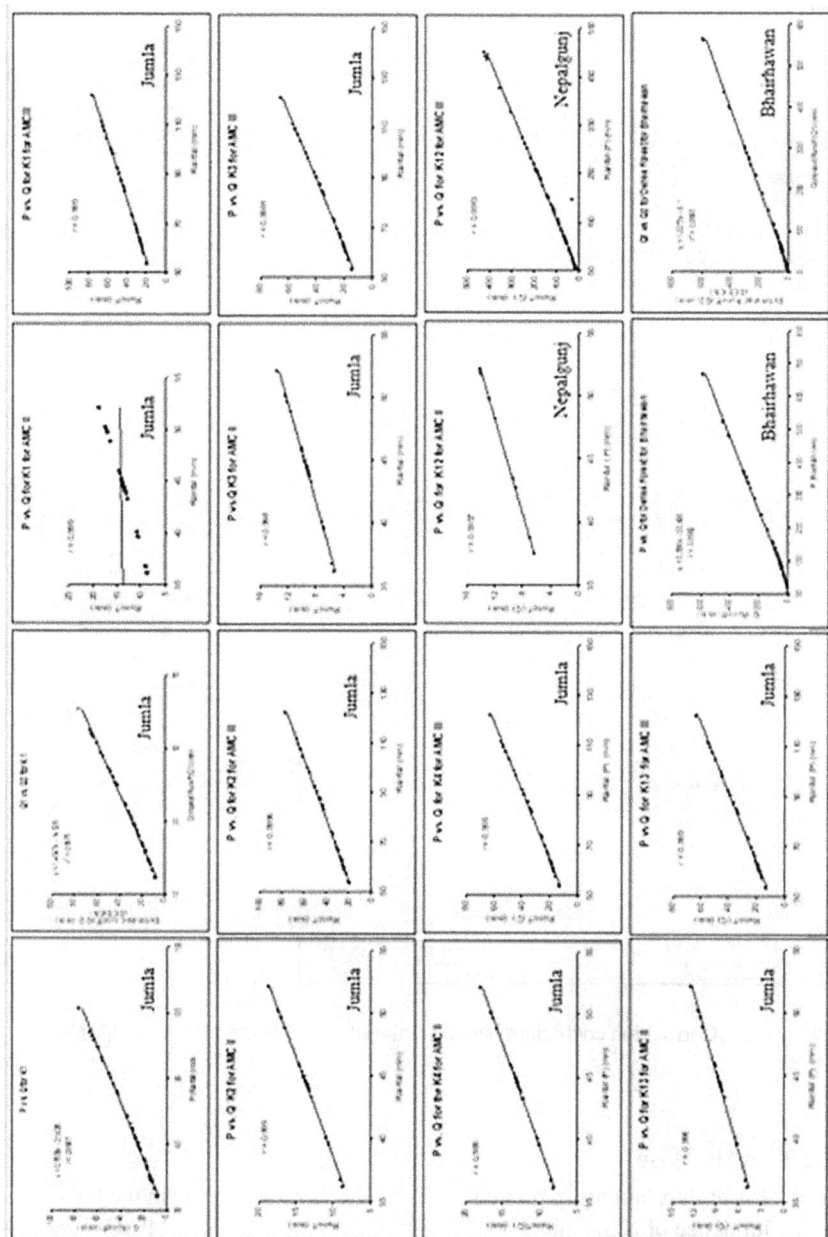

FIGURE 18.5   Rainfall versus runoff for AMCs for HRUs for rain gauze stations.

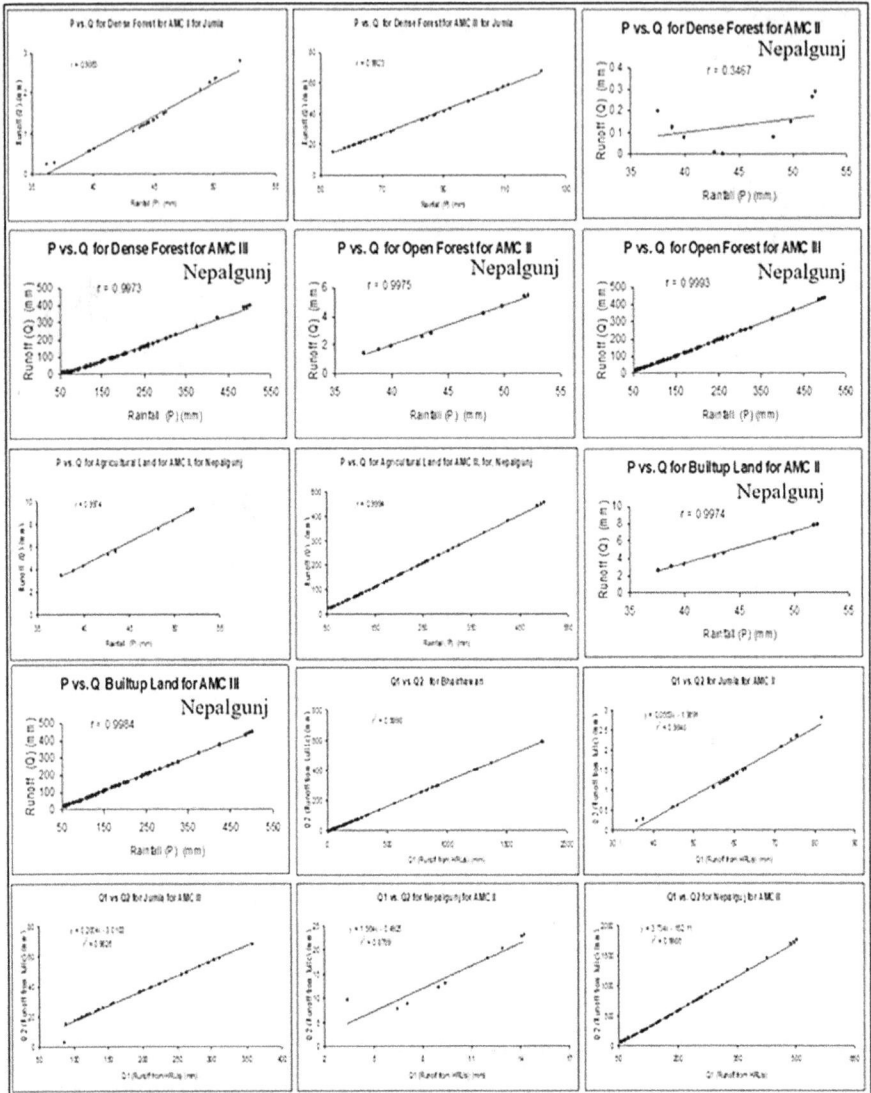

FIGURE 18.6 Correlation coefficient between rainfall and runoff for different AMCs.

18.3.2.1 AMC Wise

The rainfall-runoff relationship plotted for different categories lying in the land units under the influence of Nepalgunj and Jumla rain gauze stations in AMC II (average) and AMC III (saturated) conditions also show r values of 0.99 except in one situation, i.e., for the dense forest in AMC II condition in Nepalgunj rain gauze station where the r value is abnormally low (0.35).

18.3.3 COMPARATIVE ANALYSIS BETWEEN THE CUMULATIVE RUNOFF ESTIMATED FROM HRUs AND THAT FROM THE CORRESPONDING LULC CATEGORIES IN THE INDIVIDUAL LAND UNITS COVERED UNDER THE INFLUENCE OF EACH RAIN GAUZE STATION

There occur significantly high values of coefficient of determination ($r^2 = 0.99$) between the cumulative runoff estimated from HRUs and that from the corresponding LULC categories in the individual land units covered under the influence of each rain gauze station.

18.3.3.1 AMC Wise

Similar analyses performed for Jumla and Nepalgunj rainfall stations in AMC II and AMC III conditions exhibit significantly large values of r^2.

18.4 SUMMARY AND CONCLUSION

18.4.1 SUMMARY OF THE RESEARCH WORK

The correlation coefficient (r) determined between rainfall and runoff for each HRU shows a value of 0.99. The coefficient of determination (r^2) between the estimated runoff from the SCS-CN method and the computed runoff determined using the regression equation from the SCS-CN method is also found to be very high (0.99). The higher values of r and r^2 are also found for the individual LULC categories. A high correlation is obtained between the rainfall and runoff for AMC II and AMC III for the individual HRUs and LULC categories. Significantly very high value of r^2 is observed between the runoff estimated from all HRUs and all the LULC categories covered in a land unit falling under the influence of each rain gauze station.

18.4.2 CONCLUSIONS

The values of r and r^2 between P and Q and between $Q_{estimated}$ (SCS-CN) and $Q_{computed}$ (using the regression equation from SCS-CN), respectively, are found to be very high for the individual HRUs, and each LULC category and the same results are also found for AMC II and AMC III for both HRUs and the various LULC categories. There is also good agreement between the cumulative runoff obtained from the HRUs present in a land unit and the cumulative runoff obtained from their corresponding LULC categories falling under the influence of each rainfall station. These findings ascertain the efficacy of the SCS-CN method for the estimation of runoff in a mountainous basin.

ACKNOWLEDGEMENTS

The work was carried out in the Department of Remote Sensing, Birla Institute of Technology Mesra, Ranchi, India. The authors acknowledge the cooperation received from the Institute.

REFERENCES

Anon. 1973. A method for estimating volume and rate of runoff in small watersheds. Technical pp. 149, Soil Conservation Service, USDA- SCS, Washington, D.C.

Anbazhagan, A., S. M. Ramaswamy, and S. Das Gupta. 2005. Remote sensing and GIS for artificial recharge study, runoff estimation and planning in Ayyar basin, Tamilnadu, India. *Journal of Environmental Geology* 48: 158–170.

Daofeng, L. I., Tian Ying, Liu Changming, and H. Fanghua. 2004. Impact of land cover and climate changes on runoff of the source regions of Yellow River. *Journal of Geographical Sciences* 14 (3): 330–338.

Sarangi, A., D. K. Singh, and A. K. Singh. 2008. Evaluation of curve number and geomorphology-based models for surface runoff prediction from ungauged watershed. *Journal of Current Science* 94 (12): 1620–1626.

Seghal, J. et al. 1994. Soil Series of India, NBSSLUP, Nagpur.

Sharma, A. K., and P. K. Thakur. 2007 Quantitative assessment of sustainability of proposed watershed development plans for Kharod watershed, Western India. *Journal of Indian Society of Remote Sensing* 35 (3): 231–241.

Shi Zhi-Hua, Li-Ding Chen, Nu-Fang Fang, De-Fu Qin, and Chong-Fa Cai. 2009. Research on the SCS-CN initial abstraction ratio using rainfall runoff event analysis in the Three Gorges Area, China. *Journal of Catena* 77 (1): 1–7.

Topography of Nepal. 2009. Available via DIALOG. http://countrystudies.us

USDA-SCS. 1985. National Engineering Handbook.

Viswanadh, G. K., and M. V. S. S. Girdhar. 2007. Semi-distributed runoff model for a Semi-Arid area of Andhra Pradesh- A geomatic approach. In: *Proceedings of Water Resource Management*, ACTA.

19 Methodologies of Scenario Development for Water Resource Management
A Review

Gaurav Kumar and Rajiv Gupta

CONTENTS

DOI: 10.1201/9781003203445-22

19.1   INTRODUCTION

A thought-provoking procedure in identifying the probable situations and their effect in the near/far future is called scenario development (Schoemaker, 1995). Scenario development accounts for uncertain futures to generate ideas and business planning strategies. Organizational planners consider paradigms challenging their present knowledge to solve the short/long term future situations (Chermack et al., 2001). Managers articulate the models intellectually for making improved decisions as the concluding part of the scenario development (Martelli, 2001). Scenarios are broadly applied in diverse research fields to estimate the system's outcomes in several policy-making and strategic planning (Yoe, 2004).

19.1.1   SCENARIO PLANNING PERSPECTIVE FOR WATER RESOURCES

Over-exploitation of the water resources against their replenishment is a significant concern for future generations. Such a shift in the status of water resources results from fast economic expansion and additional human actions. Balanced management of water resources needs perfect per capita utilization estimations and a good perception of the utilization aspects. It is required to quantify ecological water requirements to balance human water needs, including environmental conditions to retain an operative ecosystem (Yuan et al., 2016; Maurya and Singh, 2021). The advantage of scenarios lies in their accountability to uncertainties associated with the climatic, economic, demographic, technical, social, and political conditions affecting the water resource system's performance, forthcoming water availability, need, and management plans. A watershed, a river basin, or groundwater can generally be aimed at managing and planning the development (Alcamo and Gallopin, 2009).

The review is focusing on the keywords scenario planning and water resources. Various methods applied in the process of scenario development were prediction and derivation methods. Since scenario development is a prediction process, primarily focusing on the prediction methods, a classification has been done among the methods, whether predictive or derivative. It is observed from the literature that the prediction methods used in scenario development of the water resources were complex and challenging to understand for a novice user. In this chapter, the accuracy and usability of the prediction methods have been compiled. Besides, it has also been reviewed whether GIS (Geographical Information System) encountered in the literature has been used as a predictive or derivative tool. Based on the review, it is recommended that: 1) prediction methods more straightforward to use/generalized with lesser technical parameters should be developed to support sustainability everywhere; and 2) GIS should be extended with the capabilities of prediction besides the derivation or analysis.

19.1.2   Basic Terminology

A brief explanation for the basic terms used in the scenario development/planning has been presented in the following subsections.

19.1.2.1   Types of Scenarios

There are various categories of the scenarios, as presented in Figure 19.1. Strategic scenarios are the primary area of interest for the modellers and researchers, where the main aim is to identify the irregularities in the methodologies applied by diverse subjects describing the modules of a sophisticated system. Here the emphasis is given to the explicit patterns, assumptions, and data selected subject wise.

Scenarios that describe the future centred on prior data of the earlier observations are considered exploratory scenarios (McCarthy et al., 2001). Therefore, exploratory scenarios are centred on the trend and pattern of the previously observed data projecting the future trends. The application of such trends is simple but does not allow identifying all the related policies affecting the future outcomes (Steinitz, 2003). Future trends of the explorative scenarios may be either projective or prospective. The projective scenarios are the projections in the future using the trends observed in earlier times, whereas prospective scenarios predict a future shift that differs substantially from its history. The anticipatory scenarios are developed to check the viability of the various solutions for future visions that may be desirable or undesirable. They are highly subjective and assume the previously happened events or the probable future situations (McCarthy et al., 2001). The policy responsive scenarios are part of an anticipatory approach where critical issues are planned to make the policy decisions targeting the future outcomes. Governmental and organizational decisions are made using policy responsive scenarios to analyze better and manage threatening situations (Baker et al., 2004). Again, policy responsive scenarios are classified into expert judgement and stakeholders' views. Future conditions are

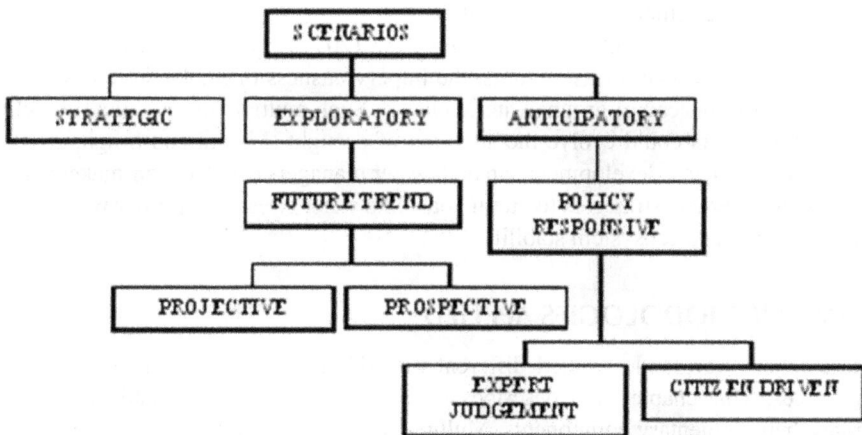

FIGURE 19.1   Scenario types.

(Source: Mahmoud, 2008).

modelled concerning field experts and science investigators who set the rules and criteria to fulfil the objectives in expert judgement. Present knowledge regarding future shift, a broad spectrum of relevant evidence, and a scientific unanimity are the advantages of such scenarios. Lack of political involvement and subjectively biased decisions are the major disadvantages of expert judgement scenarios (McCarthy et al., 2001). In citizen-driven scenarios, since the stakeholders are actively involved, there is greater political involvement and mass acceptance than expert judgement (Hulse et al., 2004). However, since only the very active citizens are involved, citizen-driven scenarios are also biased potentially.

19.1.2.2 Factors for Scenarios Planning

Tankersley (2006) stated that different factors considered in scenario planning are abbreviated as STEEPV (Social, Technological, Economic (macro), Environment, Political, and Values). General factors considered for studying the variations in water availability include climatic, anthropogenic, and socioeconomic factors (Wang et al., 2011). It is necessary to simulate the evolving pattern of these factors to predict the effect of such factors on water resources in the future. For example, inhabitants are an anthropogenic factor affecting water demand and supply issues in any study region. Climatic variables like temperature and rainfall affect the water reserves. Socioeconomic factors of land use and land cover may affect the groundwater recharge through the pervious/impervious. Such factors evolve over time and directly or indirectly affect a system. Therefore, any group of factors critical to the specific problem can be considered for scenario development and planning.

19.1.2.3 Water Management

The sources of freshwater for humankind are river flows (surface water), underground aquifers (groundwater), and meteorological rainfall (precipitation). Climate change, limited replenishment, land-use changes, and increased demand affect their potential (Kløve et al., 2014). Climate change directly impacts altered rainfall patterns and rising temperature on irrigation water demand and availability of water resources (Zhu and Ringler, 2012). Wong et al. (2011b) mentioned that due to the development of urban infrastructures, the imperviousness of the land increased the rainfall-runoff instead of groundwater recharge. Such multiple hydrological and climatic parameters could evolve the situation of drought (Mishra and Singh, 2010). Therefore, Scenario development can help water managers and decision-makers plan robust and alternate strategies to attain goals like water stress relief, improvement of water quality, and ecosystem stability.

19.2 METHODOLOGIES APPLIED

Researchers worldwide applied different methods for scenario planning of water resources. This chapter first describes the various methodologies and later illustrates their elementary components. Multiple studies from 2005 onwards have been referred to in this chapter.

Pallottino et al. (2005) implemented a multistage scenario tree for a group of hydrological scenarios which were statistically independent. The multistage scenario tree included nodes as the reservoir and demand representing storage and need at different time steps, whereas the connectors between the nodes represented inflows and outflows. The tree represented a broad time horizon as a whole, where the scenarios moved parallel until the decisions applied were the same and broke into different paths if the decision got changed.

Pulido-Velazquez et al. (2011) downscaled the historical data relevant to climate change at a basin scale to analyze the future hydrological scenarios. Future outcomes were analyzed in terms of various indices such as: 1) demand reliability index; 2) demand satisfaction index; 3) withdrawal use index; and 4) withdrawal index. Maurya et al. (2020) developed an index called Water for Development Planning Index (WDPI) which can be used for further development planning incorporating all aspects of urban water system. Kwak et al. (2016) analyzed the climate change effects and future spatial distribution of drought using time series of SPI (standardized precipitation index) evaluated from the Hadley Centre Global Environment Model (HADGEM3-RA) for RCPs (Representative Control Pathways) scenarios.

Rainfall and streamflow vary from month to month in a year, so the available data provides the monthly variations. Serinaldi and Kilsby (2012) implemented a deseasonalization scheme based on log transformation to normalize the monthly data into yearly. Their study used the Generalized Additive Model for location, scale, and shape (GAMLSS) for modeling the deseasonalized data. López and Francés (2013) developed the framework to analyze the flood frequency using the GAMLSS with non-stationary time series flood data. Suen and Lai (2013) examined the future salinity concentration of rivers using an Artificial Neural Network (ANN) since lower precipitation and lesser stream flows generate higher salinity in waterways. Similarly, Kourgialas et al. (2015) forecasted river flows to estimate the extreme flows hourly using ANN.

Since water balance is the elementary approach for water resource management, various methods accounting for its different components are applied. Griffin et al. (2013) assessed water resources availability in the future for land use and climate changes using USGS Thornthwaite water balance model. Jiang et al. (2007) tested a few other water balance models like Xinanjiang (XAJ), WatBal (WM), Vrije Universiteit Brussel (VJB), Guo (GM), and Shaake (SM), besides the Thornthwaite model producing different results for comparison. ISAT (impervious surface analysis tool) has been used to identify the impervious surface over a region. Yang et al. (2011) estimated the effective impervious area for better examination of runoff for severing networks. Rainfall-runoff models are another class of tools for water resource management affected due to rainfall and its runoff over the ground surface. Silberstein et al. (2012) estimated runoff in the catchments with rainfall-runoff models like SMARG, IHACRES, Sacramento, SIMHYD, and AWBM.

Two broad perspectives of analyses for such methodologies have been considered: top-down approach by downscaling the components at the region level and bottom-up approach by upscaling the components at the community level (Kalaugher et al., 2013). Socioeconomic vulnerabilities are the main focus in the bottom-up approach, whereas uncertainties are burst in the top-down approach generating complex decision-making. Wang et al. (2011) combined the top-down system dynamics model, bottom-up cellular automaton model, and the ANN model to simulate land use pattern changes for drought scenarios. Sahin et al. (2016) simulated proposed changes to water governance using a system dynamics model by integrating demand, supply, and asset management processes. Halmy et al. (2015) also used the Markov Cellular Automaton (Markov-CA) model to predict land-use changes. Therefore, identifying current and future risks is required to develop robust adaptation measures and strategies (Ludwig et al., 2014).

Future prediction is full of uncertainties; hence, estimating the uncertainties is also one of the research fields in scenario planning and structured in the vague or fuzzy form. Li et al. (2009) incorporated MFSQP (Multistage Fuzzy-Stochastic Quadratic Programming) method in a multi-layered scenario tree to determine uncertainties. Li et al. (2009) applied Inexact Multistage Stochastic Integer Programming (IMSIP) to facilitate the examination of several policy scenarios in terms of economic penalties if the promised goals are interrupted and the assessment of the economics of the diversion of surplus water.

Water scarcity effect on crop production has been estimated by a hydrological model tied with crop and water measurements to imitate the water cycle components like precipitation, evapotranspiration, soil moisture, deep drainage, and runoff associated with crop production. The modeling tool used was SWAT (Soil and Water Assessment Tool) which determined the crop water productivity (Huang and Li, 2010). Assessed the different responses for the climate change using three hydrological models named SWAT, PRMS (precipitation runoff modeling system), and SLURP (semi-distributed land use runoff process), applying seven distinctive PET (potential evapotranspiration) computational models (Hamon and Jensen-Haise methods for PRMS, Penman-Monteith, Granger, and Spittlehous-Black for SLURP, Penman-Monteith, Priestley-Taylor, and Hargreaves for SWAT).

19.2.1 PREDICTION/DERIVATION METHODS

This chapter reviewed various methods employed in scenario planning of water resources. There are different components for every method. Further, a description of elementary components in the methods is carried forward, focusing on their complexity and ease of use. Also, it is provided whether the component used in scenario planning is a predictive or derivative method.

19.2.1.1 System Dynamics (SD)

It is a modeling approach to comprehend complex systems' behaviour across time. It models internal feedback loops and time delays affecting the complete system's behaviour (Sterman, 2002). It can represent physical and information flows that enable us to understand the nonlinear dynamics behaviour in uncertain conditions. SD is unique from other methods since it studies complex systems, structuring them by deploying stocks and flows and feedback loops. A simple-looking system will emerge as a strange nonlinear system deploying stocks, flows, and loops. It is a complex method of prediction incorporating all the variables considered producing results and feedbacks concurrently to ensure better accuracy. There is extensive memory usage and challenging to implement over a problem domain.

19.2.1.2 Markov Model

Problems with the statistical or probabilistic frameworks are dealt with by Markov model. It generates predictions that are plausible. Markov model is a three-state system called full-up (all the components of the system are operating), intermediate (partially failed system), and full-fail (the system fails to execute its functions designed). There may be repair and failure transition paths in the model. The Markov model analyst writes the equations for the system that represent the system's time history based on the transition paths identified among the system states. The method will be more complex if the chain order of the previous events is higher, whereas the model's prediction accuracy is ultimately a hit and trial over different orders of chains.

19.2.1.3 GAMLSS (Generalized Additive Model for Location, Scale, and Shape)

Rigby and Stasinopoulos (2005) developed a statistical model called GAMLSS, which overcame few limitations of GLMs (generalized linear models) and GAMs (generalized additive models). In GLMs and GAMs, the location parameter is only the typical value describing the probability distribution of any dependent variable. In contrast, other distribution parameters are also related to the explanatory variable as scale and shape in GAMLSS, though the parameters are not limited. A typical set of distributions involving discrete, highly skewed, and kurtotic distributions can be availed in GAMLSS for the response variable distribution (Stasinopoulos and Rigby, 2007). The complexity of GAMLSS depends on the PDF (probability distribution function) for any dataset, whereas prediction accuracy depends on the data volume, i.e., the bigger the historic data size, the better the accuracy.

19.2.1.4 ANN

Imitating the functions of a human brain, ANNs (Artificial Neural Networks) are algorithms that statistically learn the behaviour/pattern of data. These algorithms can generate approximate functions for a large set of input variables (Haykin, 1999). ANNs are adaptive nonlinear networks made up of multiple distributed processing units (Principe et al., 2000). Inputs to the ANN are scaled by associated weights that adjust during the learning process through different processing units for known outputs. The learning process for the ANN is carried on the training data with no assumptions for the distribution characteristics of the data (Hagan et al., 1996). ANN requires a complex network with many trials for improved accuracy.

19.2.1.5 Water Balance Models

Hydrological models that forecast the water resource scenarios for certain predicted climatic prospects for a diverse range of hydrological situations are called water balance (Xu and Singh, 1998). Water balance models account for the water allocation among different elements of a hydrologic structure. A simple water balance model linking rainfall-runoff and climate in a single equation is Thornthwaite and Mather (1955). Whereas few other complex models like the Stanford watershed model (1966) account for the soil moisture (Schaake et al., 1996). Monthly precipitation and temperature are the standard inputs to the water balance models, while its outputs may include evapotranspiration, snow storage, soil moisture storage, runoff, and surplus. Climate variables either predicted or from climate models are input to water balance models for water resource scenario development. Therefore, water balance models are derivative models instead of predictive in scenario development of water resources.

19.2.1.6 PET (Potential Evapotranspiration) Models

The hydrological models analyzing the potential evapotranspiration in mm/month are called PET models (Thornthwaite, 1948). PET models are employed to approximate actual evapotranspiration in the ecosystem and rainfall-runoff modeling (Lu et al., 2005) as per soil moisture status (Bormann, 2011). The amount of evaporation that occurs in an adequate water resource or water demand by the atmosphere is termed PET. PET is affected by different climatic/atmospheric variables like air and surface temperature, wind speed, and insolation. PET models are also used to derive the evaporation's scenario outcomes in water resource management accounting.

19.2.1.7 ISAT (Impervious Surface Analysis Tool)

ISAT, a GIS extension tool, helps to determine the impervious area within a geographic region selected by a user for study, including municipalities, watersheds, or subdivisions. Impervious area is the area within a region covered by artificial buildings or pavements that hinder the rainfall infiltrate in the soil and instead promote evaporation (Chabaeva et al., 2009). There is a high correlation between rising impervious areas and lowering water quality within small watersheds. Therefore, ISAT is used to derive or predict the impacts of various scenarios on water quality. To apply ISAT, GIS is used to derive the water quality for the impervious area within a region instead of prediction.

19.2.1.8 Rainfall-Runoff Models

As the name determines, such models are the mathematical models explaining the relation of rainfall-runoff with a watershed, catchment area, drainage basin, or river discharge (Sun et al., 2010), including potential evaporation estimations and many other parameters (Vaze et al., 2010). Watershed area within any region is dependent on the rainfall characteristics in that region. If the size of the study area is very large, then the watershed splits into small catchments where runoff from all sub-catchments is combined based on flood routing methods. Rainfall models apply to the rainfall data, which is seasonal. Hence such models can help in short-term scenario planning, say monthly/annual rainfall. These models are again an intermediate tool deriving rainfall-runoff, which in its short regime may affect the water resource management over the long term.

19.2.1.9 SWAT (Soil and Water Assessment Tool)

SWAT model quantifies the effect of terrain management systems in vast and complicated watersheds. It is a river basin tier model available in the public domain backed by the USDA Agricultural Research Service at the Grassland, Soil, and Water Research laboratory in Temple, Texas, USA. SWAT is a watershed hydrologic transport model with the modules evapotranspiration, surface runoff, percolation, weather, return flow, reservoir and pond storage, transmission losses, groundwater flow, river routing, water transfer, irrigation and crop growth, and pesticide and nutrient loading. SWAT is used globally for hydrological modeling and still under further development (Gassman et al., 2007). Therefore, SWAT will be considered as an analysis or derivative tool.

19.2.1.10   Precipitation Runoff Modeling System (PRMS)

PRMS is a physical model that is deterministic and examines the reaction of different combinations for land use and climate on streamflow and watershed with the distributed parameters. The main objectives of PRMS are: 1) to imitate hydrological processes like evaporation, infiltration, runoff, transpiration, and interflow based on the energy and water budgetary systems of soil zone, plant canopy, and snowpack employing distributed climate data like rainfall, temperature, and insolation; 2) to simulate watershed-scale hydrological water budgets temporally for days and centuries; 3) to integrate itself with other natural resource management or other scientific models; and 4) to provide a standard library of different hydrologic process algorithms. PRMS is also a derivative system in scenario planning of water resources.

19.2.1.11   Semi-distributed Land Use-based Runoff Processes (SLURP)

It is a basin-scale model that imitates the hydrologic cycle of precipitation and runoff involving the impacts of dams, reservoirs, regulators, irrigation schemes, and water extractions (Wu et al., 2012). SLURP model examines the results of projected modifications in water resources management at the basin scale to observe the impact of climate and land cover changes on different water users. The model may use satellite images to identify the land cover, cloud cover, and vegetation indices. It uses cloud cover and vegetation indices to determine the precipitation distribution and evapotranspiration, respectively. Besides, it can also observe the snow extent to determine snow water equivalent. Full windows capability of the model makes it efficient to use inputs and generate outputs graphically. SLURP is also classified as a derivative kind of tool to produce outputs corresponding to the predicted variables in scenario planning.

19.2.1.12   Fuzzy Logic

Fuzzy logic is an alternative way to produce potential scenarios without sufficient historical data and experience (Zadeh, 1973). The scenario team applies their subjective and contextual knowledge in a fuzzy manner. Hence, values of the important variables and their numerical ranges are determined subjectively using supporting reference data that may be either historical or general but serve to assess boundary limits from the field of key variables. The fuzzy approach is not deterministic like various statistical techniques, yet suitable for scenario development. Therefore, results obtained from the fuzzy logic approach are apart from the facts, instead produce imprecise but reasonable projections dependent on the probability of views.

19.2.1.13   Water Indices

It is one of the methods to scale the state of water resource operations, e.g., water scarcity index, water stress index, etc. The problem of describing water stress is concerned with various similarly significant dimensions to water usage, supply, and shortage. The criteria to assess water can be either a policy choice or a scientific choice. The index term is subjective because water demand may vary from society to culture to region (Rijsberman, 2006). Therefore, using the indices as variables may lead to inexact evaluations. Water indices can also be considered to belong to a derivative category.

19.2.1.14 Stochastic Programming

Time series is theoretically a discrete observation of a stochastic process, which is a generalization of the concept of a random variable. The stochastic programming models have extended the linear and nonlinear programming models to the decision models where uncertain coefficients are given a probabilistic representation (Wets, 1989). GAMLSS and Markov chains are examples of the stochastic models. Such models fall in the predictive category. Accuracy depends upon the volume of observed data.

A summary of the tools discussed above is shown in Table 19.1, classified into predictive, derivative, and GIS applications.

19.3 DISCUSSION

A per the literature reviewed in the preceding section, various methodologies applied in scenario development are found as a combination of different prediction and derivation methods. Derivation methods like water balance models, rainfall-runoff models, and potential evapotranspiration modes are the core of the hydrological systems. At the same time, a set of prediction methods is used to forecast the future situation and applies to any study area or field. Soft computing and statistical analysis are the base of such predictive methods. It is observed that these predictive methods have many

TABLE 19.1

Models Classification in Predictive, Derivative, and GIS Applications

Models	Predictive	Derivative	GIS Application
System Dynamics	√	×	×
Markov	√	×	×
GAMLSS	√	×	×
Artificial Neural Network	√	×	×
Water Balance	×	√	×
Potential evapotranspiration	×	√	×
ISAT	×	√	√
Rainfall Runoff	×	√	×
SWAT	×	√	√
PRMS	×	√	√
SLURP	×	√	√
Fuzzy Logic	√	×	×
Water Indices	×	√	×
Stochastic Programming	√	×	×

technical parameters for consideration and therefore are a bit complex in their application. The purpose of the review in the chapter is to suggest reducing the complexity of such methods by developing generalized alternatives so that a novice user can use them with little or more straightforward technical terms without exploring the different parameters required, say future time steps to generate scenario outputs. It may help to avail scenario development process for sustainability everywhere, and ordinary people can also examine the future status for their different endeavours to plan efficiently.

Besides the review of prediction and derivation methods, it is also observed that GIS application is limited to the derivation purpose only. The prediction methods employed use time-series data to forecast in one dimension. In contrast, GIS is capable of presenting the results in two/three dimensions. This chapter recommends generating predictions in the form of thematic maps using time series maps by developing specific GIS tools. Therefore, GIS should be taken forward for prediction purposes as well instead of presentation and analysis purposes.

19.4   CONCLUSIONS

Scenario planning, an essential and helpful process, should be generalized so that any user can apply it to examine the future opportunities or risks for taking the proper measures beforehand. Though the prediction methods could be similar, yet generalized systems should be designed with lesser parameters so that any novice user can apply them easily with the known and non-technical dataset, e.g., time step ahead to forecast any variable of importance. Such an approach of generalization will help to make scenario development applicable everywhere, promoting sustainability all around. Application of GIS in scenario development and planning should be taken forward for prediction besides the presentation and analysis, i.e., the one-dimensional prediction process should be carried forward for two/three dimensional.

ACKNOWLEDGEMENTS

The study's idea originated from the implementation of the project entitled "Structured Dialogues for Sustainable Urban Water Management" sponsored by the Department of Science and Technology, Government of India under the grant number DST/TMD/EWO/WTI/2K19/UWS-04(C1).

REFERENCES

Alcamo, J. and Gallopin, G., 2009. Building a 2nd generation of World Water Scenarios. World Water Assessment Programme Side Publications Series. Paris, UNESCO. Accessed on 30 June 2021 https://unesdoc.unesco.org/ark:/48223/pf0000181796

Baker, J.P., Hulse, D.W., Gregory, S.V., White, D., Van Sickle, J., Berger, P.A., Dole, D., and Schumaker, N.H., 2004. Alternative futures for the Willamette river basin. *Ecological Applications* 14:313–324.

Bormann, H., 2011. Sensitivity analysis of 18 different potential evapotranspiration models to observed climatic change at German climate stations. *Climatic Change* 104(3–4):729–753.

Chabaeva, A., Civco, D.L. and Hurd, J.D., 2009. Assessment of impervious surface estimation techniques. *Journal of Hydrologic Engineering* 14(4):377–387.

Chermack, T.J., Lynham, S.A. and Ruona, W.E., 2001. A review of scenario planning literature. *Futures Research Quarterly* 17(2):7–32.

Gassman, P.W., Reyes, M.R., Green, C.H. and Arnold, J.G., 2007. The soil and water assessment tool: Historical development, applications, and future research directions. *Transactions of the ASABE* 50(4):1211–1250.

Goodarzi, M., Abedi-Koupai, J., Heidarpour, M., Safavi, H.R., 2016. Development of a new drought index for groundwater and its application in sustainable groundwater extraction, *Journal of Water Resources Planning and Management*, 142:1–12.

Griffin, M.T., Montz, B.E. and Arrigo, J.S., 2013. Evaluating climate change induced water stress: A case study of the Lower Cape Fear basin, NC. *Applied Geography* 40:115–128.

Hagan, M.T., Demuth, H.B. and Beale, M.H., 1996. Neural network design, Vol. 20. PWS Pub., Boston, USA.

Halmy, M.W.A., Gessler, P.E., Hicke, J.A. and Salem, B.B., 2015. Land use/land cover change detection and prediction in the north-western coastal desert of Egypt using Markov-CA. *Applied Geography* 63:101–112.

Haykin, S., 1999. Neural networks: A comprehensive foundation, 2nd edn. Prentice-Hall, Upper Saddle River, N.J., USA

Huang, F. and Li, B., 2010. Assessing grain crop water productivity of China using a hydro-model-coupled-statistics approach: Part I: Method development and validation. *Agricultural Water Management* 97(7):1077–1092.

Hulse, D.W., Branscomb, A., and Payne, S.G., 2004. Envisioning alternatives: Using citizen guidance to map future land and water use. *Ecological Applications* 14:325–341.

Jiang, T., Chen, Y.D., Xu, C.Y., Chen, X., Chen, X. and Singh, V.P., 2007. Comparison of hydrological impacts of climate change simulated by six hydrological models in the Dongjiang Basin, South China. *Journal of Hydrology* 336(3):316–333.

Kalaugher, E., Bornman, J.F., Clark, A. and Beukes, P., 2013. An integrated biophysical and socioeconomic framework for analysis of climate change adaptation strategies: The case of a New Zealand dairy farming system. *Environmental Modeling & Software* 39:176–187.

Kløve, B., Ala-Aho, P., Bertrand, G., Gurdak, J.J., Kupfersberger, H., Kværner, J., Muotka, T., Mykrä, H., Preda, E., Rossi, P. and Uvo, C.B., 2014. Climate change impacts on groundwater and dependent ecosystems. *Journal of Hydrology* 518:250–266.

Kourgialas, N.N., Dokou, Z. and Karatzas, G.P., 2015. Statistical analysis and ANN modeling for predicting hydrological extremes under climate change scenarios: The example of a small Mediterranean agro-watershed. *Journal of Environmental Management* 154:86–101.

Kwak, J., Kim, S., Jung, J., Singh, V.P., Lee, D.R. and Kim, H.S., 2016. Assessment of Meteorological Drought in Korea under Climate Change. *Advances in Meteorology* 1879024:1–13.

Li, Y.P., Huang, G.H., Wang, G.Q. and Huang, Y.F., 2009. FSWM: A hybrid fuzzy-stochastic water-management model for agricultural sustainability under uncertainty. *Agricultural Water Management* 96(12):1807–1818.

López, J. and Francés, F., 2013. Non-stationary flood frequency analysis in continental Spanish rivers, using climate and reservoir indices as external covariates. *Hydrology and Earth System Sciences* 17(8):3189.

Lu, J., Sun, G., McNulty, S.G. and Amatya, D.M., 2005. A comparison of six potential evapotranspiration methods for regional use in the southeastern United States. *JAWRA Journal of the American Water Resources Association* 41(3):621–633.

Ludwig, F., van Slobbe, E. and Cofino, W., 2014. Climate change adaptation and integrated water resource management in the water sector. *Journal of Hydrology* 518:235–242.

Mahmoud, M.I., 2008. Scenario development for water resources decision-making. *The University of Arizona*, USA.

Martelli, A., 2001. Scenario building and scenario planning: State of the art and prospects of evolution. *Futures Research Quarterly* 17(2):57–74.

Maurya S.P. and Singh R., 2021. Sustainable water resources. *Sustainable Resource Management*, 2021:147–162.

Maurya, S.P., Singh, P.K., Ohri, A. and Singh, R., 2020. Identification of indicators for sustainable urban water development planning. *Ecological Indicators* 108:105691.

McCarthy, J.J., Canziani, O.F., Leary, N.A., Dokken, D.J. and White, K.S. (Eds), 2001. Climate change 2001: Impacts, adaptation and vulnerability, contribution of Working Group II to the third assessment report of the Intergovernmental Panel on Climate Change, Cambridge University Press, Cambridge, United Kingdom.

Mishra, A.K. and Singh, V.P., 2010. A review of drought concepts. *Journal of Hydrology* 391(1):202–216.

Pallottino, S., Sechi, G.M. and Zuddas, P., 2005. A DSS for water resources management under uncertainty by scenario analysis. *Environmental Modeling & Software* 20(8):1031–1042.

Principe, J.C., Euliano, N.R. and Lefebvre, W.C., 2000. Neural and adaptive systems: Fundamentals through simulation. John Wiley, New York, USA.

Pulido-Velazquez, D., Garrote, L., Andreu, J., Martin-Carrasco, F.J. and Iglesias, A., 2011. A methodology to diagnose the effect of climate change and to identify adaptive strategies to reduce its impacts in conjunctive-use systems at basin scale. *Journal of Hydrology* 405(1):110–122.

Rigby, R.A. and Stasinopoulos, D.M., 2001, July. The GAMLSS project: A flexible approach to statistical modeling. In *New trends in statistical modeling: Proceedings of the 16th international workshop on statistical modeling.* pp. 337–345.

Rigby, R.A. and Stasinopoulos, D.M., 2005. Generalized additive models for location, scale and shape. *Journal of the Royal Statistical Society: Series C, Applied Statistics* 54(3):507–554.

Rijsberman, F.R., 2006. Water scarcity: Fact or fiction? *Agricultural Water Management* 80(1):5–22.

Sahin, O., Siems, R.S., Stewart, R.A. and Porter, M.G., 2016. Paradigm shift to enhanced water supply planning through augmented grids, scarcity pricing and adaptive factory water: A system dynamics approach. *Environmental Modeling & Software* 75:348–361.

Schaake, J.C., Koren, V.I., Duan, Q.Y., Mitchell, K. and Chen, F., 1996. Simple water balance model for estimating runoff at different spatial and temporal scales. *Journal of Geophysical Research. D. Atmospheres* 101:7461–7475.

Schoemaker, P.J., 1995. Scenario planning: A tool for strategic thinking. *Sloan Management Review* 36(2):25.

Serinaldi, F. and Kilsby, C.G., 2012. A modular class of multisite monthly rainfall generators for water resource management and impact studies. *Journal of Hydrology* 464:528–540.

Silberstein, R.P., Aryal, S.K., Durrant, J., Pearcey, M., Braccia, M., Charles, S.P., Boniecka, L., Hodgson, G.A., Bari, M.A., Viney, N.R. and McFarlane, D.J., 2012. Climate change and runoff in south-western Australia. *Journal of Hydrology* 475:441–455.

Stasinopoulos, D.M. and Rigby, R.A., 2007. Generalized additive models for location scale and shape (GAMLSS) in R. *Journal of Statistical Software* 23(7):1–46.

Steinitz, C, 2003. *Alternative futures for changing landscapes: The Upper San Pedro River Basin in Arizona and Sonora.* Island Press, New York.

Sterman, J.D., 2002. Systems dynamics modeling: Tools for learning in a complex world. *Engineering Management Review, IEEE* 30(1):42–42.

Suen, J.P. and Lai, H.N., 2013. A salinity projection model for determining impacts of climate change on river ecosystems in Taiwan. *Journal of Hydrology* 493:124–131.

Sun, W.C., Ishidaira, H. and Bastola, S., 2010. Towards improving river discharge estimation in ungauged basins: Calibration of rainfall-runoff models based on satellite observations of river flow width at basin outlet. *Hydrology and Earth System Sciences* 14(10):2011.

SWAT. n.d. Soil and water assessment tool. Retrieved from http://swat.tamu.edu/

Tankersley, J., 2006. Ten tips for creating more powerful future stories. FUTURETAKES (late fall 2006). *The Electronic Newsletter of the World Future Society's US National Capital Region Chapter* 5(3):19.

Thornthwaite, C. W., 1948. An approach toward a rational classification of climate. *Geographical Review* 38 (1): 55–94.Thornthwaite, C.W. and Mather, J. R. 1955. *The Water Balance, Publications in Climatology VIII(1).* 1–104, Drexel Institute of Climatology, Centerton, New Jersey.

Vaze, J., Post, DA, Chiew, F.H.S., Perraud, J.M., Viney, N.R. and Teng, J., 2010. Climate non-stationarity–validity of calibrated rainfall runoff models for use in climate change studies. *Journal of Hydrology* 394(3):447–457.

Wang, X.J., Zhang, J.Y., Shahid, S., ElMahdi, A., He, R.-M., Wang, X.-G. and Ali, M., 2011a. Gini coefficient to assess equity in domestic water supply in the Yellow River. *Mitigation and Adaptation Strategies for Global Change,* 17(1): 65–75.

Wets, R.J.B., 1989. Chapter VIII Stochastic programming. *Handbooks in Operations Research and Management Science* 1:573–629.

Wang, N.H., Jusuf, S.K. and Tan, C.L., 2011b. Integrated urban microclimate assessment method as a sustainable urban development and urban design tool. *Landscape and Urban Planning* 100(4):386–389.

Wu, L., Long, T.Y., Liu, X. and Guo, J.S., 2012. Impacts of climate and land-use changes on the migration of non-point source nitrogen and phosphorus during rainfall-runoff in the Jialing River Watershed, China. *Journal of Hydrology* 475:26–41.

Xu, C.Y. and Singh, V.P., 1998. A review on monthly water balance models for water resources investigations. *Water Resources Management* 12(1):20–50.

Yang, G., Bowling, L.C., Cherkauer, K.A. and Pijanowski, B.C., 2011. The impact of urban development on hydrologic regime from catchment to basin scales. *Landscape and Urban Planning* 103(2):237–247.

Yoe, C., 2004. Scenario planning literature review. US Army Corps of Engineers Institute for Water Resources, Alexandria.

Yuan, G., Zhu, X., Tang, X., Du, T. and Yi, X., 2016. A species-specific and spatially explicit model for estimating vegetation water requirements in desert riparian forest zones. *Water Resources Management*:1–19.

Zadeh, L.A., 1973. Outline of a new approach to the analysis of complex systems and decision processes. IEEE Transactions on Systems, Man, and Cybernetics, 3(1): 28–44.

Zhu, T. and Ringler, C., 2012. Climate change impacts on water availability and use in the Limpopo River Basin. *Water* 4(1):63–84.

Part IV

Future Algorithms in Environmental Systems

20 Process-Based Scenario Analyses of Future Socio-Environmental Systems
Recent Efforts and a Salient Research Agenda for Decision-Making

Rakesh Kadaverugu, Rajesh Biniwale, and Chandrasekhar Matli

CONTENTS

20.1 INTRODUCTION

Socio-environmental systems (SES) represent a wide range of environmental systems with significant contributions from humans (social, cultural, and economic aspects). People are an inseparable part of any SES, which governs the present state and future pathways of the systems. Mathematical modeling integrates the knowledge and perspectives of the people (or stakeholders) into concepts so that they can be computationally solved or optimized to understand more clearly how human decisions affect the environment, and probably vice-versa. The concept of scenario-based modeling of SES is ever-increasing with more practical applications related to the complex environmental systems, but with very little progress on the

DOI: 10.1201/9781003203445-24

theoretical development on the concepts associated to the scenario-based models (Tourki et al., 2013).

There is a need to synthesize the recent developments and identify the challenges in the process-based scenario analyses of SES to provide a more comprehensive understanding for the researchers and policy-makers about the multiple dimensions of SES. Current and future societal goals, including but not limited to sustainable development goals, goals related to climate change and resilience building, disaster risk reduction and mitigation, etc., involve multiple stakeholders at varying scales. Rapid economic development, emerging new markets, and international policies have consequences on human systems, which are critical to be evaluated to avoid the future pathways leading to the tipping points. In this context, the scenario-based modeling of SES can bolster evidence-based high-impact research to build more resilient human systems.

20.2 PROCESS-BASED MODELING OF SOCIO-ENVIRONMENTAL SYSTEMS

Process-based models are mathematical models (mostly computer models) that represent a system's behaviour or process, generally using the first principles or governing equations. These models generally use a set of ordinary or partial differential equations to capture the essence of the system behaviour. Some examples of process-based models include treatment wetland modeling (Kadaverugu, 2016), soil-plant system modeling (Kadaverugu, 2015), and regional weather forecasting (Kadaverugu et al., 2019, 2021b, 2021c). Even the global climate models, which operate on a continental scale, also employ the governing equations of fluid motion to simulate the meteorological variables.

Unlike the environmental systems that mainly consist of interactions and feedback among the atmosphere, lithosphere, hydrosphere, and biosphere, the socio-environmental systems have an additional dimension of human behaviour that constantly interacts and influences the SES. For example, indigenous and local people near the protected and sacred forests rely on forest produce for their sustenance (Dhyani et al., 2021; Kadaverugu et al., 2021d). People's behaviour and interaction with nature drastically change according to the policies as well. For instance, changes in oil prices will affect the prices of commodities and thereby influence the consumption pattern and demand of goods. Another example can be the traffic movement, which is also driven by the people's behaviour, which causes environmental impacts related to air and noise quality in the surroundings. Particular flora and fauna around the roadside are severely affected due to environmental pollution. The change in land use and land cover (LULC) is a classic example where people significantly modify the urban systems or regional landscape (Kadaverugu et al., 2021a).

20.2.1 Defining a Socio-Environmental System

The definition of any SES should consist of four significant aspects: 1) identification of the SES; 2) assessment of the stakeholders or people involved in the systems while decision-making; 3) defining the SES boundaries in space and time, like the geographical or spatial extent of the system, time-horizon of the developmental goals, coarseness of the information, and a clear cut understanding about the environmental setting or context of the system; and 4) information on system description, on the flow of information among sub-systems of the system, the parameters, and constraints involved in the system (a general schematic of an SES is shown in Figure 20.1).

Several authors have suggested that the SES should be looked at as a hierarchical theory of ecosystems (Musters et al., 1998). It is generally suggestive of organizing the SES as per the hierarchy of human activities, like governmental or non-governmental structures. But, such an organization should be entirely based on the description of the SES in consideration and the expected objectives. Defining a level of spatio-temporal scale at which the SES is desired is also a vital aspect. Along with that, while representing an SES, the people involved with the system (both at present and in the future) should also be known. They should be included in all debates about the SES and must reach a consensus on the structure and functioning of the SES. The steps involved in defining an SES are provided in Figure 20.2.

20.2.2   PROCESS-BASED MODELING

A process-based model represents the behaviour of a system and its components with a set of mathematical equations. Generally, the cause-and-effect relations within a system are captured through these equations representing first principles (or governing equations), but to some extent, they can be complemented by a well-defined empirical relation. Within the framework of an SES, the processes of sub-systems are represented through a set of equations, and the inputs and outputs of each subsystem are interconnected. Such a modeling paradigm generally emerges out by a set of nonlinear and coupled partial differential equations. For instance, the crop yield models, population dynamics, species dynamics modeling, dynamic forest models, etc., are well-defined examples.

Through the modeling of SES, multiple goals related to informed decision-making and system optimization for achieving sustainability targets can be planned. The scope of process-based modeling of SES has expanded to include the recent modeling techniques, including but not limited to system dynamics, agent-based modeling, dynamic stochastic equilibrium modeling, Bayesian networks, statistical simulation models, and cellular automata (Kelly et al., 2013). Several researchers argue that it is not reasonably possible to capture human behaviour in decision-making through a set of differential equations. The choices made by humans are determined to a significant extent by their previous experiences, environmental conditions, and options available so that they survive and become competitive. In this context, agent-based models and cellular automata represent human behaviour quite well at the micro-level in decision-making. Such modeling approaches complement each other in defining the complex nature of SES. All these approaches can be considered as process-based models, as far as they represent the system behaviour in a more accurate way.

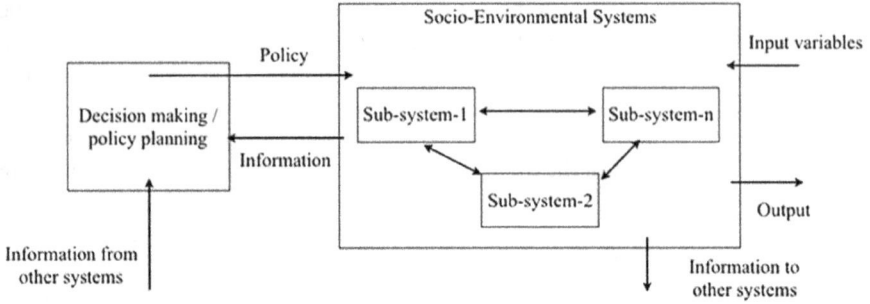

FIGURE 20.1   Schematic of a socio-environmental system with the policy interactions.
(Adapted from Musters et al., 1998).

Based on the literature review, we observed that the SESs in the studies represent a diverse range of spatial scales, including but not limited to urban centres, river catchments, protected forest zones, coastal areas, islands, districts, and varied till country level to global scale. Most of the studies have used the following models to represent the dynamics of SES, such as Random Forest, soft systems models, conceptual models, Water Evaluation and Planning Model (WEAP), artificial neural networks, agent-based discrete event models, Markov chain analysis and cellular automata, Image GLOBIO, CLIMSAVE IAP, Soil and Water Assessment Tool (SWAT), Fuzzy Cognitive Mapping software, forecasting tools, and various economic models, etc. among others. The scenarios in the studies were developed using participatory, exploratory, and normative approaches. Several studies have also integrated the future climate change scenarios, land use change scenarios, and socioeconomic scenarios developed by various international assessments.

It can be summarized from the literature that the significant drivers of change considered in SES modeling are climate variables, socio-economic and demographic variables, urban development, tourism, drought, land use changes, agricultural expansion, political will, soil and water quality, operational, behavioural factors, and others. Several studies that modelled the SES have tried to simulate the aspects of sustainable land uses, optimized benefits from natural resources, ensuring environmental quality, conservation of nature and cultural diversity, maximizing sustainable development goals, exploring low carbon pathways, and biodiversity conservation.

Several earlier studies have utilized ABMs to simulate SES to capture the interactions of autonomous agents or groups which can make decisions on their own according to the environmental conditions, competition behaviour, and resources, or other limiting criteria. These kinds of models at a higher scale can simulate the learning patterns or evolving behaviour of the whole system, which would otherwise go undetected by modeling the SES with a differential equation. Defining a topography of the network of agents is crucial in modeling the SES with the ABM approach. The ABMs also help visualize the overall system behaviour at a higher scale with respect to the changes in model parameters, environmental conditions, or decision-making constraints. Hence, ABMs are also proved to be a powerful approach in scenario-based modeling of SES. Still, mainstreaming of the workflow of using ABMs to SES is not

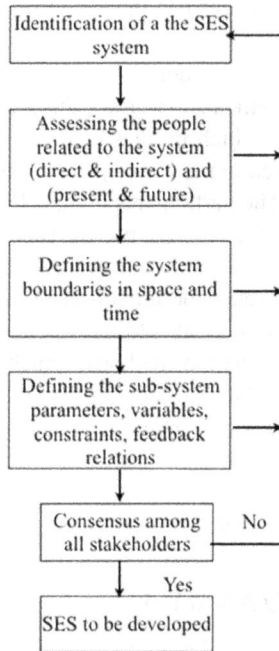

FIGURE 20.2   Schematic of the steps involved in defining a socio-environmental system.

(Adapted from Musters et al., 1998).

yet picked up as expected. Polhill et al. (2019) argue that the ABMs are still new enough to most of the researchers in the field, and it requires a multi-disciplinary approach to solve the SES objectives. In recent years, the ABMs are developed with more empiricism or experimental data, unlike with abstracted representation of reality. This has led to a more realistic picture of the reality with credible modeling outcomes (Polhill et al., 2019). Yet ABMs are still facing the challenge of validation with the actual data as it is hard to match the model performance with the natural and random world.

The regional climate dynamics have significant impacts on the decision-making of individuals and households according to the prevailing environmental conditions. They show that the cultural bonding and cooperation among different groups of a rural community through the labour-sharing agreements are impacted by the changing scenarios of water availability, drought, and other socio-environmental risks. NetLogo (https://ccl.northwestern.edu/netlogo/) is a widely used computer program for implementing various ABMs.

Game-based learning is one of the methods for teaching systems thinking and educating the stakeholders. Serious games devised based on the SES utilize the concepts of systems behaviour, which expands the stakeholders' thinking perspectives and gives a sense of interconnectedness with the system elements, which helps to achieve the system's overall goals in consideration. Papathanasiou et al. (2019) argue that this approach will encourage the participants to express more clearly and helps policymakers to make decisions based on the scientific analysis about future scenarios. In this approach, a system is represented using the Causal Loop Diagrams (CLD), which qualitatively represent the components of the system and its interconnections among various elements (Papathanasiou et al., 2019). One such model developed using CLD is SUSTAIN CLD which mimics an urban system consisting of several subcomponents like people, economic activity, environment, waste and water management, transportation, etc. The participants or stakeholders play this game to optimize the outcome – attractiveness of a city, by varying the states of the system subcomponents.

Further, it is essential to discuss the model uncertainties for informed decision-making about SES. See Elsawah et al. (2020a) for more information on different types of model uncertainty evaluation methods, which broadly include but are not limited to mental modeling, critical systems thinking, global sensitivity and uncertainty analysis, crash or stress testing of model parameters, Monte Carlo methods, Bayesian inference methods, exploratory analysis, uncertainty matrix method, automated scientific workflows, multi-model analysis, parameter estimation, identifiability analysis, model auditing, and peer review, etc.

20.3 SCENARIO-BASED ANALYSIS

The process involved in scenario-based modeling of SES stimulates the thinking about the plausible alternative futures, which might arise due to uncertainty in policy, people's behaviour, randomness, surprise, etc. The scenario analysis is not exactly a prediction about the future, which is not quite possible owing to the highly nonlinear, uncertain, and complex behaviour of an SES, but tries to divide the unknown future space into a more like a set of plausible spaces (independent and unrelated). The scenario analysis (SA) stretches our imagination to rather explore and compare a multitude of diverse conceivable futures, which might also reflect the plausible states of the world in the future and the pathways that lead to them (Börjeson et al., 2006).

Some of the earlier applications of SA include planning defence strategies and corporate practices, but now SA has percolated into many disciplines where uncertainty about the future is too much to tackle. Many scenario development exercises are based on the story-and-simulation (SAS) approach (Alcamo, 2008). Prediction about the future is crippled with uncertainty and complexity involved between human behaviour and the environment (Guivarch et al., 2017; Elsawah et al., 2020a). Generation of scenarios is not a task to be performed in isolation but rather involves a more like co-generation of scenarios with the concerned stakeholders. Some of the challenges in scenario-based modeling of SES as identified by Elsawah et al. (2020b) include: a) involvement of stakeholders in co-creation of the scenarios; b) development of more realistic storylines about the plausible alternative future scenarios at multiple spatial, temporal, and sectoral scales; c) establishing a proper linkage between the qualitative and quantitative aspects of the scenarios; d) to what extent the scenarios capture the uncertainties and surprise events; e) the diversity and consistency of the proposed scenarios; and f) how best the scenarios help in decision-making. Also, communicating the scenarios with the stakeholders and policymakers to engage them actively while planning the scenarios is an essential element in the co-development of the scenarios.

The scientific rigour in the scenario analysis has been evolving according to the needs of the study and challenges involved by using various analytical tools and models. Qualitative scenario analysis is more like a narration of the plausible future when the future is highly uncertain about imagining. In contrast, quantitative analysis is achieved through the forecast of calibrated models (like weather prediction) (Tourki et al., 2013). It can be said that these qualitative and quantitative approaches are two extremes for studying future scenarios.

Participatory scenario planning exercise has a significant role in involving local stakeholders in devising the policies that best suit the local communities with an acceptable trade-off between the development and the environment. Several studies have applied the Kesho model in participatory scenario planning, and it is widely used in devising policies on developmental actives in Africa (http://www.real-project.eu). This scenario framework, Kesho, is named after the Swahili word for 'tomorrow' or 'later', hoping to plan a better future with the help of this framework. The essence of this framework lies in bringing together all the stakeholders with divergent views and in addressing the trade-offs between them for building long-term sustainable policies. The impacts of various proposed policies are explored by what-if scenarios to bring consensus for future developmental pathways. The whole process of Kesho can be divided into four steps: the first step includes identification of system boundaries, finding out the objectives of scenario building, and the stakeholders involved will build storylines, which to some extent germinates a sense of connectedness and ownership with the outputs generated (Kariuki et al., 2021); the second step encourages the stakeholders to visualize the future socio-economic and environmental trajectories in a specific relation with livelihood and land use modifications; in the third step, experts or modellers translate the qualitative narration into quantities using models and spatial maps; lastly, these results are brainstormed among the stakeholders for feedback and valuation, so that the modeling can be amended to reflect the desired intentions of the stakeholders. Kesho framework is being applied in a multitude of SES for future planning. For instance: Capitani et al. (2019) have applied the Kesho to portray the future land use changes according to the scenarios – green economy (where REDD+ framework is implemented) and business-as-usual (REDD+ is not effectively implemented), and quantified the impacts on natural capital in Tanzania.

Exploratory scenario generation by established techniques can also help to visualize the extent of all possible outcomes in the near future. One such example shown by Kadaverugu et al. (2021a) involves the generation of plausible future scenarios of a city's land use driven by two major drives of changes consisting of (a) urbanization, and (b) commitment towards green spaces. Positioning these two drives on the x and y axes results in four quadrants with a combination of high and low of each driver. These four quadrants represent four different scenarios emerging out of the two major drivers. This approach is called as two-axes method. Several studies have applied the two-axes method for deriving future scenarios in modeling various SESs (for example, DasGupta et al., 2018; Hashimoto et al., 2018). Based on the information gathered from the literature survey and screening of the city policy documents, the storylines in each scenario have been developed, and the corresponding plausible land use modification that would result in each scenario was modelled using the InVEST proximity-based land use change simulation model. The obtained future land use categories are utilized for the quantification of the heat mitigation role of the urban green spaces (Kadaverugu et al., 2021a). This kind of exploratory scenario-based quantification of urban heat mitigation modeling linked with future land uses helps policy-makers for making the cities more climate-resilient.

It is quite possible to have differences among the stakeholders with different interests about the outcomes of an SES. For instance, a certain infrastructure development project might be of interest to a set of stakeholders (businessmen, policymakers, government, etc.), whereas the agricultural landowners might feel under-compensated when their agricultural land is diverted for developmental activity. Therefore, it is of primary concern for the policymakers to communicate the benefits and trade-offs due to a change or a scenario with all the stakeholders. In a participatory scenario building session, the stakeholders with different opinions might be asked to develop the SES (with all the attributes and interconnections or feedbacks of the sub-systems), which will help the policymakers and modellers to clearly visualize the thought process of various stakeholders. The fuzzy cognitive mapping approach is helpful in quantifying the complex interconnections of an SES with weightage and signs (positive or negative) so that the system behaviour can be put in numbers or quantified. The creative narrations of stakeholders and their abstract thoughts about the system or about the future scenarios can be given shape with this approach. The relative changes in the outcomes of a system due to subtle changes in the weights or connections can be comprehended. Mental Modeler is a free web-based software based on fuzzy cognitive mapping, which is used in scenario analysis of complex systems. For instance, Singer et al. (2017) used this software to understand the water crisis in Michigan State and demonstrated how this tool is helpful in structuring the thoughts of local people.

Results of two different scenario analyses of the same SES might lead to different results. Hence, it is essential to evaluate the validity of scenario analysis to have an objective judgement. According to Tourki et al. (2013), evaluation of SA studies can be carried out by four criteria: coverage, consistency, uncertainty assessment, and efficiency. Here, coverage includes the extent of the relevant cases considered in the SA, such as various strategic options considered and the range of environmental factors describing an SES and their variables states considered in the SA. The consistency aspect deals with the logical agreeability of a scenario at all stages, and it is a necessary condition to check for the possibility of a scenario without considering its probability. Consistency check has more relevance, especially where scenarios are generated by an algorithm by varying the environmental variables of an SES. Ensuring the consistency of the scenarios leads to an efficient study (Scholz et al., 2000). Also, understanding uncertainty in SA is a necessity in appreciating the context and limitations of the study. Majorly, uncertainty in the SES modeling and SA stems from: lack of understanding about the SES; errors in model development, data, and system variables; biased judgements; inappropriate system assumptions; ambiguity in defining the system behaviour; inadequacy in accounting for human behaviour; and due to other unaccounted reasons (Mahmoud et al., 2009).

20.4   RESEARCH AGENDA FOR DECISION-MAKING AND THE WAY FORWARD

Future research agenda should try to fulfil the environmental concerns through a fresh perspective on SES modeling using SA. Important contributions from basic science and traditional or local knowledge should be fused to solve societal problems in general. The participation of all stakeholders in co-design is to be ensured for trust-building, and the management policies should be flexible and people-centric. The case studies and the methodologies discussed in this chapter highlight the need for realizing the full potential in solving the socio-environmental issues with a creative mindset to confront the global challenges of humanity and ecosystems.

Some of the challenges identified in modeling SES with SA include co-design, co-learning, appreciating the diversity of knowledge, translation of concepts into models, knowledge about varying spatial and temporal scales, the flow of data across the scales in the system, scenario development, effective communication, effective policymaking, public participation, building trust and connectedness, effective project management and monitoring, which needs to be overcome to ensure the effectiveness of the SES modeling.

Although the need for scenario-based thinking is well recognized and utilized in better framing the policies on SES, the connection between the scenarios and the actual policy implementation on practice is quite weak. Several studies have noted this mismatch, which could be attributed to not accounting for the evolving needs of society and policy (Elsawah et al., 2020b). Some of the main challenges that cripple this mismatch can be attributed to lack of stakeholder engagement or participation in co-development of scenarios, unaddressed scale issues and surprise events, poor communication of the scenarios across the social sectors (according to Elsawah et al., 2020b).

Unified models on SES that can perform scenario-based future assessments are needed to be developed with process-based or robust models describing the system as building blocks. Especially, the focus of these models should be on capturing the multi-scale factors in understanding the climate-energy-ecosystem nexus with the involvement of policies. The literature review suggests that the current research trends are more focused on mitigating the impacts due to the climate variability at multiple spatial and temporal scales. Designing resilient urban systems and agricultural systems is of prime concern that has direct connections with the people and food security. However, the need for the protection of biodiversity and conserving pristine forests should not be undermined.

Integrating process-based models (or robust machine learning models) and human factors is needed, which requires sophisticated coordination and data flow across different tools, scales, and knowledge structures. Several efforts have been underway in this direction. For instance, a web-based model developed by Belete et al. (2019) illustrates the dynamics of energy, climate change, and the economy. The framework includes several toolkits consisting of an integrated assessment model (GCAM), an equilibrium model (EXIOMOD), and an ABM (BENCH). Applications of soft computing methods and hybrid modeling paradigms involving, Game theory, Bayesian systems, Markov chain, Monte Carlo simulations, cellular automata, genetic algorithms, etc., and their integration with the robust process-based models describing the SES should be fully explored.

Inclusion of Indigenous and Local Knowledge (ILK) and traditional practices and customs of local people or stakeholders into scenario planning will ensure greater success in effective scenario development and implementation. Integrating knowledge from multiple disciplines describing the SES components and translating the conceptual wisdom into quantitative connections is a challenge for SES modellers and policy-makers.

Uncertainty analysis is largely missing in the description of SES and SA. Research on SES modeling should embrace more concepts on complexity, adaptiveness, and resilience of better understanding the evolution patterns of the systems. Widlok et al. (2012) argue that the modeling of SES should move away from the typical cause-effect paradigm and should adapt theories on resilience and adaptive cycles of a system, inspired by Gunderson and Holling et al. (2002). Further, they mention that integrated SES modeling should address the trade-offs between short-term gains and long-term resilience or sustainable development goals across multiple spatial and temporal scales. Concepts on the collapse of a system due to abrupt events or slow decay should also be integrated into future studies on SES modeling.

ACKNOWLEDGEMENTS

The authors sincerely thank the Director of CSIR National Environmental Engineering Research Institute, and Director of National Institute of Technology for providing the facilities to work on this chapter. Also the services of Knowledge Resource Center of NEERI are acknowledged for processing this chapter for similarity report having a reference number of CSIR-NEERI/KRC/2021/JULY/CTMD/1.

Declarations: The authors declare that they have no known competing financial interests or personal relationships that could have appeared to influence the work reported in this paper.

REFERENCES

Alcamo, J. (2008). Chapter six the SAS approach: Combining qualitative and quantitative knowledge in environmental scenarios. In *Developments in integrated environmental assessment* (Vol. 2, pp. 123–150). Elsevier. https://doi.org/10.1016/S1574-101X(08)00406-7

Belete, G. F., Voinov, A., Arto, I., Dhavala, K., Bulavskaya, T., Niamir, L., Moghayer, S., & Filatova, T. (2019). Exploring low-carbon futures: A web service approach to linking diverse climate-energy-economy models. *Energies*, *12*(15), 2880. https://doi.org/10.3390/en12152880

Börjeson, L., Höjer, M., Dreborg, K.-H., Ekvall, T., & Finnveden, G. (2006). Scenario types and techniques: Towards a user's guide. *Futures*, *38*(7), 723–739. https://doi.org/10.1016/j.futures.2005.12.002

Capitani, C., van Soesbergen, A., Mukama, K., Malugu, I., Mbilinyi, B., Chamuya, N., Kempen, B., Malimbwi, R., Mant, R., Munishi, P., Njana, M. A., Ortmann, A., Platts, P. J., Runsten, L., Sassen, M., Sayo, P., Shirima, D., Zahabu, E., Burgess, N. D., & Marchant, R. (2019). Scenarios of land use and land cover change and their multiple impacts on natural capital in Tanzania. *Environmental Conservation*, *46*(1), 17–24. https://doi.org/10.1017/S0376892918000255

DasGupta, R., Hashimoto, S., Okuro, T., & Basu, M. (2018). Scenario-based land change modeling in the Indian Sundarban delta: An exploratory analysis of plausible alternative regional futures. *Sustainability Science*. https://doi.org/10.1007/s11625-018-0642-6

Dhyani, A., Kadaverugu, R., Nautiyal, B. P., & Nautiyal, M. C. (2021). Predicting the potential distribution of a critically endangered medicinal plant Lilium polyphyllum in Indian Western Himalayan Region. *Regional Environmental Change, 21*(2), 30. https://doi.org/10.1007/s10113-021-01763-5

Elsawah, S., Filatova, T., Jakeman, A. J., Kettner, A. J., Zellner, M. L., Athanasiadis, I. N., Hamilton, S. H., Axtell, R. L., Brown, D. G., Gilligan, J. M., Janssen, M. A., Robinson, D. T., Rozenberg, J., Ullah, I. I. T., & Lade, S. J. (2020a). Eight grand challenges in socio-environmental systems modeling. *Socio-Environmental Systems Modeling, 2*, 16226. https://doi.org/10.18174/sesmo.2020a16226

Elsawah, S., Hamilton, S. H., Jakeman, A. J., Rothman, D., Schweizer, V., Trutnevyte, E., Carlsen, H., Drakes, C., Frame, B., Fu, B., Guivarch, C., Haasnoot, M., Kemp-Benedict, E., Kok, K., Kosow, H., Ryan, M., & van Delden, H. (2020b). Scenario processes for socio-environmental systems analysis of futures: A review of recent efforts and a salient research agenda for supporting decision-making. *Science of The Total Environment, 729*, 138393. https://doi.org/10.1016/j.scitotenv.2020.138393

Guivarch, C., Lempert, R., & Trutnevyte, E. (2017). Scenario techniques for energy and environmental research: An overview of recent developments to broaden the capacity to deal with complexity and uncertainty. *Environmental Modeling & Software, 97*, 201–210. https://doi.org/10.1016/j.envsoft.2017.07.017

Gunderson, L. H., & Holling, C. S. (2002). *Panarchy: Understanding transformations in human and natural systems*. Island press.

Hashimoto, S., DasGupta, R., Kabaya, K., Matsui, T., Haga, C., Saito, O., & Takeuchi, K. (2018). Scenario analysis of land use and ecosystem services of social-ecological landscapes: Implications of alternative development pathways under declining population in the Noto Peninsula, Japan. *Sustainability Science*. https://doi.org/10.1007/s11625-018-0626-6

Kadaverugu, R. (2015). Framework for mathematical modeling of Soil-Tree system. *Modeling Earth Systems and Environment, 1*(3). https://doi.org/10.1007/s40808-015-0017-2

Kadaverugu, R. (2016). Modeling of subsurface horizontal flow constructed wetlands using OpenFOAM®. *Modeling Earth Systems and Environment, 2*(2), 55.

Kadaverugu, R., Gurav, C., Rai, A., Sharma, A., Matli, C., & Biniwale, R. (2021a). Quantification of heat mitigation by urban green spaces using InVEST model–A scenario analysis of Nagpur City, India. *Arabian Journal of Geosciences, 14*(2), 82. https://doi.org/10.1007/s12517-020-06380-w

Kadaverugu, R., Matli, C., & Biniwale, R. (2021b). Suitability of WRF model for simulating meteorological variables in rural, semi-urban and urban environments of Central India. *Meteorology and Atmospheric Physics*. https://doi.org/10.1007/s00703-021-00816-y

Kadaverugu, R., Purohit, V., Matli, C., & Biniwale, R. (2021c). Improving accuracy in simulation of urban wind flows by dynamic downscaling WRF with OpenFOAM. *Urban Climate, 38*, 100912. https://doi.org/10.1016/j.uclim.2021.100912

Kadaverugu, R., Dhyani, S., Dasgupta, R., Kumar, P., Hashimoto, S., & Pujari, P. (2021d). Multiple values of Bhitarkanika mangroves for human wellbeing: Synthesis of contemporary scientific knowledge for mainstreaming ecosystem services in policy planning. *Journal of Coastal Conservation, 25*(2), 32. https://doi.org/10.1007/s11852-021-00819-2

Kadaverugu, R., Sharma, A., Matli, C., & Biniwale, R. (2019). High resolution urban air quality modeling by coupling CFD and mesoscale models: A review. *Asia-Pacific Journal of Atmospheric Sciences*. https://doi.org/10.1007/s13143-019-00110-3

Kariuki, R. W., Munishi, L. K., Courtney-Mustaphi, C. J., Capitani, C., Shoemaker, A., Lane, P. J., & Marchant, R. (2021). Integrating stakeholders' perspectives and spatial modeling to develop scenarios of future land use and land cover change in northern Tanzania. *PLOS ONE, 16*(2), e0245516. https://doi.org/10.1371/journal.pone.0245516

Kelly, R. A., Jakeman, A. J., Barreteau, O., Borsuk, M. E., ElSawah, S., Hamilton, S. H., Henriksen, H. J., Kuikka, S., Maier, H. R., Rizzoli, A. E., van Delden, H., & Voinov, A. A. (2013). Selecting among five common modeling approaches for integrated environmental assessment and management. *Environmental Modeling & Software, 47,* 159–181. https://doi.org/10.1016/j.envsoft.2013.05.005

Mahmoud, M., Liu, Y., Hartmann, H., Stewart, S., Wagener, T., Semmens, D., Stewart, R., Gupta, H., Dominguez, D., Dominguez, F., Hulse, D., Letcher, R., Rashleigh, B., Smith, C., Street, R., Ticehurst, J., Twery, M., van Delden, H., Waldick, R., ... Winter, L. (2009). A formal framework for scenario development in support of environmental decision-making. *Environmental Modeling & Software, 24*(7), 798–808. https://doi.org/10.1016/j.envsoft.2008.11.010

Musters, C. J. M., de Graaf, H. J., & ter Keurs, W. J. (1998). Defining socio-environmental systems for sustainable development. *Ecological Economics, 26*(3), 243–258. https://doi.org/10.1016/S0921-8009(97)00104-3

Papathanasiou, J., Armenia, S., Pompei, A., Scolozzi, R., Barnabè, F., & Tsaples, G. (2019). Sustainability literacy through game-based learning. 5470–5474. https://doi.org/10.21125/iceri.2019.1313

Polhill, J. G., Ge, J., Hare, M. P., Matthews, K. B., Gimona, A., Salt, D., & Yeluripati, J. (2019). Crossing the chasm: A 'tube-map' for agent-based social simulation of policy scenarios in spatially distributed systems. *GeoInformatica, 23*(2), 169–199. https://doi.org/10.1007/s10707-018-00340-z

Scholz, R. W., Mieg, H. A., & Oswald, J. E. (2000). Transdisciplinarity in groundwater management–Towards mutual learning of science and society. *Water, Air, and Soil Pollution, 123*(1), 477–487.

Singer, A., Gray, S., Sadler, A., Schmitt Olabisi, L., Metta, K., Wallace, R., Lopez, M. C., Introne, J., Gorman, M., & Henderson, J. (2017). Translating community narratives into semi-quantitative models to understand the dynamics of socio-environmental crises. *Environmental Modeling & Software, 97,* 46–55. https://doi.org/10.1016/j.envsoft.2017.07.010

Tourki, Y., Keisler, J., & Linkov, I. (2013). Scenario analysis: A review of methods and applications for engineering and environmental systems. *Environment Systems & Decisions, 33*(1), 3–20. https://doi.org/10.1007/s10669-013-9437-6

Widlok, T., Aufgebauer, A., Bradtmöller, M., Dikau, R., Hoffmann, T., Kretschmer, I., Panagiotopoulos, K., Pastoors, A., Peters, R., Schäbitz, F., Schlummer, M., Solich, M., Wagner, B., Weniger, G.-C., & Zimmermann, A. (2012). Towards a theoretical framework for analyzing integrated socio-environmental systems. *Quaternary International, 274,* 259–272. https://doi.org/10.1016/j.quaint.2012.01.020.

21 From Quantitative to Qualitative Environmental Analyses

Translating Mental Modeling into Physical Modeling

Ramesh Singh, Chaitanya Nidhi, and Satya Prakash Maurya

CONTENTS

21.1 INTRODUCTION

The junction of environmental requirements, institutional expectations, and scientific instruments such as proper analytical tools and Decision Support Systems is critical to the future of environmental management. The cognitive features of

DOI: 10.1201/9781003203445-25

our collaborative attempts to understand and tackle these challenges more than ever for a sustainable system, as professionals face a barrage of environmental problems that are longer in duration, larger in scope, more complicated, and less understood.

For the first time in 2006, the synthesis of quantitative research appears to have been adopted for environmental management, and it has been evolving since then. (Roberts et al., 2006; Pullin and Stewart, 2006). Systematic reviews of quantitative research show that, while they can help us answer important questions regarding the magnitude of an impact or effect, they can't help us answer other policy and practice-related concerns. (Dalton et al., 2017; Barnett-Page and Thomas, 2009). Furthermore, a simple aggregation will merely obscure crucial differences among studies on the influence of environmental action or exposures, rather than the complexity of the studies themselves. In the context of environmental management, a synthesis of qualitative research is supported by the argument that useful input to policy and practice on: 1) appropriateness and feasibility intervention (i.e., how people benefit and their attitudes towards conservation intervention); 2) adoption or acceptability of the intervention (i.e., what is the degree of conservation intervention adoption, and what are the enablers and impediments to its acceptance?); 3) subjective experience (i.e., what are the challenges and priorities of local community?); and 4) outcome heterogeneity (i.e., what values do people place on different outcomes, and why didn't an intervention work?) can be provide qualitative research. (Hannes et al., 2013; Dalton et al., 2017; Heyvaert et al., 2017; Adams and Sandbrook, 2013; Dalton et al., 2017; Heyvaert et al., 2017; Adams and Sandbrook, 2013).

Environmental issues are difficult because they involve an extreme degree of uncertainty, spatial and temporal scale fluctuation, irreversible potential damage, and cross-sectoral and multi-level reaching. It's also well-connected to both ecological and social processes. Furthermore, environmental interventions are made up of many interconnected components that are introduced into and reliant on social systems for their execution. (Macura et al., 2019; Kirschke and Newig, 2017).

21.2 BACKGROUND

Various researches have revealed that in order to obtain a comprehensive view of the ecosystem's activity, it is important to combine the key ecological components and processes. We believe that the ecosystem functions as a unit to some extent. Modeling is required to investigate the basic characteristics of complex environmental concerns in a methodical and collaborative manner. Ecological models are simplified representations of real ecosystems that are used in environmental management with the goal of improving ecosystem health. The diagram below shows a simple simplification of ecosystem interconnection, ecological models, and environmental management (Jørgensen, 1994) (Figure 21.1).

Ecological models are used to create ecological theories and to comprehend the features of ecosystems. This is deduced from the model that: 1) for ecosystems,

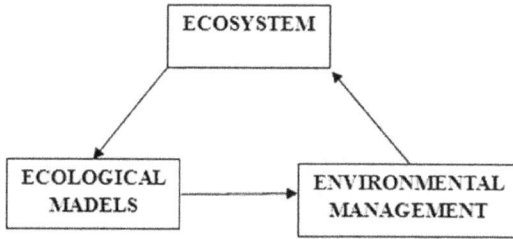

FIGURE 21.1   Simplification of ecosystem interconnectivity (Jørgensen, 1994).

quantum mechanics holds valid due to their complexity; 2) direct effects are often less important than indirect effects; 3) ecosystems have a hierarchy of regulation mechanisms and feed backs; and 4) ecosystems are soft systems (due to point 3), and their softness can be described using goal functions with correlations. Ecosystem properties, on the other hand, cannot be measured because it is an irreducible system (Jørgensen, 1994).

Socio-Environmental System (SES) modeling is used to analyze complicated problems obtained from interactions among environmental, ecological, economic, and social systems by integrating quantitative and qualitative techniques and datasets on system components, interactions among components, and their responses to changes in exogenous and endogenous drivers (Levin et al. 2013). The emergence of data sources, new techniques and computational power in one arm, and the raising challenges for sustainability on the other, posed grand challenges in modeling. SES modeling enables a scientific platform for all stakeholders and consolidates knowledge to share understanding for an alternate line of plan for action. Researchers worked hard to identify and formulate the issues and challenges of SES modeling through a literature analysis. The eight grand challenges are as follows:

1. Bridging theory of knowledge across all disciplines.
2. Integrated treatment of modeling uncertainty.
3. Integrating the qualitative and quantitative methods as well as data sources.
4. To deal with scales and its scaling.
5. Capturing systematic changes in SES.
6. Integrating the human dimension.
7. Raising the acceptance of models related to SES and impacts on policy.
8. Leveraging new data types and sources, were identified by researchers and experts of an interdisciplinary team.

Figure 21.2 depicts the eight big difficulties of SES modeling and their underlying issues.

Through a process of eliciting and sharing mental model process, academicians have come up with several techniques to address these difficulties in order to strengthen planning and support decision-making. However, the lack of long-term on-the-ground results is still perceived as a planning-implementation gap.

Disciplinary training

Limited adoption of integrated uncertainty

Limited communication of uncertainty to decision makers

Disciplinary perceptions of methods and data

Implementing mixed-methods in practice

Determining the right balance between quantitative and qualitative aspects of data collection and model building

Lack of standard collaboration norms

2.Integrated assessment in practice

2.Integrated treatment of modeling uncertainty

3. Combining qualitative and quantitative methods and data sources

Modeling phenomena across multiple scales

Representing and matching scales in SES models

Institutional gate keeping practices

1.Bridging epistemologies across disciplines

4. Dealing with scales and scaling

Different levels of knowledge and data about the social and environmental subsystems at various scales

Ambiguity about what constitutes data

8. Leveraging new data types and sources

Grand Challenges in SES Modeling

5. Capturing systemic changes in SES

Dealing with emerging ethical issues

7. Elevating the adoption of SES models and impacts on policy

6. Integrating the human dimension

Limited methods for modeling systemic changes

Methodological issues around data collection and use

Lack of understanding of the inevitable uncertainty in complex SES modeling

Measuring the impact of SES modeling on decision making

Limited funding for social science

Scaling up of outcomes from participatory modeling across multiple scales

Lack of knowledge and data for the fundamental processes that drive systemic shifts in social systems

Inherent difficulties in gathering data and representing the process of actual decision-making in models

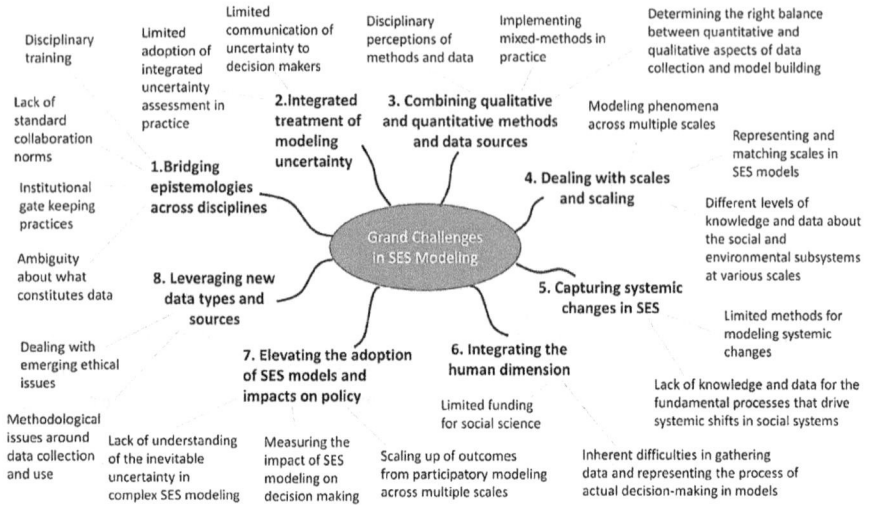

FIGURE 21.2 The grand challenges for Socio-Environmental System Modeling and their underpinning issues (Elsawah et al., 2020).

21.3 MENTAL MODEL VS PHYSICAL MODEL

21.3.1 MENTAL MODEL

Mental models are cognitive frameworks for interpreting and comprehending the world based on an individual's knowledge pattern (Carley and Palmquist, 1992; Bower and Morrow, 1990). Mental model is defined as "Internal representation of an event or process that works as a structural equivalent". The mental model is a dynamically changing picture of the environmental problem situation that is formed in working memory. Mental models are fundamental structures of cognition that are used to describe things in day-to-day life, the way, how the world works, the states of affairs, sequences of events and social and psychological acts. The mental model for any environmental problems is internal representations formed either consciously or unconsciously which is a combination of knowledge of a person, his skills, competency, his social and environmental beliefs and values, as well as his common sense, roles and actions, and help them to solve problems and make decisions.

Environmental mental model mapping can help people better understand environmental issues and sustainability, as well as learn to make decisions in difficult situations. The goal of studies attempting to find mental models of environmental problems is to learn how different decision-makers view the components of environmental problems and what changes occur in a robust manner in individual as well as in shared mental models in the event of action. The majority of the obstacles and limitations are related to the constraints of isolating and researching mental models, as well as the capture and validation of mental models. The majority of studies use qualitative analysis techniques and analytical categorization schemes to elicit mental model information. In theory, mental models are accepted as cognitive structures in addition to existing other eco-cognitive explanations of distributed cognition.

In a proposed mental model for any environmental problems, it is assumed a complete internal representation of the environment has not been made; rather, it is manipulated and even the create the environment as a representation in order to make room for new cognitive opportunities that are not immediately available. Mental models carefully filter and interpret the overwhelming information due to the limitations of the human intellect (Sabatier and Jenkins-Smith, 1999).

21.3.2 PHYSICAL MODEL

The resulting description is a physical model when the claims of the theory are concerned with a simplified and idealized physical system or phenomena. The physical models fully realize the theoretical/mental model's potential.

Because the relationship between reality and the physical model is complex, when images or "visualizations" are mentioned in the context of physical models, they should be taken in their broadest sense, rather than as a pictorial relationship in which each element of the model corresponds to an element.

21.4 CORE PLANNING-IMPLEMENTATION GAPS OF MODELING

21.4.1 DYNAMISM OF ENVIRONMENTAL MODEL

Even within environmental science, mixing ideas, information, and methods from many disciplines necessitates the construction of bridges, but connecting social and natural science demands the integration of distinct worlds. The primary concept of social science is to comprehend the opinion or connection of a certain instance or time frame, whereas environmental modellers treat dynamics as an axiomatic consideration.

21.4.2 AMBIGUITY ABOUT WHAT CONSTITUTES DATA

Variations related to epistemological backgrounds may create uncertainty about the data validation, its quality, and interpretation and the methods used to acquire and analyse it.

21.4.3 LACK OF STANDARD COLLABORATION NORMS

In all sectors of science, it has been observed a growing acceptance of the absence of data and model code sharing and transparency norms (Baker, 2016), although with minimal success to yet. The proper sharing of well-documented data and model code is not yet customary across all fields (Janssen, 2017; Stodden et al., 2018).

21.4.4 MODELING UNCERTAINTY

The most common sources of uncertainty include data, model structure, and model parameters, although uncertainty sources include every modeling option, activity, and intermediate and final result obtained across the modeling procedure. The assessment of uncertainty has primarily been found to be quantitative and limited to a restricted investigation of uncertainty surrounding model parameters, inputs, and other data. Nevertheless, uncertainty is found at many stages of the modeling process, which may include defining objectives and the purpose, problem formulation, and in conceptualizing model for investigating the proper challenge, boundary conditions, in validating data and verification of parameters, qualitative aspect, model code, and modeling workflows (Jakeman and Jakeman, 2016). There has recently been a surge in interest in taking a more comprehensive approach to addressing uncertainty. Table 21.1 summarizes the various strategies used to investigate the probable sources of modeling uncertainties.

TABLE 21.1
Overview of the Modeling Uncertainties Examined through Various Methods (Elsawah et al., 2020)

Approach	Rationale	Subjectivity	Sources
Mental modeling with stakeholders	To identify and combine different stakeholders with problem framings	Framework development	Voinov et al., 2016
Sensitivity analysis	To identify model inputs, model parameters and their significant effect on the model outputs	Sensitivity of parameters and inputs	Constantine and Diaz, 2017
Convergence of modeling strategies	Establishing limitations or deficiency of a model	Data and parameters, and also structure when considering alternatives	Hrachowitz and Clark, 2017
Uncertainty analysis	To calculate distribution of outputs based on sampling a distribution of inputs, on emergence, and on stochasticity of complex processes	Parameters	Railsback and Volker, 2011
Data-driven modeling	To collaboration between data-driven and physical model which improve ability to model	Mainly data and parameter distributions; model structure (and hence conceptualization) rarely considered	Mount et al., 2016
Exploratory analysis	To search for scenarios that lead to good, poor and intermediate outcomes, or specified objectives such as robustness metrics	Parameters and model structure	Moallemi and Malekpour, 2018
Surrogate models for uncertainty quantification	To approximate the response surface of a model with a simpler, faster running model to obtain sensitivity indices and undertake various types of uncertainty analyses	Model parameters and inputs which measure uncertainty	Sudret et al., 2017
Model-independent comparison	To illustrate the impact of automatically selects output features	empirical selection of output features	Fachada et al., 2017

(Continued)

TABLE 21.1 (CONTINUED)

Approach	Rationale	Subjectivity	Sources
(automated) Scientific workflows	To capture and automatically run model experiments, allowing for reproducing results and model's transparency	Communication about all sources of uncertainty, including structure, parameters, and data	Yilmaz et al., 2016
Practice documentation or logbooks	To capture the rationale and details of the methodological choices made throughout the modeling process	Identifying sources of uncertainty	Lahtinen et al., 2017
Uncertainty assessment	Assessment of uncertainty to assurance of the quality of the climate change modeling process	uncertainty identification in adaptation strategies	Refsgaard et al., 2013

21.4.5 INTEGRATION OF QUANTITATIVE AND QUALITATIVE APPROACH AND DATA SOURCE

Modellers have both obstacles and opportunities when combining qualitative and quantitative data. Many of science's major challenges are qualitative rather than quantitative, however qualitative data is frequently being portrayed as relatively inferior in comparison to quantitative data for modeling uses. Models generated from qualitative data are likewise seen to be ambiguous and so it is very difficult and complex to validate and prove (Di Baldassarre et al., 2015). Furthermore, qualitative data, which is sometimes presented as narratives, provides in-depth opinions on persons and their response with the natural and socio-environmental systems.

21.4.6 ADVANCEMENT IN SCALES AND SCALING

Socio-environmental system models include a wide range of spatial and temporal scales, and the social dimension is represented at several levels, such as household, community, or society, state or province, and country. Modeling of such a complex, multiscale system necessitates a clear understanding of scale representations and the matching of a conceptual representation of variables and processes to their data (Scholes et al., 2013; Robinson et al., 2018). Moreover, because the geographical, temporal and spatial domains of socio-environmental systems must be matched, progress in the scaling system is required.

21.4.7 HUMAN DIMENSIONS

For the sake of convenience and to avoid qualitative ideas, models employed in socio-environmental systems for framework and policy assistance are influenced by three general techniques to expressing the human dimensions: scenario, statistical models, and equilibrium models. Formalisms, e.g., agent-based models, as well as interactive techniques in which individuals and representatives of organizations and institutions make decisions on the human dimension, better describe social behaviour.

21.5   THE TRANSLATIONAL UNDERSTANDING: A MENTAL MODEL TO A PHYSICAL MODEL

The existence of working models in the mind of the person who understands is at the heart of understanding. Understanding a scientific theory would necessitate the creation of mental models that could be converted into physical models in the mind of the person seeking to comprehend it. If comprehension of discourse necessitates the creation of mental models, then comprehension of scientific theory statements, and by extension, comprehension of their physical models, which are a particularization of these assertions, will necessitate the creation of mental models as well. These mental models, in turn, will shape how people perceive experiences, and thus the types of mental models that need be developed to forecast or explain those same occurrences.

To comprehend an environmental phenomenon, the first step is to create mental models that will allow the individual to comprehend the statements that make up the semantic structure of the theory. Simultaneously, it is required to alter one's perception of phenomena by creating mental models that will allow a system to evaluate (as true or false values) the descriptions that the theory is composed of. If the above twofold process flow is achieved with the specific phenomenon in question, and the consequences (i.e., predictions and explanations) are comparable to scientifically acknowledged facts, then the individual has developed an adequate mental model of the physical model of the theory (Figure 21.3).

A physical model consists of mathematical equations and textual descriptions, for which an instruction or interpretation may be given. These texts contain claims about physical phenomena that the theory was designed to explain using dictionary notations. The translating issues of a mental model to a physical model are illustrated in Figure 21.4.

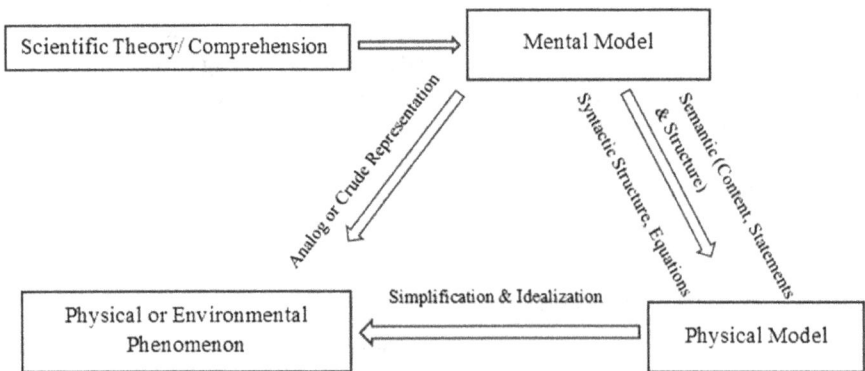

FIGURE 21.3 Conceptual representation of scientific theory, mental model, environmental phenomenon, and physical model (Greca and Moreira, 2002).

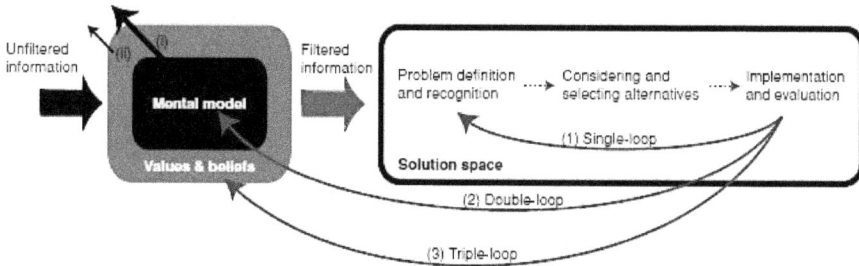

FIGURE 21.4 Translational challenges to implement a mental model to a physical model (Biggs et al., 2011).

In Figure 21.4, a single, double, or triple looped learning mental model is shown, which is informed by values and beliefs (filter) and gets data from the real world. Within a solution space, this model with filtered data ensures a decision-making process (i.e., problem definition, recognition, weighing, selection of options, execution, and evaluation). Due to incompatibility, single loop learning often functions within the existing model (less incoming information is recognized or evaluated for decision-making), whereas double-looped learning expands the model (as the amount of incoming information that is unrecognized and unassimilated is reduced). Individuals' values and beliefs adaptability (when questioned and changed, the mental model is transformed to suit the new learning), and ideological attitudes are all part of a triple looping learning process. This learning ensures the most inclusive approach to the greatest solution space (stakeholder values and views are actively questioned and investigated). However, the semantics of physical theory cannot be referred to as objects, events, or systems, and a physical model must be used to establish a relationship between theory and reality. As a result, a physical model must include the following features:

1. Methods for resolving biases and uncertainty by identifying and integrating model structure sources of uncertainty.
2. The uncertainty's qualitative aspect.
3. Finding the correct combination of quantitative and qualitative data collecting and model construction.
4. Comparative and reflective studies to assess the impact of alternative designs.
5. Evaluation and comparison of various scale-related methodological choices.
6. Representing vertical interactions within the social subsystem and cross-scale processes using social models at various scales.
7. Evolutionary mechanisms and other AI methodologies, as well as tales and visions drawn from participatory environments, to allow the production of new formal and informal rules.
8. Better correlation between theory and facts informing social decision-making.
9. Going beyond arbitrary assumptions or simplified notions that govern human behaviour.

10. In-depth knowledge of participatory modeling components.
11. A better understanding of the political process that underpins decision-making.
12. Incorporation of ethical and equity concerns.

21.6 CONCLUSION

Models of socio-environmental systems function over a wide range of spatial and temporal scales, with varying degrees of social dimension. Quantitative research could answer critical concerns regarding the degree of an effect or influence, but it couldn't help with important policy and practice issues. The mental model technique helped stakeholders comprehend and accommodate a wide range of values, perceptions, and beliefs, enhancing the likelihood of successful implementation. However, the link between social and natural science and mental models for a long-term solution to environmental problems has not been adequately addressed. We hope to adapt ways that can bridge the gap between mental and physical models by integrating qualitative and quantitative methods and data on system components, interactions among components, and their responses to changes in external and endogenous factors:

1. Recognition of stakeholders' internal assumptions and values, and how they relate to others.
2. Recognition of stakeholder commonality.
3. The formation of a shared vision for action based on the co-construction of a shared mental model that allows for an empowered and unified commitment to achieving long-term results.
4. Improved social assessments.

REFERENCES

Adams, W. M., & Sandbrook, C. (2013) Conservation, evidence and policy, *Oryx*, 47:329–333.
Baker, M. (2016) Is there a reproducibility crisis? *Nature*, 533, 452–454.
Barnett-Page, E., & Thomas, J., (2009) Methods for the synthesis of qualitative research: A critical review. *BMC Medical Research Methodology*, 9:59.
Biggs, D., Abel, N., Knight, A. T., Leitch, A., Langston, A., & Ban, N. C. (2011). The implementation crisis in conservation planning: Could "mental models" help?. *Conservation Letters*, 4(3), 169–183.
Bower, G. H., & Morrow, D. G. (1990) Mental models in narrative comprehension. *Science* 247, 44–48.
Carley, K., & Palmquist, M. (1992) Extracting, representing and analysing mental models. *Social Forces* 70, 601–636.
Constantine, P. G., & Diaz, P. (2017). Global sensitivity metrics from active subspaces. *Reliability Engineering & System Safety*, 162, 1–13.
Dalton, J., Booth, A., Noyes, J., & Sowden, A. J. (2017) Potential value of systematic reviews of qualitative evidence in informing user-centered health and social care: Findings from a descriptive overview, *Journal of Clinical Epidemiology*, 88, 37–46.
Di Baldassarre, G., Viglione, A., Carr, G., Kuil, L., Yan, K., Brandimarte, L., & Blöschl, G. (2015). Debates—Perspectives on socio-hydrology: Capturing feedbacks between physical and social processes. *Water Resources Research*, 51(6), 4770–4781. https://doi.org/10/f3n3p5

Elsawah, S., Filatova, T., Jakeman, A. J., Kettner, A. J., Zellner, M. L., Athanasiadis, I.N., Hamilton, S. H., Axtell, R. L., Brown, D.G., Gilligan, J.M., Janssen, M.A., Robinson, D.T., Rozenberg, J., Ullah, I. I.T., & Lade, S. J. (2020) Eight grand challenges in socio-environmental systems modeling. *Socio-Environmental Systems Modeling*, 2, 16226. https://doi.org/10.18174/sesmo.2020a16226

Fachada, N., Lopes, V. V., Martins, R. C., & Rosa, A. C. (2017). Model-independent comparison of simulation output. *Simulation Modeling Practice and Theory*, 72, 131–149.

Greca, I. M., & Moreira, M. A. (2002). Mental, physical, and mathematical models in the teaching and learning of physics. *Science Education*, 86(1), 106–121.

Hannes, K., Booth, A., Harris, J., & Noyes, J. (2013) Celebrating methodological challenges and changes: Reflecting on the emergence and importance of the role of qualitative evidence in Cochrane reviews. *Systematic Reviews*, 2(1), 84–84.

Heyvaert, M., Hannes, K., & Onghena, P. (2017) Using mixed methods research synthesis for literature reviews, Sage, 4. London.

Hrachowitz, M., & Clark, M. P. (2017). HESS Opinions: The complementary merits of competing modeling philosophies in hydrology. *Hydrology and Earth System Sciences*, 21(8), 3953–3973.

Jakeman, A. J., & Jakeman, J. D. (2016, November). An overview of methods to identify and manage uncertainty for modeling problems in the water–environment–agriculture cross-sector. In *Forum "Math-for-Industry"* (pp. 147–171). Springer, Singapore.

Janssen, M. A. (2017). The practice of archiving model code of agent-based models. *Journal of Artificial Societies and Social Simulation*, 20(1), 2. https://doi.org/10.18564/jasss.3317

Jørgensen, S. E., (1994) Models as instruments for combination of ecological theory and environmental practice, *Ecological Modeling*, 75/76 5–20.

Kirschke, S., & Newig, J. (2017) Addressing complexity in environmental management and governance. *Sustainability Science* 9:983.

Lahtinen, T. J., Hämäläinen, R. P., & Liesiö, J., 2017. Portfolio decision analysis methods in environmental decision making, *Environmental Modeling and Software*, 94:73–86.

Levin, S., Xepapadeas, T., Crepin, A. S., & Norber, J. (2013) Social-ecological system as complex adaptive systems: Modeling and policy implications, *Environmental and Development Economics*, 18(2), 111–132. https://doi.org/10.1017/s1355770x12000460

Macura, B., Suškevičs, M., Garside, R., Hannes, K., Ree, R., & Rodela, R. (2019) Systematic reviews of qualitative evidence for environmental policy and management: An overview of different methodological options, *Environmental Evidence*. https://dol.org/10.1186/s13750-019-0168-0

Moallemi, E. A., & Malekpour, S. (2018). A participatory exploratory modeling approach for long-term planning in energy transitions. *Energy Research & Social Science*, 35, 205–216.

Mount, N. J., Maier, H. R., Toth, E., Elshorbagy, A., Solomatine, D., Chang, F. J., & Abrahart, R. J. (2016). Data-driven modeling approaches for socio-hydrology: Opportunities and challenges within the Panta Rhei Science Plan. *Hydrological Sciences Journal*, 61(7), 1192–1208.

Pullin, A. S., & Stewart, G. B., (2006) Guidelines for systematic review in conservation and environmental management, *Conservation Biology*, 20, 1647–1656.

Railsback, S., & Volker, G., (2011). Agent-based and individual-based modeling: A practical introduction, Princeton University Press, New Jersey, ISBN: 0691136742.

Refsgaard, J. C., Arnbjerg-Nielsen, K., Drews, M., Halsnæs, K., Jeppesen, E., Madsen, H., ... & Christensen, J. H. (2013). The role of uncertainty in climate change adaptation strategies—A Danish water management example. *Mitigation and Adaptation Strategies for Global Change*, 18(3), 337–359.

Roberts, P. D., Stewart, G. B., & Pullin, A. S. (2006). Are review articles a reliable source of evidence to support conservation and environmental management? A comparison with medicine. *Biological Conservation*, 132, 409–423.

Robinson, D. T., Di Vittorio, A., Alexander, P., Arneth, A., Barton, C. M., Brown, D. G., Kettner, A. J., Lemmen, C., O'Neill, B. C., Janssen, M., Pugh, T. A. M., Rabin, S. S., Rounsevell, M., Syvitski, J. P. M., Ullah, I., & Verburg, P. H. (2018). Modeling feedbacks between human and natural processes in the land system. *Earth System Dynamics*, 9, 1–47. https://doi.org/10.5194/esd-2017–68

Sabatier, P. A., & Jenkins-Smith, H. C. (1999) The advocacy coalition framework: An assessment. Pages 117–169 in *Theories of the policy process*. P.A. Sabatier, editor. Westview Press, Boulder, CO.

Scholes, R. J., Reyers, B., Biggs, R., Spierenburg, M. J., & Duriappah, A. (2013). Multi-scale and cross-scale assessments of social–ecological systems and their ecosystem services. *Current Opinion in Environmental Sustainability*, 5(1), 16–25. https://doi.org/10.1016/j.cosust.2013.01.004

Stodden, V., Seiler, J., & Ma, Z. (2018). An empirical analysis of journal policy effectiveness for computational reproducibility, *PNAS*, 115, 2583–2589. https://doi.org/10.1073/pnas.1708290115

Sudret, B., Marelli, S., & Wiart, J. (2017, March). Surrogate models for uncertainty quantification: An overview. In *2017 11th European conference on antennas and propagation (EUCAP)* (pp. 793–797). IEEE.

Voinov, A., Kolagani, N., McCall, M. K., Glynn, P. D., Kragt, M. E., Ostermann, F. O., … & Ramu, P. (2016). Modeling with stakeholders–next generation. *Environmental Modeling & Software*, 77, 196–220.

Yilmaz, L., Chakladar, S., & Doud, K., (2016). The goal-hypothesis-experiment framework: A generative cognitive domain architecture for simulation experiment management. In *Proceedings of the 2016 Winter Simulation Conference*, IEEE, pp. 1001–1012.

22 An Interdisciplinary Modeling Approach for Dynamic Adaptive Policy Pathways

Vijay P. Singh, Satya Prakash Maurya, Ramesh Singh, and Akhilesh Kumar Yadav

CONTENTS

22.1 INTRODUCTION

Environment is the combination of different natural elements, including water, air, and soil, and the existence of living beings is dependent on environmental conditions. Owing to human intervention to make life comfortable, the proportions of natural resources consumed or exploited have compromised the environment. Meanwhile, attempts have been made to monitor and model the consequences of the compromising which is complex due to the nesting of problems. Generally, modeling of an environmental system involves a trade-off among social, economic, technical, and ecological criteria. But, with the identification of issues and evolution of multilevel systems, social and ecological elements interoperate through regular interchangeable interactions and feedback loops (Folke, 2006). They are characterized by complex and dynamic interdependencies between social and ecological sub-systems (Liu et al., 2015), which are less than well understood. However, understanding the dynamics of complex environmental system interactions is essential for sustainable resources management and ecosystems (Gain et al., 2019). Failure to identify such interdependencies and dynamics leads to even more complex environmental problems and

DOI: 10.1201/9781003203445-26

development challenges, such as degraded natural resources, biodiversity loss, natural habitat loss, and climate change.

Significant progress has been reported in the last three decades on analysis of environmental problems using various systems approaches. These approaches have emerged as prominent analytical frameworks to investigate the sustainability of natural resources considering anthropogenic activities (Rockström et al., 2009). A number of frameworks have been developed to study and address environmental issues. Improving the understanding of complex dynamics has included quantitative and qualitative approaches and their adaptability for assessing human-ecological interactions. Such approaches include system dynamics modeling, multi-criteria analysis, integrated assessment, decision support systems coupled model frameworks, indicator-based aggregation, agent-based modeling (Lippe et al., 2019).

Along with the present state-of-art, there are still enough active domains for analysis of conceptual frameworks to develop methods and tools to verify sustainable environmental development. The disjointed elements of a complex system cannot be evaluated with traditional methods. Rather interdisciplinary methods may be helpful to meet such sustainability challenges.

This chapter envisages interdisciplinary approaches to solving complex environmental problems under uncertainties like climate change and natural resource degradation, including air and water. Dynamic adaptation of some existing conceptual frameworks will be discussed and reveal the challenges for its real-time implementation.

22.2   BACKGROUND

Globally, the climate change rate seems to have heightened due to various factors, while the usable natural resources continue to deplete. Many models, including the multi-stakeholder management system, attempted to model natural resources more sustainably (Basco-Carrera et al., 2017). Still, due to heterogeneities of domains, such as spatial scales, stakeholder uses, standards for various uses, measures to control pollution, over-consumption against availability, different institutional characteristics, and political interference in environmental planning, the governance make the least representation of a precise management system. Sometimes it may be misguiding the ultimate goal. To address the above issue of diverse domains with various and even conflicting objectives, a decision support system (DSS) to manage natural resources is presented by Ahmadi et al. (2020), based on Social Choice Theory (SCT). SCT is the study of collective decision procedures that aggregate multi-inputs into collective outputs and close to a multi-criteria decision-making system. Here, an attempt is made to apply systemic Stakeholder Analysis (SA) outcomes and insights of Social Network Analysis (SNA).

Different researchers developed various theoretical and policy frameworks for understanding the qualitative and quantitative behaviour at various scales, including the social-ecological system framework (SESF) (McGinnis and Ostrom 2014), the vulnerability framework (Turner II et al., 2013), the driver-pressure-state-impact-response (DPSIR) framework (Lewison et al., 2016) among others. However, when these frameworks associate with a single goal like sustainability or climate change, they are decoupled to handle different domains, including adaptation constraints. Analyzing to measure the impact of these interventions, Lim and Spanger-Siegfried (2004) revealed a framework as a series of tasks (Figure 22.1): 1) climate change; 2) socio-economic trends and risks; 3) natural resources environmental trends; and 4) adaptation barrier and opportunities.

Task 1 describes the potential future climate risk and opportunities associated with it, i.e., a set of future climate change scenarios and an analysis of associated risk. Task 2 is to characterize socio-economic trends, risks, and opportunities for adaptation strategies for the unknown hazards of future climate change (the kind of priority system adaptation will take). Task 3 is to characterize natural resources environmental trends that deal with the growth in natural resources consumption raise important issues regarding the vulnerability to future climate risks. Certain environmental scenarios may influence adaptive capacity, which may need to be developed, including important feedback exacerbating climate risk. Task 4 is to characterize adaptation barriers and opportunities of current development, and environmental policy which is essential for assessing potential barriers to adaptation. Especially important are recent or planned state reforms for economic development (e.g., privatization and liberalization of trade). Policies and programmes related to the priority system should be evaluated for their potential to support effective adaptation to climate change in the context of sustainable development.

When these tasks are comprehensively translated to an adaptation policy framework (APF), plans, processes, and key development policies are essential in an

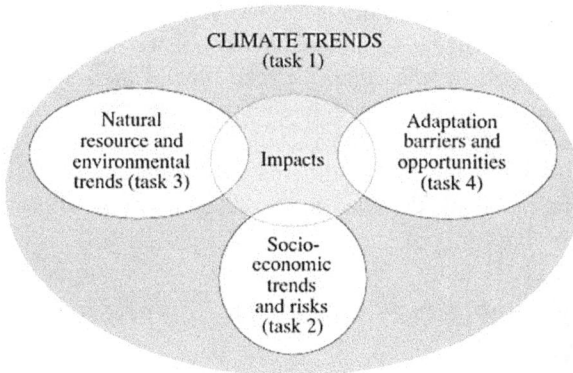

FIGURE 22.1   Conceptualization of tasks for component assessment.

[Reprinted from Lim and Spanger-Siegfried (Spanger-Siegfried et al., 2004) with permission from The Cambridge University Press].

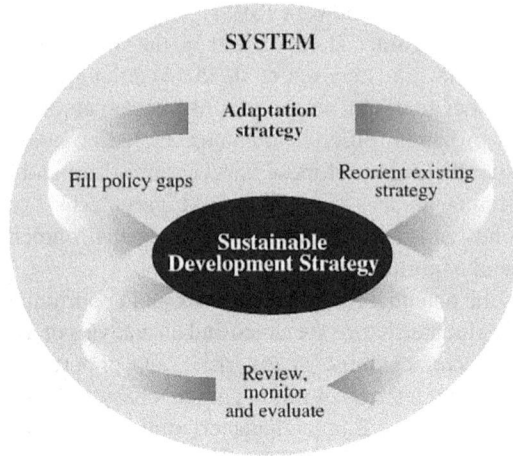

FIGURE 22.2 Conceptualization of sustainable adaptation processes for system approach.

[Reprinted from Lim and Spanger-Siegfried (Spanger-Siegfried et al., 2004) with permission from The Cambridge University Press].

adaptation strategy. The expected benefits of an adapted strategy are reducing vulnerability from current stresses, enhancing development capacity, and resources improvement. However, integrating interdisciplinary approaches may result in an efficient implementation of the policy framework. More precisely, climate change adaptation, an underdeveloped policy, may be unlikely to succeed within less integrated with well-established objectives under competitive policymaking. Figure 22.2 illustrates the activities and feedback loops in this APF component.

It suggests two paths in which (i) to make the existing policies and practices more responsive; the countries can re-orient themselves to the change in vulnerability caused by variability change (e.g., disaster management), and (ii) climate risk policy gaps may be addressed by enhancing the resilience of the priority system. These interventions remove existing barriers to adopting policies that are sensitive to the impacts of climate change, including variability.

Thus, it is essential to incorporate different disciplines to remove the subjectivity of these complex problems identified in the implementation of basic principles of environmental systems.

22.3 INTERDISCIPLINARY MODELING APPROACH FOR POLICY PATHWAYS

Earlier, the adaptation pathways identified assets of operations towards achieving goals that may be interdependent of each other, a problem simulated to identify strategies for sustainable water management in river deltas (Haasnoot et al., 2011). Furthermore, the concepts of adaptive policy making, adaptation tipping points, and adaptation pathways may be collaboratively presented to exercise the Dynamic

Adaptive Policy Pathways (DAPP) methodology (Haasnoot et al., 2013). The DAPP may achieve dynamic robustness, considering (i) an alternative policy is implementation date where the previous policy starts to perform poorly, and (ii) facilitating the end-user to monitor real-world contexts with important policy adaptation process initiation that leads to choosing an appropriate adaptation plan dynamically.

22.3.1 PERSPECTIVES AND OPPORTUNITIES FOR AIR POLLUTION

A major goal is to foster and drive new international research efforts where progress will not be possible without a collaborative interdisciplinary approach. It aims to capitalize on existing capabilities, covering topical specialties and extensive geographies with multiple groups contributing unique expertise, tools, and capacities. The adaptation pathways approach is summarized (Figure 22.3) (Haasnoot et al., 2013).

This section highlights perspectives and opportunities for air pollution, which seeks to exploit and drive forward.

22.3.2 DEVELOPING AN INTERDISCIPLINARY APPROACH

Understanding the mechanisms and processes controlling atmospheric pollution requires more coordinated observations of atmosphere dynamical processes, ocean-ice-atmosphere interactions, atmospheric boundary layer processes, and observations of atmospheric composition. Air pollution will facilitate coordinating atmospheric pollution observations and modeling studies with other international meteorology and climate research programmes (Amann et al., 2020; Vitolo et al., 2018). Air pollution will encourage the coordination of new observational capabilities and field campaigns related activities (Multidisciplinary drifting Observatory for the Study of Air

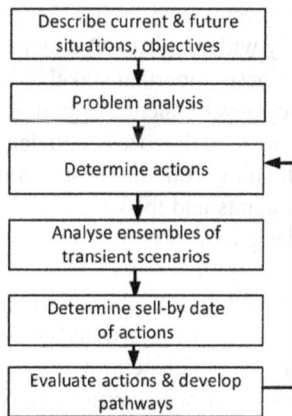

FIGURE 22.3   Stepwise policy analysis to construct adaptation pathways.

[Reprinted from Haasnoot (Haasnoot et al., 2013) with permission from The Elsevier B.V].

pollution). Existing long-term observatories also offer an opportunity for interdisciplinary questions to be addressed through already sustained and potentially expanded observations of chemical and physical characteristics of the atmosphere, cryosphere, and terrestrial ecosystems. Global Atmosphere Watch Programme (GAW) focal areas are aerosols, greenhouse gases, selected reactive gases, ozone, UV radiation and atmospheric deposition (Schultz et al., 2015; Naitza et al., 2020).

Community-based monitoring is a highly promising means by which observational networks can expand into under-sampled regions and seasons, and communities can engage in knowledge exchange frameworks with atmospheric scientists. Community-based monitoring programmes exist across the circumpolar region, although formal monitoring programmes are most developed in North America (Cohen et al., 2021). While few community-based monitoring programmes currently engage community members in air pollution monitoring, several emerging programmes have significant potential to contribute to air pollution. This emerging initiative presents a strong opportunity for collaboration that should be developed further (Boone et al., 2020).

22.3.3 FIELD MISSIONS AND LONG-TERM MONITORING

The air pollution initiative will work to develop coordinated international field missions that address the science challenges. The remoteness of the particular need for vertical information on pollutants means that airborne and ground-based in-situ measurements will necessarily play an important role. The goals of this study will be to: 1) challenge model performance regarding long-range transport, vertical distribution, deposition, and impacts on climate; 2) improve the understanding of chemical and physical processes, especially deposition; 3) better evaluate local sources, both natural and anthropogenic, and their impacts on air pollution and ecosystems; and 4) improve the understanding of social and economic interactions between air pollution and local populations and ecosystems. Future measurement programmes will need to collaborate closely with residents to understand the societal interactions with air pollution and engage these important stakeholders in understanding the science and participation in policy development. Air pollution programmes develop and exploit a new understanding of model process deficiencies and use these to target new observational mission planning from the outset. Improved understanding of relevant processes affecting pollutants and their impacts will require different tailored approaches to experimental design. In particular, a future field programme is likely to involve a more sustained effort comprised of repeated flight profiles in multiple seasons and years. Vertically resolved measurements over extended periods are needed to constrain model simulations of long-range transport to the receptor. Such observations would help determine model deficiencies associated with emissions, advective transport, and removal processes or with vertical transport and deposition within environmental surroundings.

Satellite data provide useful information about the spatial and temporal coverage of a limited number of trace gases, as well as certain aerosol properties, in regions where in-situ sampling is limited (Leon et al., 2021; Romakkaniemi et al., 2017; Ruuskanen et al., 2021). These data can be used to evaluate models, often in conjunction with in-situ data that provide much-needed vertical information (Ge et al., 2020; Jiang et al., 2021; Zhang et al., 2021). They used satellite observations to demonstrate consistent bias in simulated NO_2 among several models in high latitude regions dominated by fire emissions, where in-situ sampling was unavailable (Emmons et al., 2015; Lin et al., 2021). Satellite observations of CO can also be used to probe large-scale pollution outflow from continental regions and latitudinal gradients (Monks et al., 2015; Pope et al., 2016); while collaborations between modellers and groups making campaign observations are traditionally well-established, opportunities for collaboration between modellers and long-term monitoring observations are not always fully exploited due, in some cases, to data products which are not readily usable or have poorly documented limitations for their use (Starkweather and Uttal, 2016). Introducing more modellers into these discussions provides an opportunity to increase the relevance and accessibility of long-term observations to model evaluation and development.

22.4 COMPUTATIONAL INTERVENTIONS IN POLICY PATHWAYS

It is always important to utilize computational aspects and develop tools based on a framework that can suggest a roadmap for future sustainable development in advancing computer science. On the contrary, a few models developed with the combinatorial approach of DAPP methodology (Haasnoot et al., 2013) and the trans-disciplinary framework (i) STatistical approximation-based modEl EMulator (STEEM) (Papadelis and Flamos, 2018), which is a two-step approach to run a "black-box" emulator. Step 1, inputs and outputs simulated in the original model are given to the emulator for its training, also called calibration (learns the correlation between inputs and outputs). Step 2, the calibrated model, is further utilized to estimate the original model's outputs providing new inputs. (ii) An Agent-based Technology Adoption Model (ATOM) (Stavrakas et al., 2019), which allows the user to explicitly quantify uncertainties about agents' preferences and decision-making under different supporting policy schemes (iii) the Adaptive Policymaking Model (AIM) which (a) investigates the conditions under-, and the timeframe beyond-which a policy starts to deviate from the set targets, (b) visualizes a map of DAPP, and (c) sets up a monitoring system for real-world policy adaptation (Michas et al., 2020).

Apart from the above model's exercises, the complexities must be addressed through a well-established framework, including adaptive strategy. A dynamic adaptation approach-based framework has been proposed in Figure 22.4.

This framework is based on three pivotal entities: 1) institutions, 2) actors, and 3) processes that cover aspects, viz. problem identification, constraints and boundary conditions, implementation challenges, discipline interventions, evaluation process, policy adaptation capacity, and outcome evaluations. These aspects guide decision contexts, such as rules, knowledge, intelligence, and adaptation policy. The decision contexts will play a vital role in policy adaptation simulation. Modern technological streams of computer science such as machine learning and advanced expert system based on reinforcement learning or deep learning may help guide policy adaptation decisions.

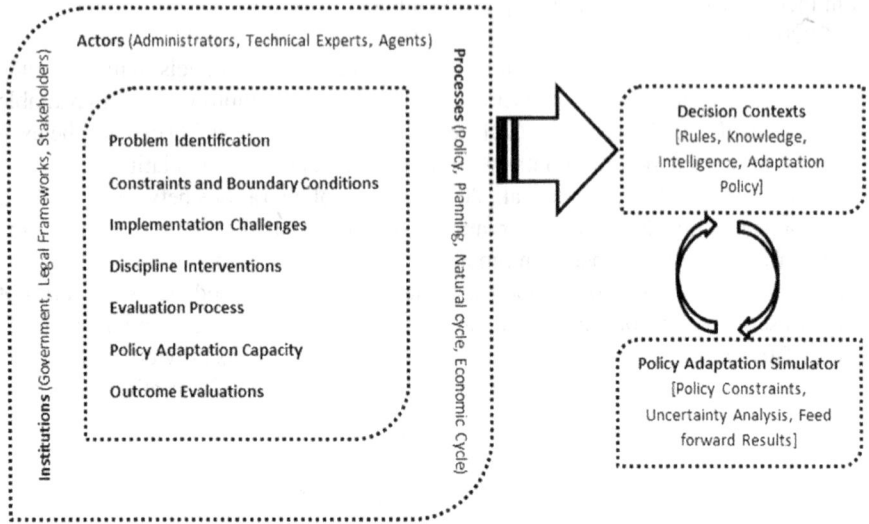

FIGURE 22.4   Conceptual framework of trans-disciplinary dynamic adaptation approach.

22.5   MOVING FORWARD AND CONCLUSIONS

This chapter has outlined how natural resources play a central role in a complex set of environmental and societal issues, driven by regional and global climate interactions and socio-economic responses to adverse environmental issues. Important goals of environmental sustainability and climate change are the provision of robust scientific knowledge to policy-makers, and engagement with local communities, to present findings and explore risks and benefits to communities while at the same time examining sustainable pathways in a changing environment. The climate change initiative aims to tackle key gaps in our understanding across the range of issues outlined by developing new focused actions and creating new collaborative efforts between observational and modeling groups, social science researchers, and local communities. Modern computational approaches and interdisciplinary models are enough exercised in literature, whereas a few models are also implemented, but the dynamic adaptation of such models is still under consideration for common framework development. A dynamic adaptation framework with possible processes has been recommended for policy pathways.

However, along with the complexities of a problem, specifically air and water under natural resources management, we make the following key overarching recommendations:

1. Advancement in resource management research should take an interdisciplinary approach. It should exploit collaborative platforms for observations across linked aspects of the earth system (atmosphere, cryosphere, ocean, land surface, society), enable community-based monitoring approaches, and consider societal and economic drivers and responses to pollution change

impacting climate and air quality and ecosystems. It should be carried in a collaborative framework linking with existing and planned programmes.

2. Improved process understanding requires further capacity in terms of regular long-term monitoring and intensive field observations, both at the surface and throughout the troposphere (in-situ and satellite). Community-based monitoring would enhance spatial and temporal coverage of surface observations while engaging local populations in science and decision-making.

3. Improved predictive capability is needed across various scales to diagnose wider impacts of sustainable resources on regional and global climate and earth system and local air quality and ecosystems.

The future models must target uncertain processes and regions to be probed by observations and observation activities. These recommended efforts should build on and link with existing programmes, ranging from networks making highly valuable observations to initiatives tackling closely related issues in different world regions.

REFERENCES

Ahmadi, A., Kerachi, R., Skardi, J. M. E., Abdolhay, A. (2020) A stakeholder-based decision support system to manage water resources, *Journal of Hydrology*, 589, 125138.

Amann, M., Kiesewetter, G., Schopp, W., et al. (2020). Reducing global air pollution: the scope for further policy interventions. *Philosophical Transactions of the Royal Society A: Mathematical, Physical and Engineering Sciences* 378 (2183):20190331.

Boone, C. G., Pickett, S. T. A., Bammer, G., et al. (2020). Preparing interdisciplinary leadership for a sustainable future. *Sustainability Science* 15 (6):1–11.

Basco-Carrera, L., Beek, E. V., Jonoski, A., Benítez-Ávila, C., Guntoro, F. P. (2017) Collaborative modeling for informed decision-making and inclusive water development, *Water Resources Management* (2017) 31:2611–2625. doi. 10.1007/s11269-017-1647-0

Cohen, A., Matthew, M., Neville, K. J., Wrightson, K. (2021). Colonialism in community-based monitoring: Knowledge systems, finance, and power in Canada. *Annals of the American Association of Geographers*:1–17.

Emmons, L. K., S. R. Arnold, S. A. Monks, et al. (2015). The POLARCAT Model Intercomparison Project (POLMIP): Overview and evaluation with observations. *Atmospheric Chemistry and Physics* 15 (12):6721–6744.

Folke, C. (2006). Resilience: The emergence of a perspective for social-ecological systems analyses. *Global Environmental Change* 16(3):253–267. https://doi.org/10.1016/j.gloenvcha.2006.04.002

Gain, A. K., Ashik-Ur-Rahman, M., Benson, D. (2019). Exploring institutional structures for tidal river management in the Ganges-Brahmaputra Delta in Bangladesh. *Die Erde* 150(3):184–195. https://doi.org/10.12854/erde-2019-434

Ge, X., Schaap, M., Kranenburg, R., et al. (2020). Modeling atmospheric ammonia using agricultural emissions with improved spatial variability and temporal dynamics. *Atmospheric Chemistry and Physics* 20 (24):16055–16087.

Haasnoot, M., Middelkoop, H., Beek, E. V., Deursen, W. P. A. V. (2011). A method to develop sustainable water management strategies for an uncertain future. *Sustaianable Development*, 19(6):369–381.

Haasnoot, M., Kwakkel, J. H., Walker, Warren E., ter Maat, Judith (2013). Dynamic adaptive policy pathways: A method for crafting robust decisions for a deeply uncertain world. *Global Environmental Change* 23 (2):485–498.

Jiang, F., Wang, H., Chen, J. M., et al. (2021). Regional CO_2 fluxes from 2010 to 2015 inferred from GOSAT XCO2 retrievals using a new version of the Global Carbon Assimilation System. *Atmos. Chem. Phys.* 21 (3):1963–1985.

Leon, J. F., Akpo, A. B., Bedou, M., et al. (2021). $PM_{2.5}$ surface concentrations in southern West African urban areas based on sun photometer and satellite observations. *Atmospheric Chemistry and Physics* 21 (3):1815–1834.

Lewison, R. L., Rudd, M. R., Al-Hayek, W., Baldwin, C., Beger, M., Lieske, S. N., Jones, C., Satumanatpan, S., Junchompoo, C., Hines, E. (2016) How the DPSIR framework can be used for structuring problems and facilitating empirical research in coastal systems, *Environmental Science & Policy*, 56, 110–119 doi. 10.1016/j.envsci.2015.11.001

Lin, H., Jacob, D. J., Lundgren, E. W., et al. (2021). Harmonized Emissions Component (HEMCO) 3.0 as a versatile emissions component for atmospheric models: application in the GEOS-Chem, NASA GEOS, WRF-GC, CESM2, NOAA GEFS-Aerosol, and NOAA UFS models. *Geoscientific Model Development* 14 (9):5487–5506.

Lim, B., Spanger-Siegfried, E. (2004). *Adaptation Policy Frameworks for Climate Change: Developing Strategies, Policies and Measures*, The Press Syndicate of The University of Cambridge, United Kingdom, ISBN 052161760X.

Lippe, M., Bithell, M., Gotts, N., Natalini, D., Barbrook-Johnson, P., Giupponi, C., Hallier, M., Hofstede, G. J., Le Page, C., Matthews, R. B., Schlüter, M., Smith, P., Teglio, A., Thellmann, K. (2019). Using agent-based modeling to simulate social-ecological systems across scales. *GeoInformatica*. https://doi.org/10.1007/s10707-018-00337-8

Liu, J., Mooney, H., Hull, V., Davis, S. J., Gaskell, J., Hertel, T., Lubchenco, J., Seto, K. C., Gleick, P., Kremen, C., Li, S. (2015). Systems integration for global sustainability. *Science* 347(6225). https://doi.org/10.1126/science.1258832

McGinnis, M. D., Ostrom, E. (2014). Social-ecological system framework: initial changes and continuing challenges, *Ecology and Society*, 19 (2):30.

Michas, S., Stavrakas, V., Papadelis, S., Flamos, A. (2020) A transdisciplinary modeling framework for the participatory design of dynamic adaptive policy pathways, *Energy Policy*, 139, 111350.

Monks, S. A., Arnold, S. R., Emmons, L. K., et al. (2015) Multi-model study of chemical and physical controls on transport of anthropogenic and biomass burning pollution to the Arctic. *Atmospheric Chemistry and Physics* 15 (6):3575–3603.

Naitza, L., Cristofanelli, P., Marinoni, A. (2020). Increasing the maturity of measurements of essential climate variables (ECVs) at Italian atmospheric WMO/GAW observatories by implementing automated data elaboration chains. *Computers & Geosciences* 137:104432.

Papadelis, S., Flamos, A. (2018). An application of calibration and uncertainty quantification techniques for agent-based models. In: *Understanding risks and uncertainties in energy and climate policy: Multidisciplinary methods and tools for a low carbon society*. Springer, pp. 79–95. https://doi.org/10.1007/978-3-030-03152-7

Pope, R. J., Richards, N. A. D., Chipperfield, M. P., et al. (2016). Intercomparison and evaluation of satellite peroxyacetyl nitrate observations in the upper troposphere–lower stratosphere. *Atmospheric Chemistry and Physics* 16 (21):13541–13559.

Rockström, J., Steffen, W., Noone, K., Persson, Å., Chapin, F. S., Lambin, E. F., Lenton, T. M., Scheffer, M., Folke, C., Schellnhuber, H. J., Nykvist, B., de Wit, C. A., Hughes, T., van der Leeuw, S., Rodhe, H., Sorlin, S., Snyder, P. K., Costanza, R., Svedin, U., Falkenmark, M., Karlberg, L., Corell, R. W., Fabry, V. J., Hansen, J., Walker, B., Liverman, D., Richardson, K., Crutzen, P., Foley, J. A. (2009). A safe operating space for humanity. *Nature* 461 (7263):472–475. https://doi.org/10.1038/461472a

Romakkaniemi, S., Maalick, Z., Hellsten, A., et al. (2017). Aerosol–landscape–cloud interaction: signatures of topography effect on cloud droplet formation. *Atmospheric Chemistry and Physics* 17 (12): 7955–7964.

Ruuskanen, A., Romakkaniemi, S., Kokkola, H., et al. (2021). Observations on aerosol optical properties and scavenging during cloud events. *Atmospheric Chemistry and Physics* 21 (3):1683–1695.

Schultz, M. G., Akimoto, H., Bottenheim, J., et al. (2015). The global atmosphere watch reactive gases measurement network. *Elementa: Science of the Anthropocene* 3.

Starkweather, Sandra, Uttal, Taneil (2016). Cyberinfrastructure and collaboratory support for the integration of arctic atmospheric research. *Bulletin of the American Meteorological Society* 97 (6):917–922.

Stavrakas, V., Papadelis, S., Flamos, A., (2019). An agent-based model to simulate technology adoption quantifying behavioural uncertainty of consumers. *Applied Energy* 255, 113795. https://doi.org/10.1016/j.apenergy.2019.113795

Turner II, B. L., Kasperson, R. E., Matson, P. A., McCarthy, J. J., Corell, R. W., Christensen, L., Eckley, N., Kasperson, J. X., Luers, A., Martello, M. L., Polsky, C., Pulsipher, A., Schiller, A. (2013). A framework for vulnerability analysis in sustainability science, *Proceedings of National Academy of Science United States of America*, 100 (14) 8074–8079; https://doi.org/10.1073/pnas.1231335100

Vitolo, C., Scutari, M., Ghalaieny, M., Tucker, A., Russell, A. (2018). Modeling air pollution, climate, and health data using bayesian networks: A case study of the english regions. *Earth and Space Science* 5 (4):76–88.

Zhang, B., Liu, H., Crawford, J. H., et al. (2021). Simulation of radon-222 with the GEOS-Chem global model: emissions, seasonality, and convective transport. *Atmospheric Chemistry and Physics* 21 (3):1861–1887.

Index

For Product Safety Concerns and Information please contact our EU
representative GPSR@taylorandfrancis.com
Taylor & Francis Verlag GmbH, Kaufingerstraße 24, 80331 München, Germany